A series of student texts in

CONTEMPORARY BIOLOGY

General Editors :

Professor E. J. W. Barrington, F.R.S.

Professor Arthur J. Willis

Professor Michael A. Sleigh

The Biology of Parasitism: an introduction to the study of associating organisms.

Philip J. Whitfield
Ph.D.

Lecturer in Zoology
King's College, University of London

University Park Press
Baltimore

First published in Great Britain by
Edward Arnold (Publishers) Limited, London

First published in the United States of America by
University Park Press, 233 East Redwood St., Baltimore,
Maryland 21202

Library of Congress Cataloging in Publication Data

Whitfield, Philip J.

Biology of parasitism.

(Contemporary biology)

Bibliography: p.

Includes index.

1. Host-parasite relationships 2. Symbiosis
3. Parasitology I. Title [DNLM: 1. Parasites
2. Symbiosis QX4.3 W595b]

QL757.W47 1979 574.5′24 79-4952

ISBN 0-8391-1459-1

Printed and bound in Great Britain

Contents

Preface

This book attempts to discuss parasitism and symbiosis using, as far as possible, a unified language. The ability to use a common vocabulary while considering these two types of species association is a consequence of a growing consensus in the biological literature. This emphasizes the ubiquity of the basic interactions that occur throughout the range of intimate associations between organisms. I believe that approaching host-parasite and symbiotic associations together generates a whole framework of insights and cross-references that are not provided by the single-minded study of either category of association alone.

The common ground of association biology has a significance both for the active scientist and the student. For the former, working on hosts, parasites or symbiotes, it is evident that the parallels and mirror images provided by adjacent categories of associations are a prolific source of ideas for hypotheses and experimental techniques. This reciprocal stimulation of disciplines is nowhere better exemplified than in the proceedings of a recent symposium of the Society for Experimental Biology[158] on organism associations. This volume happily accommodates together recent research findings on the evolutionary origins of mitochondria, termite symbiotes, tapeworms, fungal rusts, parasitoid bacteria and cleaner fish. It emphasizes, by practical demonstration, the underlying unity of approach in most investigations on closely associating organisms.

The student, on the other hand, is faced with the gruelling job of finding comprehensible patterns in the ever-enlarging bulk of the research literature. It is our experience, teaching introductory courses at King's College, that the unified principles which emerge from a holistic approach to association biology are an enormous aid. They certainly assist the unravelling of the crucial from the peripheral in a voluminous subject matter.

The comprehensive examination of all types of intimate association has guided at least one excellent introductory textbook. Clark Read's *Parasitism and Symbiology*[284] in many ways crystallized the methodology of the broad approach and demonstrated its validity. Since 1970, however, the flood of new results has inevitably continued. In many of the areas discussed by Read, these new findings have provided completely novel explanations or twisted old explanations into new configurations. The present book tries to incorporate these new facts and ideas into the conceptual framework constructed by Read and others. As a result, almost 70% of the sources which have been cited were produced in the interval 1970–77, that is, after the period covered by Read.

The Biology of Parasitism treats species associations using what has come to be called the topic approach. Major themes like nutrient flow between associates, the ecology of associations and the behavioural interactions within associations are discussed in turn across the range of association types. It is assumed in these discussions that the reader will have a reasonable working knowledge of invertebrate, vertebrate and plant diversity. A group-by-group, taxonomic treatment is not particularly appropriate for an introductory text on association biology as it would hinder the presentation of the unified concepts discussed above. In addition, several modern and detailed texts already exist which handle host-parasite systems in a taxonomically ordered way. The topic approach used here has necessitated harsh pruning of the range of specific organism associations dealt with in detail. It was judged best to concentrate on the associations which have been subjected to extensive experimental investigation and to attempt to provide a reasonably thorough picture of our knowledge of these examples. In this way, it was hoped that the more fragmentary findings on more obscure systems could be fitted into a firm experimental context. This philosophy has meant that some very intriguing associations have had to be dismissed in a summarizing sentence or two and the reader might at some points imagine that association biology only concerned itself with *Hymenolepis*, *Schistosoma*, the human immune system, lichens and termites!

One last characteristic of the present book requires some explanation. There exists in academic legends a fairy story which goes like this: Once upon a time a student, sitting in a lecture theatre, had a horrible shock. His lecturer said, without warning, in the middle of a lecture, that there were two diametrically opposed theories concerning the interpretation of some scientific facts. The student was terrified and thrown into such a state of indecision about the two theories that his work suffered badly. Luckily, the kind and wise lecturer understood his problem and explained that theory A was absolutely correct and theory B was pitifully inadequate. The student was overjoyed at this explanation and studied happily ever after.

Like all fairy stories this one has a kernel of sense, but carrying its philosophy to extremes in a text, all too often produces a book that gives a finished and absolute impression that is at variance with reality. In all evolving sciences nothing is finished or absolute. Scientific fields develop by the refutation of testable hypotheses. At any moment the expanding fringe of a science will be inhabited by interesting interpretations and hypothesis awaiting refutation or partial proof and many of these will be contradictory. In this book, where matters are being considered which are the centres of productive controversy, opposing views are considered together without any unnecessary decision being made between their respective total merits.

Finally, it is salutatory to consider the immense practical importance of intimate organism associations. Far from being an esoteric, unworldly field of biology, association biology abuts some of the most pressing practical problems in the world today. Man and his domesticated animals and crops are attacked by a multifarious army of viral, rickettsial, bacterial, fungal, protozoan, helminth and arthropod diseases. Each disease is the manifestation at the level of the individual host of an intimate host-parasite relationship. The maintenance of much of the angiosperm segment of the plant world, including many crops, depends on pollination associations between plants and animals. Those same crops, including forest trees, are often linked in vital symbiotic associations with microorganisms. Both nitrogen-fixing bacteria and mycorrhizal fungi have a global significance in plant productivity. Chemical and biological control measures used against parasitic diseases and pests are usually attempts to adjust the equilibrium population levels within intimate associations so as to reduce the population of the injurious species.

In all these, and other areas, man needs to learn how to avoid damaging beneficial associations and how to adjust harmful ones to his advantage without catastrophic ecological consequences. This ability must involve an understanding of the minutely regulated interactions that go on within the associations themselves. Those interactions are the subject matter of this book.

It is a pleasure to acknowledge the considerable assistance given by many people during the writing and production of this book. I could not have attempted to illustrate the text appropriately without generous help from many fellow biologists. They gave me permission to reproduce their previously published figures or photographs and provided me with new illustrations. At all times I was genuinely surprised by the ungrudging and positive way in which help of this sort was provided wherever it was requested. Such cooperation provided one of the greatest pleasures associated with the authorship of the book.

Many friends and colleagues have discussed portions of the text with me or critically examined manuscripts at different stages. For this invaluable help I am particularly grateful to Professor D. R. Arthur, Professor A. E. Bell, Professor F. E. G. Cox, Professor J. D. Smyth, Dr. R. M. Anderson, Mr. D. A. P. Bundy, Dr. G. A. Cross, Dr. K. Hudson, Dr. C. A. Mills and Dr. A. Purvis. Equal thanks must go to the biological students of King's College over the past eight years, whose probing questions and enthusiasm for association biology have stimulated much of the book.

Lastly, I must gratefully acknowledge the patient and skilled help from the staff at Edward Arnold and the constant encouragement and wise advice that I have received from the General Editor of the Contemporary Biology series, Professor E. J. W. Barrington, F.R.S.

Figure Acknowledgements

I am grateful to the authors, editors and publishers concerned for permission to reproduce figures that have been previously published elsewhere. These include: 1.1b, 2.20 and 5.22 from *Parasitology*; 2.5 from *Journal of Ultrastructure Research*; 2.6 from *Canadian Journal of Microbiology*; 2.7 from *Proceedings of the Royal Society of London B*; 2.11, 2.13, 4.15a and b from *Protoplasma*; 2.14 from *Journal of Cell Science*; 2.26 from *Experimental Parasitology*; 2.28 from *Ophelia*; 3.4 from *Journal of Protozoology*; 3.7 from *Journal of Cell Biology*; 3.20 from *New Phytologist*; 3.21, 3.22 and 4.13 from *Symposia of the Society for Experimental Biology*, eds D. H. Jennings and D. L. Lee; 4.15c from *Cytobios*; 5.3 and 5.21 from *Zeitschrift für Parasitenkunde*; 6.3a from *Journal of Parasitology*; 6.3b from *Journal of Animal Ecology*.

London
1979

P.J.W.

1

Intimate Associations between Organisms

INTRODUCTION

Every organism exists and evolves within a web of selective pressures that are imposed both by physical aspects of the environment and the other organisms that share its ecosystem. Because the adaptations and specializations of all organisms must be partly shaped by the latter inter- and intra-specific influences, there can be no such entity as an autonomous organism. Even the apparent independence shown by autotrophic plants and top predators must be incomplete. If this is so, what reasons are there for considering predator-prey and herbivore-crop relationships to be different from associations given names like parasitism, symbiosis and commensalism?

It must be admitted that some of the reasons are arbitrary. They are centred on historical accidents of terminology or the effects of different specialist trainings for alternative scientific disciplines. A scientist who calls himself a parasitologist will regard the existence of organisms called parasites as self-evident and a very personal matter indeed! Other reasons exist, though, which deserve more careful attention.

Firstly, the concept of the 'differentness' of associations like parasitism and symbiosis from relationships like that existing between predators and their prey persists in language and thought. This very persistence demonstrates the concepts' contemporary usefulness, no matter how difficult it may be to define such associations rigorously. Secondly, and independent of all attempts to put definitive bounds on them, the set of associations that will be called parasitism, symbiosis and commensalism in this book (see page 14 for working definitions)

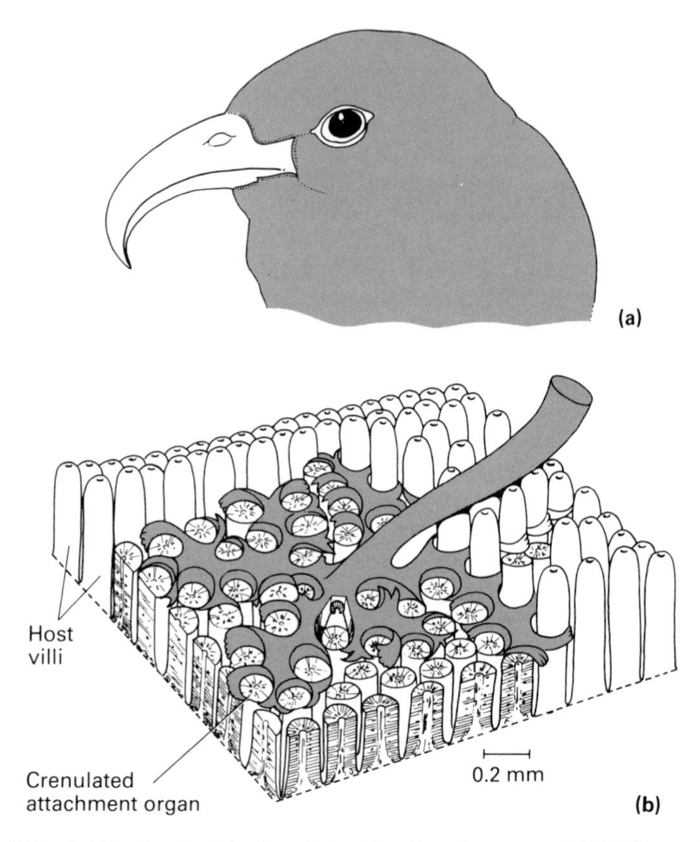

Fig. 1.1 (**a**) The head of the Everglades kite, *Rostrhamus sociabilis*, illustrating the long down-curved bill used in manipulating the molluscan food. (**b**) The attachment of the tapeworm, *Phyllobothrium piriei*, to the gut villi of its elasmobranch host. (From Williams.[378])

do have a perceivable qualitative difference from other interspecific relationships. The organisms in such associations all show an aspect of niche segregation that is uncommon in other relationships. They demonstrate a level of adaptive commitment to another species of organism which is rare or absent in the predator-prey relationship and others like it. In other words, organisms like parasites have patterns of specialization that are only satisfactorily comprehensible in the context of the partner organism (associate). These specializations can be morphological, physiological or behavioural and tend to be specific in their appropriateness to a narrow range of potential associates.

These types of specialization, of course, do occur in associations other than parasitism, symbiosis and commensalism. In a few instances predators have become committed to a single and narrow prey category. If this occurs, explanation of their patterns of specialization will require detailed reference to the biology of the prey. So, adaptive commitment to another species of organism is as well illustrated by the Everglades kite, *Rostrhamus sociabilis*, as it is by the tapeworm, *Phyllobothrium piriei* (Fig. 1.1). The latter (see Chapter 6, page 215) is an obligate parasite that lives in the intestines of elasmobranch rays. It is exquisitely structurally adapted for attachment to a particular section of the gut in a narrow range of those fish hosts.[378–9] In a parallel fashion, the Everglades kite has developed a long, down-curved beak which is unusual among the hawks. With this it can cut the columella muscle that holds the body of its snail prey, *Pomacea*, into its shell. The kites feed exclusively on this gastropod genus and have evolved novel behaviour patterns which enable them to extract the snails from their shells.[136] Thus the kites have become irrevocably adapted in morphology and behaviour to a single prey organism just as the tapeworm has become committed to a host. Such commitment, rare in predator-prey interactions, is almost ubiquitous in associations like parasitism.

DIFFICULTIES CONNECTED WITH THE DEFINITION OF ASSOCIATIONS

Man's attempts to categorize associations between organisms will never produce a system of non-overlapping groups. To begin with, these relationships can never be regarded as static for particular pairs of associates. The nature of the association can alter with changing external conditions or the developmental phases of the organisms themselves. Mycorrhizae are a good example of the former type of modulation. Many angiosperm plants have fungal mycelia intimately connected to the cortex of their roots and extending out into the soil externally. These root-fungus associations are termed mycorrhizae. In such associations the fungal partner definitely obtains organic nutrients from the plant. In some soil conditions, with optimal mineral salt concentrations, it is difficult to demonstrate any benefit accruing to the plant as a result of the association. In such circumstances, fungal parasitic aspects of the relationship appear to predominate. In poorer nutrient conditions, however, at least one class of mycorrhizae, the so-called vesicular-arbuscular (v-a) mycorrhizae, can be shown to enhance plant growth in comparison with similar plants of the same species which are not infected with the fungus.[348] This effect has been most conclusively demonstrated for phosphate nutrients. If available phosphate in the soil is at an extremely low concentrate, but there is a relatively insoluble 'buffer-store' of phosphate-containing material present, mycorrhizal roots seem to be able to mobilize the latter resource. Enhanced growth is the result. The mobilization of phosphate may be partly due to phosphatases in the hyphae, but principally appears to be due to an interesting physical effect. The hyphae which extend out from the roots act like extra root hairs in their uptake of soil phosphate and its transport back into the root. Phosphate is probably transferred from fungus to plant in the arbuscules which give the mycorrhizae their name. They are branching, tangled masses of fine hyphae within individual cortical cells of the root (Fig. 1.2a). Tinker[348] has shown that in some ways the

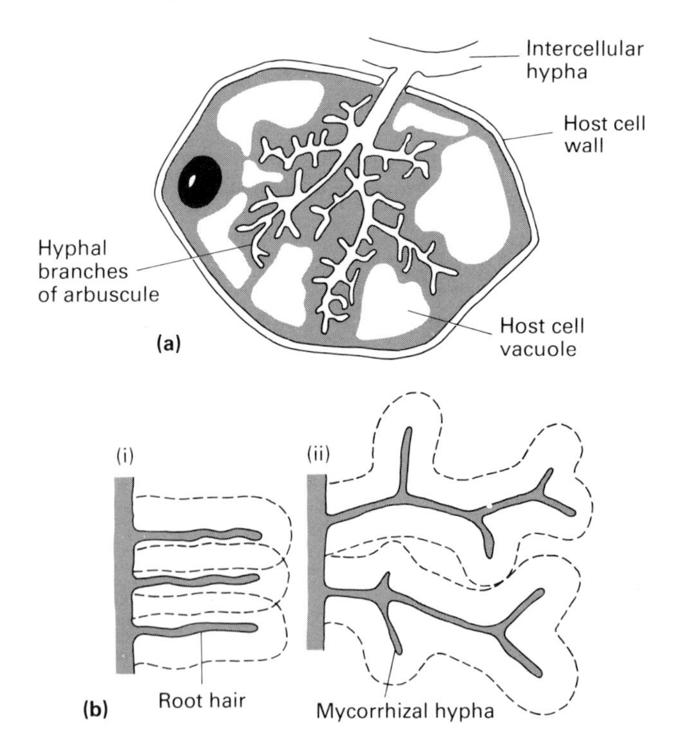

Fig. 1.2 Vesicular-arbuscular mycorrhizae. (**a**) Diagram of the relationship of an arbuscule of a v-a mycorrhiza with a cortical cell of a plant root. (**b**) Diagram to demonstrate the relative volumes and degrees of overlap of the depletion shells (dashed lines) of short root hairs (i) and long, branched mycorrhizal hyphae (ii) in the soil.

mycorrhizal soil hyphae are more efficient in their diffusive uptake of phosphate than the root hairs themselves. The latter, being short and closely packed, are somewhat inefficient at this uptake because their individual phosphate 'catchment areas' (depletion shells) overlap one another to a considerable extent. The external mycorrhizal hyphae, which are more dispersed and can stretch several centimetres out from the root, have much larger, relatively non-overlapping depletion shells. They can therefore make efficient use of the phosphate in larger soil volumes than could a root equipped with root hairs alone (Fig. 1.2b). So, in external conditions of low phosphate availability, the possession of the mycorrhizal fungus is an advantage to the plant and the balance of benefit in the relationship points to a symbiotic type of association. In different soil conditions the association has a different character.

Some helminth life cycles provide excellent examples of relationships altering in character with different developmental phases of the organisms concerned. The nematode hookworm, *Ancylostoma duodenale*, causes serious disease as an intestinal parasite of man. The male and female worms live attached to the villi of the small intestine. They abrade villi with teeth situated within their buccal cavities and ingest blood. In infected human populations the parasite is a proven cause of morbidity and mortality. Eggs of *Ancylostoma* pass out in the faeces of a host and embryonate in the soil. Here L_1 (rhabditiform) larvae hatch from the eggs, grow and moult their cuticles to become L_2 larvae. These in turn themselves moult to become larger L_3 (strongyliform) larvae enclosed within the separated cuticle (sheath) of the L_2 form. L_1 and L_2 larvae feed actively on bacteria in the faecally contaminated soil, whereas the ensheathed L_3 forms are non-feeding and can reinfect man by direct penetration of the skin. Thus in a single nematode life cycle one finds, at different developmental phases, parasitic worms ingesting human blood, free-living worms feeding on soil bacteria and non-feeding infective forms that can make the abrupt transition from a free-living existence to a parasitic mode of life.

So alterations in external conditions and patterns of development can change the objective nature of organism associations. In addition, though, our criteria for separating the associations will always be conditional upon our ability to analyse the interactions between associates that constitute the dynamics of the associations. As our abilities to monitor these interactions alter, so will our conceptions of the relationships. To this extent the categories can never be mutually exclusive pigeon-holes in which particular relationships must be placed. A pigeon-hole image of the relationship between the range of actual associations and our categorizations is often conceptually inadequate. In practice, the definitions of the categories evolve with new experimental methods and theoretical insights. If the definitions are this fluid, it will not be surprising if relationships exist that shuttle in time between two or more of our categories. Equally unexceptional will be associations that can simultaneously exhibit some of the characteristics of more than one category.

The 'spectrum' image is often used[330] to obviate some of the difficulties in the pigeon-hole concept. It suggests that association types are distributed along an axis of interactions. This alteration in imagery is helpful in that it incorporates the possibility of intermediate association types. One further change makes the image even more flexible. If we substitute a multi-dimensional space for the interaction axis, each dimension can stand for a particular aspect of inter-associate interaction. This model removes the implication that association types replace one another along a single, linear array. In a multi-dimensional association space, symbiosis, parasitism, commensalism and the like do not have to be placed in one sequence of overlapping types. Association types that are widely and discontinuously separate in one dimension can have mutually overlapping zones of intermediate forms in a different dimension. This holistic view of associations corresponds in many ways to that recently expounded by Starr.[335]

If the definitions of association categories are seen as utterly dependent on our everchanging understanding of the processes occurring in them, a new and useful perspective is reached. It enables one to see that a primary concern of association biology (or symbiology as Read[284] has called it) must be the phenomenological investigation of the processes, rather than any overriding concern about the way in which definitions can be formed around sets of processes.

This book attempts to provide a process-orientated description of the overlapping classes of association called parasitism, symbiosis and commensalism. Each of the following five chapters deals with different categories of interactions within these associations. The interactions concerned, whether they are structural (Chapter 2), nutritional (Chapter 3), physiological (Chapter 4), behavioural (Chapter 5) or ecological and evolutionary (Chapter 6), have all been used in the past as crucial differentiating features between types of association.

PREVIOUS TERMINOLOGY

It will be obvious, from the attitude taken above to the description of associations between organisms, that this book will not attempt any inviolable definitions of associations. It will be necessary, though, to discuss the unfortunate confusion of terms which already exists in this area to perplex both student and research worker alike. The coexisting sets of contradictory definitions have been described by Starr[335] as 'confusing, parochial and highly imprecise' and by Henry[138] as 'pure chaos' and a 'quagmire'. Although many reviewers[21,62,284,330,335] have bravely navigated through this particular quagmire before, it must unfortunately be crossed once more in this chapter. Previous definitions will be examined and a set of working definitions erected.

Symbiosis

This word can mean two utterly different things in the context of organism associations. As originally defined by de Bary about 100 years ago it meant the 'living together of dissimilarly named organisms'. Thus it was an all-embracing term including the vast majority of interspecific associations between organisms. De Bary did not in any way constrain the breadth of his concept by reference to the way in which the organisms lived together or to the effects, harmful, beneficial or otherwise, that the associates had on one another. Unhappily this beautifully general term has been rendered almost unusable without further qualification because it came to be used for a more specific area of associations (for a discussion of this transition in meaning, see Hertig *et al*[139]). Many workers, with Caullery[62] being an influential example, have stabilized this second usage, namely that symbiosis is a closely integrated, spatial association between organisms of different species which is mutually beneficial.

As if this semantic anarchy were not sufficient, there exists another controversy within the first. Should the associates in a symbiosis (however defined!) be called symbionts or symbiotes? Different authorities use each of the two terms with the same apparent sense of justification which has resulted in the deadlocked coexistence of both. Etymological experts suggest that symbiote is the purer Greek, but this knowledge has not shifted the verbal log-jam.

Commensalism

Starr[335] has recently made an inventory of the variety of meanings which have been attached to the form of associate called a commensal. The original meaning, coined in 1876 by van Beneden, invoked a relationship in which one organism benefited nutritionally from another, without at the same time harming its benefactor. This concept has been the progenitor of many other definitions, most of which are not really compatible with the original. In these new definitions, benefits other than nutritional ones have been encompassed and commensals are allowed to harm their hosts on occasion. Many newer expositions of the commensal idea have discarded the central theme of a trophic link completely and concentrate on the continued physical proximity of the two associates.

Parasitism

If one is prepared to use extremely inexplicit terms it is easy to provide a widely acceptable classical definition of parasitism. The following definition is a typical example, 'Parasitism refers to associations in which there is overt exploitation of one associate by the other, leading ultimately to severe injury or death'.[138] Simply stated, parasites benefit at the expense of their hosts. Unfortunately these types of definition beg very many questions. How does one measure exploitation? What is the difference between overt exploitation and other types? Are concepts like injury, harm and benefit quantifiable? Should injury be assessed at the level of the individual or of a population? Questions like these have encouraged attempts to pin down parasitism in more specific terms. Two very fruitful attempts have centred round, on the one hand, new knowledge concerning parasite metabolism[330], and on the other, the application of population dynamic ideas.[8, 81–2]

Smyth[330] has emphasized the importance of the concept of metabolic dependence levels in organism associations, and moved attention away from the central idea of 'harm' in host-parasite relationships. A parasite is seen, in this new light, as an organism showing varying degrees of metabolic dependence on its host. This metabolic dependence can occur in the areas of nutritional materials, developmental stimuli or the control of maturation.

Crofton[81–2] sought to express the fundamentals of the host-parasite relationship in population terms. He postulated that in such associations the following three population relationships will exist:

(i) The parasite individuals are distributed in an aggregated, over-dispersed fashion (see Chapter 6) among the individuals of the host population.

(ii) For each host-parasite relationship there must exist a parasite population density in an individual host (the lethal level) which kills that host.

(iii) A parasite species always has a higher reproductive potential than that of the host species which it parasitizes.

Thus Crofton specified that 'harm' in his definition of parasitism can be equated with the potential for host death that follows as a consequence of a sufficiently high parasite population density.

Mutualism

Those modern workers who utilize the word symbiosis in the broad de Bary sense often find that they still require a sub-category to describe associations which involve mutual benefit. Such associations are termed mutualism, and the members of the associations, mutuals.[284] Thus mutualism is in almost every way equivalent to symbiosis in its restricted sense.

Phoresy[138] (Phoresis[330])

This type of relationship, as usual, is the subject of a cluster of somewhat differing definitions. All of them accept, however, that phoretic relationships are at the more temporary, least integrated end of any spectrum of interspecific relationships. One organism transports, shelters or supports another. There is usually no implication of metabolic dependence in the relationship.

Most of the recorded examples of phoretic relationships concern interactions involving insects or acarines[137]. The borborid fly, *Limosina sacra*, is often transported on the backs of dung beetles, and trichopterans can carry out their larval development in gelatinous capsules attached to the bodies of chironomids. Parasitic insects can use phoretic partners to transport their eggs to a host. Female botflies (*Dermatobia hominis*) attach their eggs to the abdomen and legs of mosquitoes and other dipterans which bite man. When the insects are taking their blood meal eggs become detached, releasing botfly larvae which bore into human skin to carry on their development. In each of these cases the phoretic relationship simply involves attachment and transportation. No metabolic dependence is implicated.

Inquilinism[138]

Inquilinism is a particular grouping of commensal and/or symbiotic (narrow sense) relationships. They share a common feature, namely that one inquiline shares the nest, hole or home of the other.

Many of the known examples involve marine animals that share a burrow or hole, or terrestrial insects that share a nest. Shrimps of the genus *Alpheus* enter into symbiotic inquiline relationships with different species of gobiid fish in many tropical seas.[167] On flat sandy bottoms the shrimps, mainly living in pairs, dig burrows. These holes are used by a fish as a resting place at night and a temporary shelter in the daylight hours. The shrimps use the fish as a 'sentry' and a complex communication system exists between the animals. Shrimps only leave the burrow when the fish is at its mouth and they keep antennal contact with the fish's tail region whilst feeding. At the approach of a predator, either tail-flicking by the fish or its movement towards the tunnel mouth cause the shrimp to return (see[121] for a discussion of such relationships).

The nests of many social insects contain inquilines.[137] Those of ants' nests are termed myrmecophilous animals.[18] Army ants in Central America, for instance, share their nests with at least three species of myrmecophilous thysanuran bristle tails. These primitive insects usually remain inside the nests and feed there on debris, scraps of food and corpses of ants. Many insect inquilines that live with social insects respond in similar ways to their nest-building associates to pheromone signals used to integrate the insect society. Pheromone trail-following behaviour has been demonstrated in the inquiline cockroach, *Attaphila fungicola*, which lives in the nests of ants of the genus *Atta*.[245]

Close integration of the biology of host and inquiline of this sort seems to have led in evolution to ever more complex relationships that can no longer be regarded as simple inquilinism. In these, the associations come to have distinct nutritional and population dynamic consequences for both host society and inquilines. Two myrmecophilous examples, *Anergates* and *Maculinea*, will illustrate this complexity. The ant, *Anergates atratulus*, lives as an inquiline in the large nests of another ant, *Tetramorium caespitum*.[104] Inquiline males are wingless and pupal in appearance and females when virgin are winged but after copulation become apterous and bloated with developing eggs. Neither males nor females can feed themselves and depend entirely on food given to them by workers of *T. caespitum*. The presence of *Anergates*, however, signifies the collapse of that particular *Tetramorium* nest because such colonies always come to consist of sterile *T. caespitum* workers only. It is believed that the invading females kill the resident queens and the population of *Tetramorium* workers then devotes itself to the rearing of the offspring of the social parasites.

Equally complex is the relationship between the 'honey' caterpillars in the lepidopteran family, Lycaenidae, and their ant partners. A considerable amount of work has been carried out recently on the population dynamics of one British member of the family, namely the large blue, *Maculinea* (*Lycaena*) *arion*. This work has been summarized by Dempster.[99] This butterfly has always been a very local species in Britain but in the last two decades it has declined disastrously in numbers until there is now only one small colony surviving in S.W. England. The adult butterflies emerge in June and lay eggs on the flower buds of thyme (*Thymus drucei*). The young caterpillars at first eat the developing seeds in the flower heads. In August they leave the plants and begin to secrete a sugary substance from a dorsal pair of glands on the eleventh segment. This secretion is attractive to ants in the genus *Myrmica* which take the caterpillars back

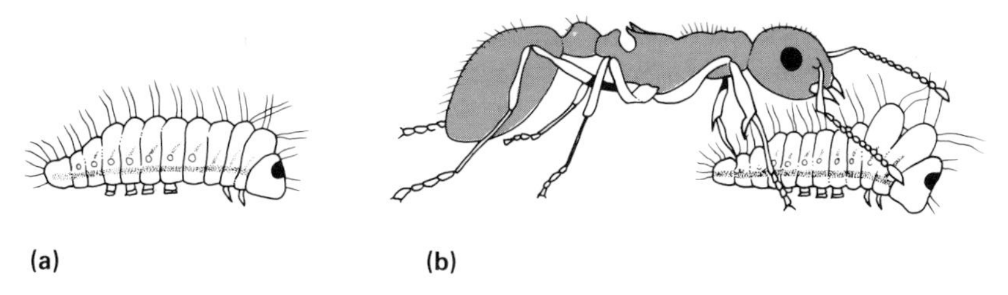

Fig. 1.3 The myrmecophilous caterpillar, *Maculinea* (*Lycaena*) *arion.* (**a**) In normal configuration. (**b**) In 'hunched' configuration, producing a dorsal projection by which it will be carried back to the underground nest of the worker ant (*Myrmica* sp.). (Redrawn from Donisthorpe.[104])

to their underground nests (Fig. 1.3). Here the caterpillars spend the rest of their larval life feeding on ant grubs. When fully grown, the larvae pupate near the surface of the nest, the adult butterflies crawling to the surface of the nest to escape from it in June. The reasons for the decline of this intriguing species are complex and include reduction in the number of suitable butterfly habitats, cannibalism among caterpillars on flower beds when flower production is low and changes in the reproductive status of the ant colonies utilizing and being utilized by the butterflies.

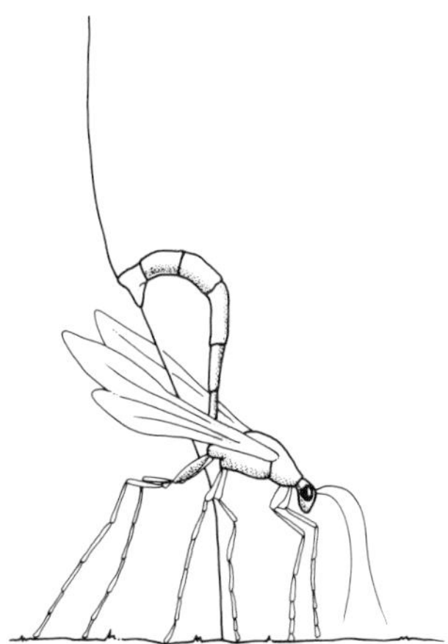

Fig. 1.4 An ichneumenoid hymenopteran parasitoid showing the method of oviposition used by species like *Pseudorhyssa alpestris,* which parasitizes the larvae of the wood wasp *Xiphydria camelus.* In such forms the long ovipositor is braced between the coxae of the hind legs and the ovipositor sheath is held vertically after the initial stages of drilling through wood to locate the larva. (Redrawn from Askew.[18])

Parasitoid relationships[18,361]

Within some taxa of insects there exist examples of a particular form of predatory behaviour which has parasitic overtones. Such insects, and the other animals that show this mode of existence, are called parasitoids. Parasitoid insects introduce their larvae, in a variety of ways (Fig. 1.4), into host insects or other invertebrates. Typically the parasitoid larvae develop and grow in the host individual, using the

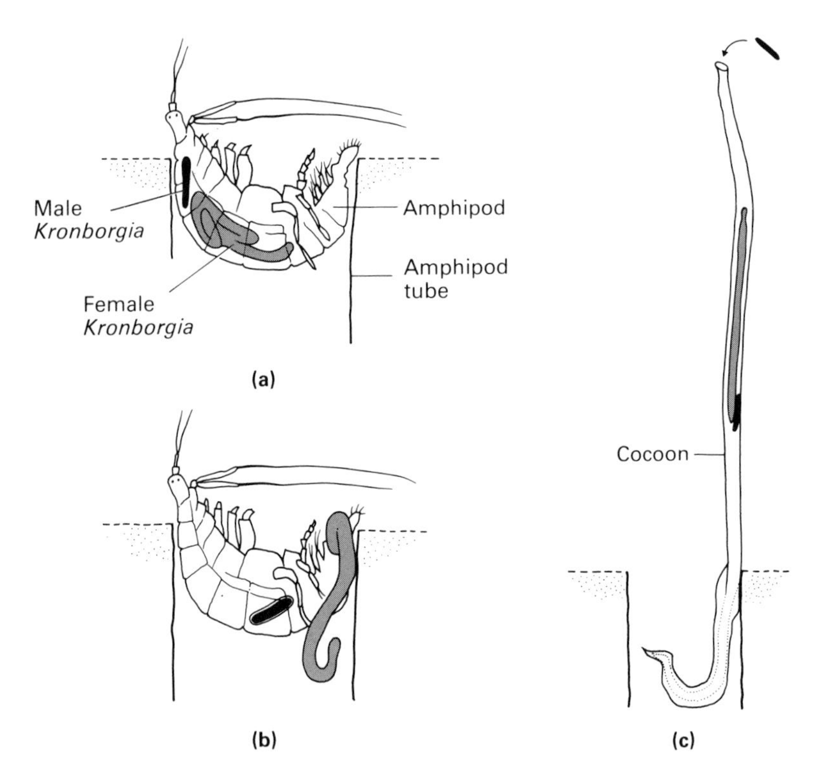

Fig. 1.5 A rhabdocoele platyhelminth parasitoid, *Kronborgia amphipodicola*. (**a**) Gutless, ciliated males and females develop in the haemocoele of the tube-dwelling, marine ampeliscid amphipod, *Ampelisca macrocephala*. (**b**) The worms release themselves from the crustacean, killing it in the process. (**c**). The female worm then secretes a long tubular cocoon around herself which is attached to the tube wall of the deceased host. Free swimming males enter the open distal end of the cocoon and fertilize the female's eggs. She then spawns and dies leaving the cocoon filled with developing eggs each of which eventually contains two ciliated larvae. These hatch and swim to new hosts, penetrating their cuticles directly and completing the life cycle. (Redrawn from Christensen and Kanneworf[72])

biomass of the host in a slow, controlled manner as food to sustain its maturation. In most examples this utilization ultimately kills the host. This fatal end-result means that in reality the association is a temporally elongate predator-prey relationship. The long period of association and the controlled feeding, however, show some similarities to host-parasite relationships. Although most research work on the host-parasitoid relationship has concentrated on insect examples, others do exist. The rhabdocoele platyhelminth *Kronborgia*, for instance, has been shown to be a parasitoid of marine amphipod crustaceans (Fig. 1.5).

INTEGRATED APPROACHES IN DEFINING ORGANISM ASSOCIATIONS

Several modern workers have tried to achieve unitary, integrated approaches to the definition of organism associations. Starr's attempt[335] has been the most ambitious. He has set up a series of 'criterional continua' and hopes that by specifying the position of a relationship on each of these continua, a characteristic pattern of continuum values will identify the uniqueness of that relationship. At present there are nine such criterional continua, and there is an implication that as new, important aspects of relationships are recognized, they would be added as new continua to the scheme. The approach has the advantages of open-endedness. It allows for intermediate categories of associations and does not emphasize one aspect of the relationship above others. In this respect it resembles a non-weighted numerical taxonomic strategy. Unfortunately, some of the continua suggested are not as yet continua in any quantifiably assessable sense. This stricture applies to those continua based on 'criteria pertaining to harmful or beneficial effects on each associate' and on 'integrational criteria'. The attempt to classify associations in this way provides classifications which do not telescope down to easily handled labels. Thus, *Bdellovibrio*, which is a vibrio bacterium that has a parasitoid-like relationship with other bacteria, would have a partial description in Starr's scheme as follows:[335]

> 'a symbiosis-competent *Bdellovibrio* behaves like an obligately to facultatively dependent, persistent, fairly specific, biotrophic becoming necrotrophic, antagonistic (possibly physically harmful, toxigenic and pathogenic), generally (intramurally) inhabiting and possibly exhabiting symbiont of other bacteria. The bdellovibrios are benefited nutritionally and possibly regulatively'.

One important aspect of a classificatory scheme must be communicatory convenience. So although Starr's approach is of great importance in terms of specifying the dynamic interactions that occur in associations, it cannot supplant the old 'labels' like parasite. These names, no matter how 'fuzzy' their definitions may be, are essential for concise scientific communication.

Smyth[330] has made the practically convenient step of linking parasitism and symbiosis (narrow sense) in a category of 'intimate associations'. Such associations share the characteristic of the metabolism of one species being dependent to some degree on a persistent association with another species. On this view symbiosis (narrow sense) is recognized merely as a special sub-category of parasitism in which some metabolic products of the parasite are of value to the host.

WORKING DEFINITIONS

As mentioned above, the approach of this book does not admit the possibility of strictly defined association categories. We shall need, however, to have some nomenclatural ground rules concerning such categories. These working definitions will be used, purely for convenience, whilst the interactions occurring in associations are discussed.

The phrase '***intimate association***' will be used broadly in Smyth's sense,[330] to include parasitism and symbiosis (narrow sense). The latter relationship will, on all subsequent occasions, be referred to simply as symbiosis. In all intimate associations we will expect to see varying degrees of adaptive commitment by one associate in respect of the other. Such commitment will have large elements of species-specificity. The associates will tend to demonstrate spatial proximity (usually contact) of some persistence. The associations will also exhibit varying levels of metabolic dependence of one associate on the other.

Parasitism will include all those intimate associations that meet Crofton's ecological population criteria (see page 8). It must also encompass intimate associations that show patterns of detriment to the host other than 'all-or-nothing' host death induced by a lethal level parsite density. In these instances, at a population level, one would expect to find parasites depressing the host's reproductive rate or increasing its death rate in a gradual density dependent manner[8]. The terms ***ectoparasite*** and ***endoparasite*** will be used to describe the topological relationship of the parasite to its host.

Symbiosis will encompass those associations in which the metabolic dependence of associates is to any degree mutual. The associates will be termed ***symbiotes***. If there is a large discrepancy in the sizes of the symbiotic associates they will be called ***macro-*** and ***microsymbiotes***.

Commensalism will be used, in a very particular sense, to describe all those relationships which appear to be intimate associations as described above, but which on presently available evidence cannot be incorporated into the parasitic or symbiotic categories. Many of these associations show minimal metabolic dependence. The term 'host' seems inappropriate for the larger, more passive, member of commensal relationships and the phrase ***substrate organism*** will be used here. The presence of commensals will not be expected to influence either the death or reproductive rates of the substrate organisms.

The proviso concerning presently available evidence built into the working definition of commensalism has a central relevance in all attempts to label associations. In so many cases, close physical coexistence and mechanisms for maintaining this closeness show us that a relationship is more than a casual association, but we know next to nothing about the nature of the interactions occurring between the associates.

The oligochaete annelid, *Chaetogaster*, provides a thought-provoking example of these labelling problems. It is difficult to make a large collection of limnaeid freshwater snails in Britain without finding that several of them have annelids of this genus in their mantle cavity. The worms are rarely found off their hosts. They possess fans of chaetae with recurved, hooked tips for grasping the mucus-covered lining of the mantle cavity. They seem therefore to be in an intimate association with the snails. In most works they have been considered to be commensals[333] as they appear to feed on algae which enter the mantle cavity. One recently acquired piece of information, however, confuses this simple categorization. *Chaetogaster* has been shown to eat miracidial[174] and cercarial[300] larvae of digenean platyhelminth parasites that use the snails as intermediate hosts. Miracidia moving into the mantle cavity to infect the snail and cercariae passing out of the snail *en route* for the next host in the life cycle can both be killed by the snail's annelid associate. Hence, there would seem to be elements of mutual benefit in the association. Snail cilia bring algal food to the annelids and the mantle cavity provides a refuge from predators. On the other hand, the annelids are a potentially depressive influence on the reproductive success of digenean parasites of the snail. Should we therefore regard the two animals as symbiotes? If so, is the relationship still symbiotic when the snail population concerned is not infected with digeneans? An example of this sort, which is far from unusual in its complexity, serves to show how one new piece of experimental evidence can change our understanding of an association. It also demonstrates the impossibility of maintaining static association definitions.

2

Structural Aspects of the Association Interface

INTRODUCTION

The associations between organisms that are the subject matter of this book are, in the main, close physical ones. In both symbioses and host-parasite interactions the organisms concerned usually touch each other. The contacts are often of some temporal permanence and exceptionally intimate.

The morphology of the interface between associating organisms is the structural framework within which nutritional, physiological and regulatory interactions occur. These intrinsically dynamic areas of association biology are analysed in the next two chapters. The present chapter will provide the morphological milieu for its successors.

Of course, this conceptual dichotomy of structure and function is an abstraction, because at the level of molecular structure the morphology of the interface and its activities become different aspects of the same entity. In practice the greater part of this section will concern ultrastructural interpretations of association contacts. This is because it has been findings from transmission and scanning electron-microscopical studies that have done most in recent years to engender what must be called a new understanding of the physical basis of associations. The range of structures which these techniques have revealed are obviously diverse, as the range of organisms scrutinized extends from viruses to vertebrates. Despite this multiplicity, unifying general concepts recur. For instance, ultrastructural specializations for attachment and the inter-associate transport of materials are common at interfaces. In addition, some general physical principles are equally true for interfaces in wildly disparate types of association.

Two of these generalizations may be stated as postulates that will be constantly confirmed concretely in this chapter. Firstly, surfaces which must sustain high levels of transmembranous molecular transport will often be greatly amplified in area. Secondly, adjacent cell membranes which interact adhesively with one another will either do so by direct molecular bridging or at a distance from one another of from 6–30 nm, the 'secondary minimum' gap predicted by the Derjaguin-Landau-Verwey-Overbeck (DLVO) theory of colloid adhesion (Fig. 2.1).[92]

There have been two recent reviews which have considered the general principles of association interfaces, in addition to the cell

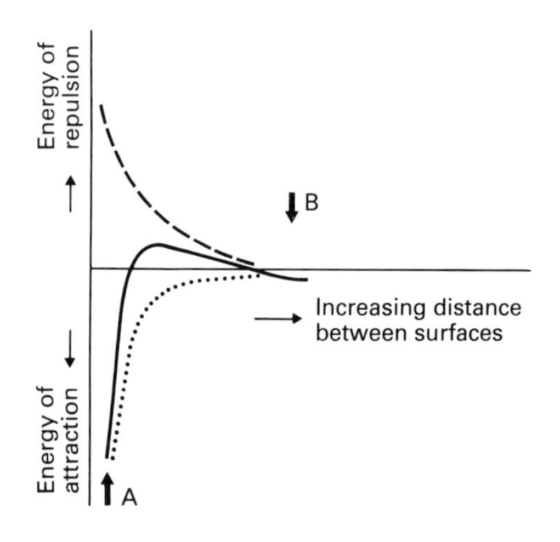

Fig. 2.1 Potential energy diagram for the interaction of two surfaces (in this instance two membranes) according to the DLVO theory (see text). Dashed line indicates electrostatic energy of repulsion; dotted line Van der Waals energy of attraction, and the solid line the resultant of the two. Distances A and B indicate the primary and secondary minima of repulsive energy respectively. The secondary minimum occurring with a gap between the membranes in normal physiological conditions of between 6 and 30 nm. (Redrawn from Curtis.[92])

biological analysis of Curtis.[92] Nachtigall[253] has investigated the biomechanical options for mutual attachment open to associating organisms. Smyth[329] has studied, more specifically, the interface phenomena which occur between parasitic protozoa and platyhelminths and their hosts. He has introduced a convenient categorization of intimate interface types. They may be membrane-to-membrane, cytoplasm-to-membrane or cytoplasm-to-cytoplasm in

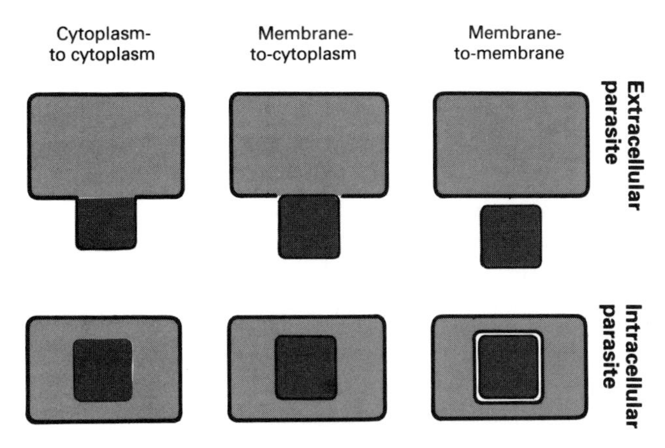

Fig. 2.2 Diagram of the categories of associate contact described by Smyth[329]. The large cell in each case is the host cell, the smaller cell the parasitic cell.

character (Fig. 2.2). While this chapter considers the fine structure of specific association interfaces, these subdivisions will be utilized when relevant.

There are a number of associating organisms which, because they are in such favour with experimental biologists, will make numerous appearances in this book. While their interface phenomena are being described in this chapter, the opportunity will be taken of describing their basic biological characteristics.

BACTERIOPHAGE ASSOCIATIONS[39,105]

Viruses are the smallest organisms, whose structure and biology are at all well understood, that are implicated in intimate parasitic relationships. Any precise categorization of the relationships between the non-cellular viruses and their host cells is very difficult because the relationships are different in kind from those that exist between cellular organisms.

A virus is an organized assemblage of proteins and an RNA or DNA genome which, within the correct host cell, can reproduce to form many new virions (the general term for infective virus particles). The metabolic and synthetic biochemical machinery of the host cell is utilized to effect this reproduction and the host cell is lysed and destroyed by the eventual release of the new generation of virions. The organization of the virion is concerned with the protection of the viral genome and its passage into the cytoplasm of the appropriate host cell.

The nature of the association interface in viral relationships is best characterized in the case of the bacteriophages which use prokaryotic bacterial cells as hosts.[39,105] Bacteriophage (phage) virions of several different morphologies are known. The tail-less or minute phages such as ΦX174, f2 and Qβ are parasites of *Escherichia coli* and other Gram-negative bacteria. They are the smallest of the discovered viruses and are small spheres only 22–25 nm in diameter with single-stranded nucleic acid genomes. In the tail-less virions we perhaps approach the physical limit of specializations for host-parasite attachment. Within the group we know most about phage ΦX174, thanks to the epic investigations of Sanger and his collaborators[305] who recently elucidated the complete DNA genome sequence for this virus in its 5375 nucleotide entirety! This coding information complements earlier results concerning the nature and functions of the nine protein types for which the virus carries a nucleic acid blueprint. Proteins F (M.W. 48 000), G (M.W. 19 000) and H (M.W. 37 000) are all components of the protein capsid or coat which surrounds the viral genome. Protein F is the main capsid subunit and proteins G and H form the spikes which protrude from the coat (Fig. 2.3). Of this latter

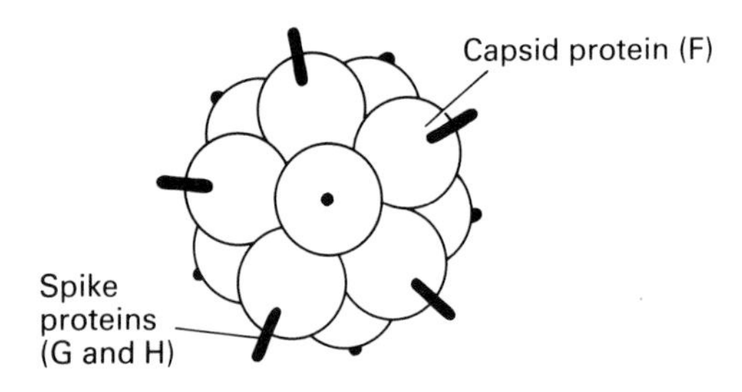

Fig. 2.3 The virion of the bacteriophage ΦX174.

pair, H seems to be the most likely candidate to be a recognition and attachment protein which binds to the cell membrane of *E. coli* and initiates the phage-bacterium interaction. Its amino-acid sequence at the N-terminal end is very rich in hydrophobic residues and if these were located near the tips of the spikes they might be well placed to interact with the hydrophobic regions of the bacterial membrane.

Without doubt, the best understood phage is the tailed coliphage (bacteriophage of coliform bacteria) T2 which infects *E. coli* cells (Fig. 2.4a). It is one of the most complex virions, possessing specialized multimolecular components for firstly attachment to the bacterial cell, then secondly the injection of its double-stranded DNA into the bacterial cytoplasm (Fig. 2.4b).

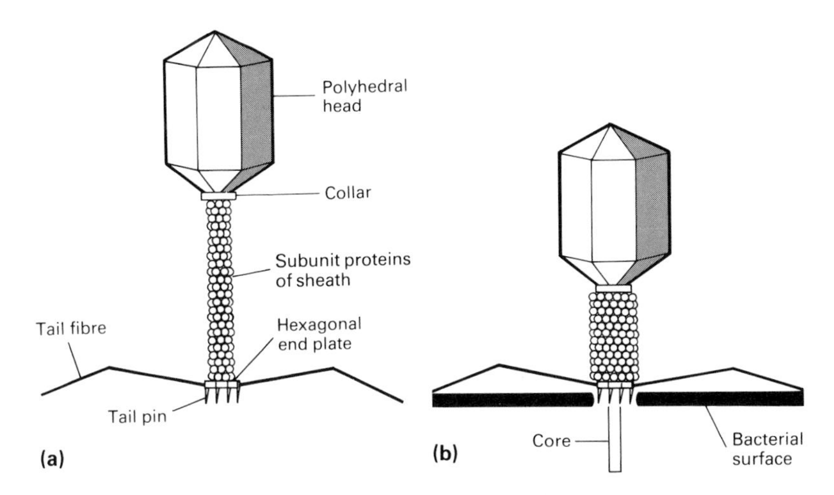

Fig. 2.4 The virion of the tailed coliphage T2. (**a**) Free, unattached configuration. (**b**) Configuration when attached to a bacterial surface. The sheath has contracted, forcing the core through the bacterial surface.

Each virion consists of a polyhedral head about 100 nm long which contains the DNA genome. The latter is in the form of a single double helix more than 50 μm long, stored in a manner which enables it to leave the head in a linear fashion. The head is connected to a tail, 100 nm in length, via a collar piece. The tail has a complex morphology of its own. It consists of two concentric tubes, the outer sheath and an inner core, with a fine lumen down the centre of the latter. At the distal end of the tail is a hexagonal end plate, each vertex of which bears a long jointed tail fibre which shows every appearance of having a flexible connection to the plate. Also present at each vertex is a tail pin, a distally pointing short projection.

Attachment of phage to bacterium is termed absorption. It is an extremely host-specific process and appears to depend initially on the interaction of distal parts of the tail with particular receptor molecules in the bacterial cell membrane. In one case, these receptors have been isolated and shown to be high molecular weight lipoproteins.

The T2 virions are moved constantly in an aqueous environment by Brownian motion. When a tail fibre touches the *E. coli* surface it adheres and soon all the other five fibres adhere also. The tail plate then engages with the surface, probably by the involvement of the tail pins. A remarkable active process then occurs. The tail sheath, which consists of 144 helically arranged protein molecules, abruptly changes its configuration by altering from a helix of 12 turns of 12 molecules to one of half the original length consisting of 6 turns of 24 molecular subunits. This energy-requiring activity is 'driven' by ATP molecules which are stored in the tail. The sheath protein molecules and the ATP appear to exist in a 1 : 1 ratio, each of the 144 subunits requiring a single ATP molecule to accomplish its own particular conformational change. The result of the contraction of the sheath (see Fig. 2.4b) is to force the core tube through the bacterial cell wall into the cytoplasm. This breach of the bacterial surface is possibly aided by the action of enzymes in the core canal which may weaken the outer layers of the host cell immediately under the tail plate. Once the distal end of the core is within the *E. coli* cell the long viral DNA molecule passes into the bacterium down the core lumen. In the host cytoplasm it catalyses its own reproduction and initiates the synthesis of the various phage proteins which undergo self-assemblage to form many new virions around the multiple DNA copies.

BACTERIAL ASSOCIATIONS

As with many associations, one component of which is a microorganism, the relationships between bacteria (or rickettsiae) and other organisms are difficult to classify. Overt disease with pathogenesis, caused by the bacterial utilization of host tissues and the production of toxins, must be regarded as a manifestation of a host-parasite relationship. At the other extreme of a spectrum of association types are those which include bacteria that have been demonstrated to perform vital metabolic roles for their macrosymbiotes. A huge literature exists on the endosymbiotic bacteria of insects,[176,252] the gut bacteria of vertebrates[153] and the nitrogen-fixing bacteria associated with plants.[186] In all these instances, distinct metabolic advantages are enjoyed by the macrosymbiote as a result of the activities of the microsymbiotes. There exist, however, many associations between bacteria and other organisms where the effects of the association on the putative macrosymbiote or host are not so easy to determine. In such cases, the ultrastructural nature of the interface between bacterium and associate can often provide clues about the nature of the relationship. For instance, it is often tempting to lean in the direction of a symbiotic interpretation when the associates show signs of mutual physical integration at their areas of contact. A consideration of the interface morphology in a range of bacterial relationships will illustrate the value of such structural evidence.

Extracellular symbiotic bacteria are found on the outer surfaces of animals and plants, and on the surfaces of internal spaces as well.

Specific strains of bacteria associated with restricted areas of human skin surface produce micro-environments inimical to pathogenic bacterial colonization. These and most of the important gut dwelling bacterial symbiotes are extracellular forms. The often-observed species specificity of such relationships has always suggested that intimate and integrated physical contacts must exist between such bacteria and the specific epithelia with which they are associated. Only recently though have such interfaces been examined rigorously. An excellent example of such an analysis is that of Brooker and Fuller[53] on the lactobacilli that are found attached to the crop epithelium of the chicken. In a normal chicken, the distal surfaces of the stratified squamous epithelial cells of the crop are almost completely overlain with a continuous mat of lactobacilli cells. Other bacterial types are found associating with different gut epithelia in the same host. There is good evidence that several species of the genus *Lactobacillus*, which are found attached to the internal epithelia of vertebrates, are playing symbiotic roles. They produce acid excretory products during their utilization of host carbohydrates that can induce significant local lowerings of pH values. Such acidic conditions act to regulate the composition of other coexistent microflora, specifically excluding many pathogenic forms. *L. acidophilus* (Döderlein's bacillus), for instance, is attached to the human vaginal epithelium during the post-pubertal, pre-menopausal reproductive years, and produces pH values as low as 4 in the vaginal lumen. This pH is rapidly lethal for many bacterial pathogens. The chicken crop *Lactobacillus* is almost certainly performing a similar service for its macrosymbiote, and is of importance in regulating the composition of the intestinal flora. Electron-microscopical studies wth cytochemical techniques that demonstrate acidic carbohydrate-rich macromolecules have shown that layers of such material coat both the *Lactobacillus* cells and the outer surfaces of the epithelial cells.[53] Adhesion between the bacteria and the epithelium appears to be due to intimate association between these two layers (Fig. 2.5). The adhesion is a mutual and specific interaction between the two cell types. Thus, lactobacilli of a strain (strain 59 belonging to biotype B) isolated from chicken crop walls adhere strongly to the crop epithelial cells of germ-free chicks when fed to them. In contrast, a closely related strain, *Lactobacillus acidophilus* NCTC 1723, isolated from rat faeces, does not.

The prokaryotic symbiotes in the bed bug, *Cimex lectularis*, provide a model example for almost all symbioses which involve intracellular bacteria or rickettsiae. As in many insects, the symbiotes of *Cimex* are found within the cells of specialized symbiote-containing organs called mycetomes and their presence seems to be obligatory for normal

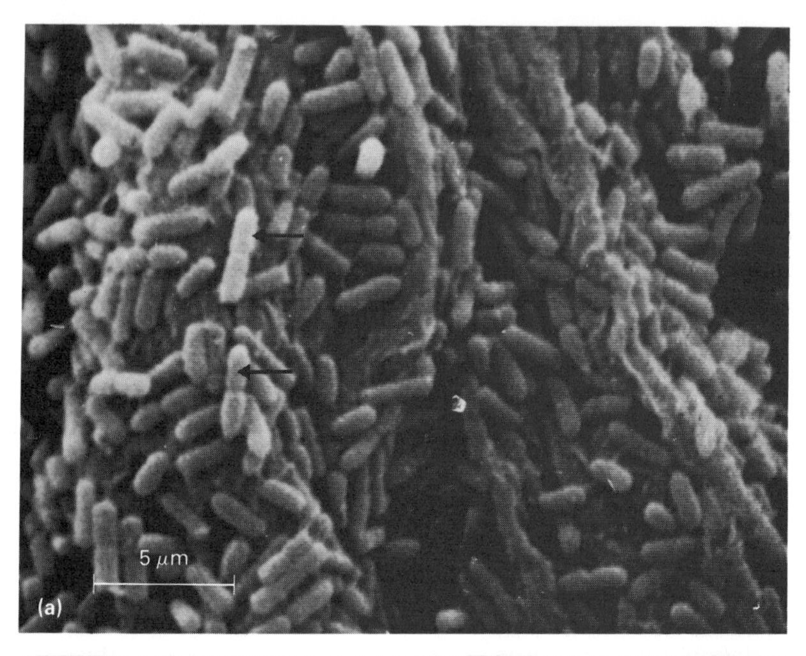
5 μm
(a)

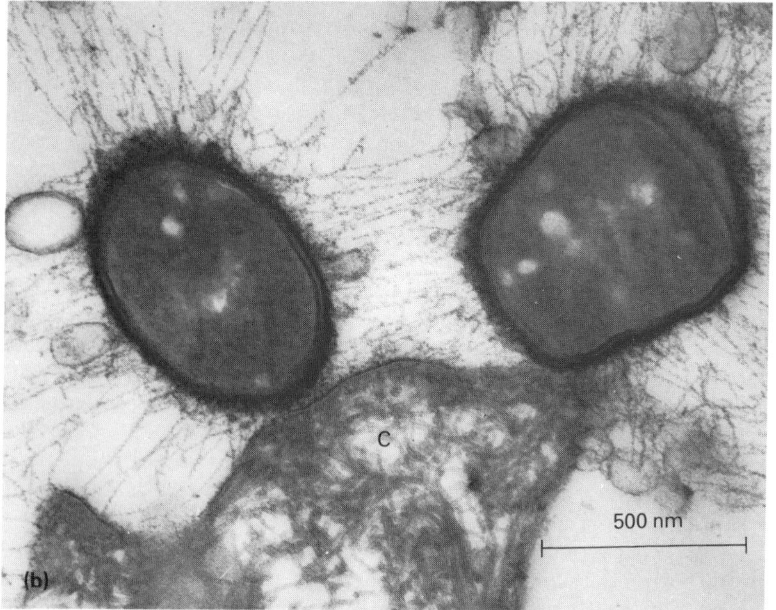
C
500 nm
(b)

reproduction (see Chapter 3, pp. 74–5 for a discussion of the nutritional aspects of such relationships).

Although earlier workers thought that insect mycetomal symbiotes were fungi or bacteria, modern opinion considers that many are probably symbiotic rickettsiae.[63,65] The cells of the mycetomes of *Cimex* form multinucleate syncytia and two types of intracellular symbiote can be distinguished within them (Fig. 2.6). The first type, rod-shaped symbiotes, are about 1 μm long and 0.3–0.4 μm broad. They are found in both cytoplasmic and nuclear situations within the mycetomal syncytia. The complex wall of the symbiote is of Gram-negative ultrastructure and is directly in contact with the cytoplasm or nucleoplasm of the mycetomal syncytium. No *Cimex*-produced plasma membrane is interposed between the membrane of the symbiote and the mycetomal cytoplasm. The second type of rickettsia, the pleomorphic symbiote, has been termed the primary symbiote. It can assume a variety of shapes, being spherical, ovoid or elongate in different parts of a mycetome. The spherical forms are about 0.5–0.8 μm in diameter and all the variants of the primary symbiote are found exclusively in cytoplasmic mycetomal locations. In contrast to the rod-shaped types the primary symbiotes do not have a direct contact with *Cimex* cytoplasm. Either singly or in clusters they are surrounded by a membrane of *Cimex* origin and are thus enclosed in a macrosymbiote-produced vacuole (see Fig. 2.6b).

The two rickettsiae found in *Cimex* elegantly summarize the differences between two of the interface strategies possible for intracellular parasites or symbiotes. In the terminology of Smyth,[329] the rod-shaped type has a membrane-to-cytoplasm interface whereas the pleomorphic primary symbiotes exhibit the more indirect membrane-to-membrane mode of contact.

In the *Lactobacillus* and rickettsial examples described above, the nature of the symbiotic interaction between the associates is, to some extent, understood and the close physical approximation of the organisms can be interpreted in the light of this understanding. In many other bacterial associations, however, the ultrastructural information concerning the association interface stands almost alone as evidence, when little or nothing is known directly about nutritional

Fig. 2.5 The attachment of symbiotic lactobacilli to the chicken crop. (**a**) Scanning electron micrograph demonstrating the layer of bacteria almost completely covering the epithelium of the crop. Arrows indicate dividing bacteria. (**b**) Transmission electron micrograph demonstrating the apparent involvement of strands of acidic carbohydrate-rich macromolecules in the attachment of the bacteria to one another and to the crop epithelium. C, apical portion of a crop epithelial cell. (Both from Brooker and Fuller.[53])

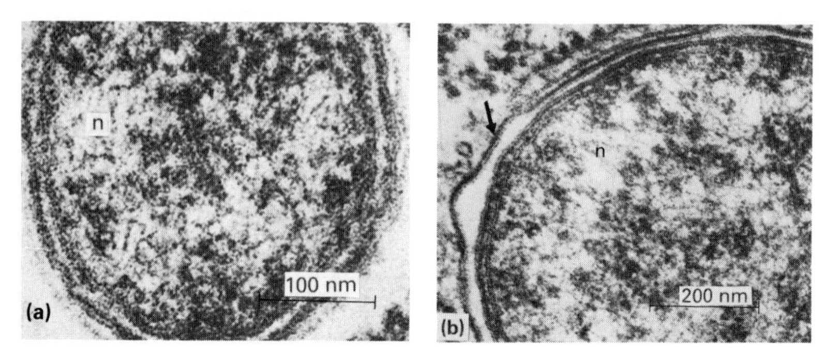

Fig. 2.6 The mycetomal symbiotes of the bed bug, *Cimex lectularis.* (**a**) Transmission electron micrograph of the rod-shaped symbiote showing the absence of an enclosing vacuole membrane of host origin. The membranous layers around the symbiote are parts of its complex cell wall. n, lightly stained nuclear body of symbiote. (**b**) Transmission electron micrograph of the pleomorphic symbiote which is surrounded by a membrane of host origin. n, lightly stained nuclear body of symbiote, arrow indicates host membrane. (Both from Chang and Musgrave.[65])

or physiological interactions between bacterium and eukaryotic associate. Many of the relationships between protozoans and bacteria fall into this category. Perhaps the most extraordinary are those in which spirochaetes or bacilli are attached to the surfaces of flagellates and ciliates. Cleveland and Grimstone[77] showed that symbiotic spirochaetes were adherent to specific attachment brackets on the surface of the flagellate *Mixotricha paradoxa* from the gut of the primitive termite *Mastotermes* (Fig. 2.7). These spirochaetes by their intrinsic motility appeared to act as 'pseudo-flagella', helping to propel the protozoan forward with a steady gliding motion. Each attachment bracket in this association also bears yet another symbiote, an adherent rod-shaped bacterium. Not surprisingly the discoverer of this strange flagellate identified the spirochaetes as cilia and the bacteria as basal bodies! More recent studies[36] on similar spirochaetes on the surface of *Pyrsonympha vertens*, a flagellate from the gut of the termite *Reticulitermes flavipes*, have shown once more the staggering complexity of the physical relationship between protozoan and spirochaete in this type of symbiosis. The flagellate possess attachment brackets which contain striated rootlets whose morphology is remarkably similar to those attached, in more orthodox circumstances, to the basal bodies beneath cilia or flagella. The attachment region of the spirochaete is no less specialized, with an attachment plaque and connecting fibres in the periplasmic space between the spirochaete cell

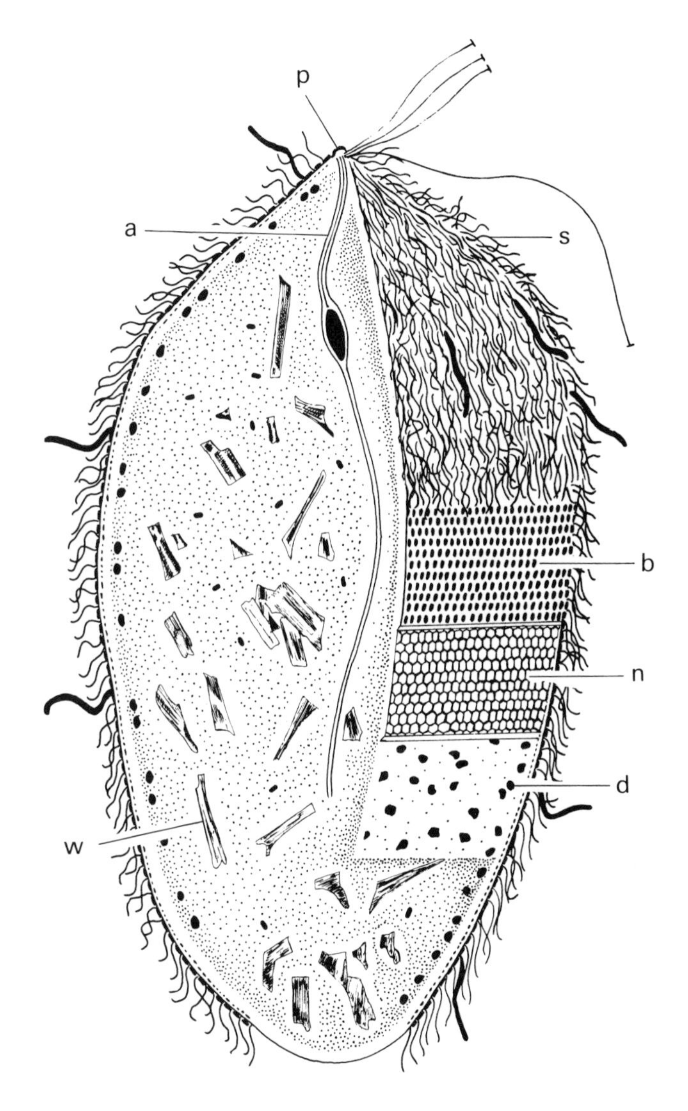

Fig. 2.7 A diagram of the spirochaete-covered flagellate *Mixotricha paradoxa*, a symbiote in the gut of the termite *Mastotermes dawiniensis.* The protozoan is drawn in optical section on the left. On the right, surface structures are shown at progressively deeper levels passing down the diagram. The spirochaete attachment brackets are not shown. a, axostyle; b, bacterium; d, dictyosome (Golgi body); n, fibrous network; p, papilla; s, spirochaete; w, ingested wood particle. (From Cleveland and Grimstone.[77])

wall and cell membrane. In the area of contact between the bracket and the spirochaete the two surfaces are cemented together with a dense transition zone (Fig. 2.8). Thus, not only do the attached symbiotes function in a manner analogous to flagella, but they are also anchored to the protozoan's surface by complex organelles synthesized by *Pyrsonympha* which mimic those that would be present beneath an actual flagellum. Even if there were not good, independent indications that the spirochaetes have a locomotory function for the protozoan partner, this intimacy of ultrastructural integration between the organisms would point most strongly to some form of symbiotic relationship.

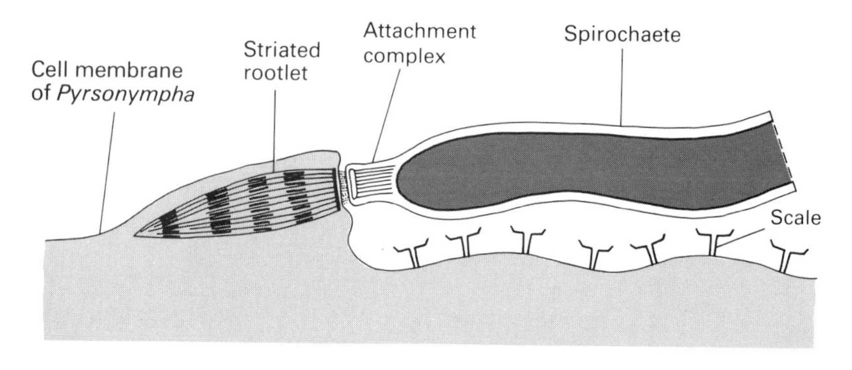

Fig. 2.8 The attachment of a spirochaete to the surface cell membrane of *Pyrsonympha.* (Based on the electron micrographs of Bloodgood, Miller, Fitzharris and McIntosh[36].)

UNICELLULAR ALGAL ASSOCIATIONS[223,320]

Single-celled algae exist in symbiotic associations with a number of protozoan and metazoan animals. The role of the autotrophic microsymbiote has been a matter of some dispute but there is now good evidence that organic photosynthate produced by the algae can be utilized by the animal partner in at least some of these relationships. It certainly occurs in the case of the green alga, *Chlorella*, in association with the hydrozoan cnidarian, *Hydra viridis*,[251] and in the many situations where dinoflagellate symbiotes coexist with invertebrates such as corals,[249] sea anemones and giant clams.

The prompt movement of photosynthate from algae to animal cells is made possible by the close contact between the associates. Most of

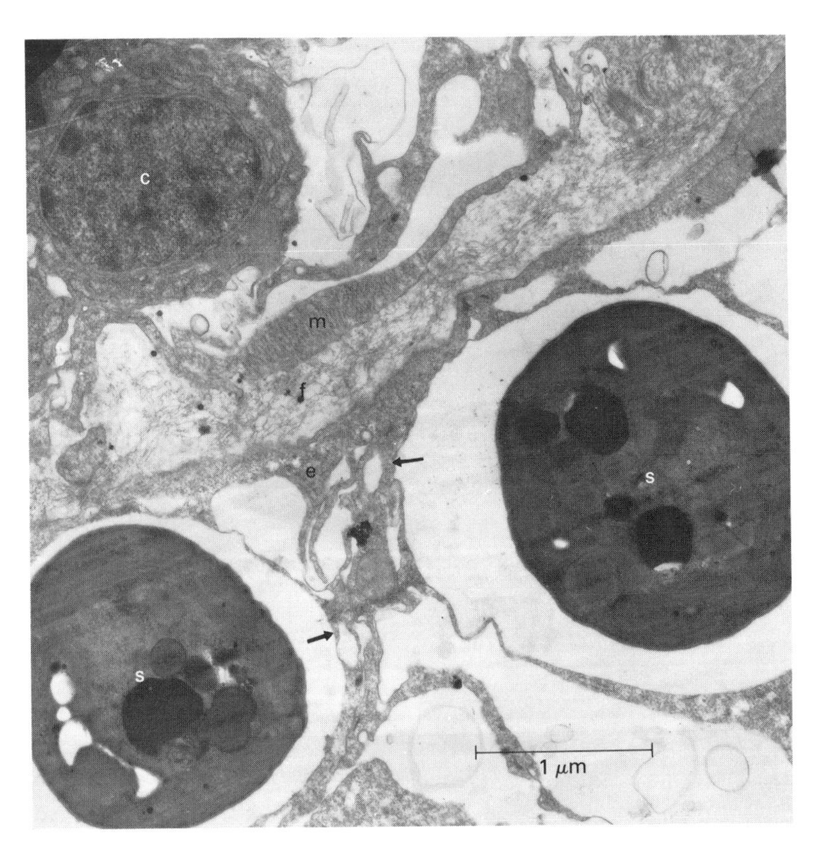

Fig. 2.9 Electron micrograph of a transverse section through the mesogloeal zone of the green hydra, *Hydra viridis*, showing the perialgal vacuoles in an endoderm cell. c, young cnidoblast (nematoblast) in ectoderm; e, cytoplasm of the basal region of an endodermal cell; f, fibrous mesogloea; m, muscle strand in the base of an ectodermal myoepithelial cell; s, *Chlorella* cell; arrows indicate the bounding membranes of perialgal vacuoles.

the algae in such associations exist within intracellular perialgal vacuoles bounded by membranes of animal origin. The ciliate, *Paramecium bursaria*, for instance, normally harbours several hundred symbiotic cells of *Chlorella*.[166] The green symbiotes are situated within individual vacuoles and these algal-vacuole complexes grow and divide as units. In this way the vacuoles never contain more than a single algal cell. The algal association survives through mitosis and conjugation with their consequent nuclear and cytoplasmic rearrangements. Similar perialgal vacuoles (Fig. 2.9) surround *Chlorella* symbiotes in the endodermal cells surrounding the enteron of *Hydra viridis*,[250] and enclose blue-green algal cells, probably of the genus *Synechococcus*, which are symbiotes in the cytoplasm of the testate amoeba *Paulinella*.[175] A non-dividing *Paulinella* possesses two sausage-shaped algae, or cyanelles as they are called, each in its own perialgal vacuole. When the sarcodine divides each daughter *Paulinella* cell receives one of these algal vacuole complexes. Before the next division each algal cell divides just once to restore the initial symbiote quota (Fig. 2.10). The surface membrane of the cyanelle is a

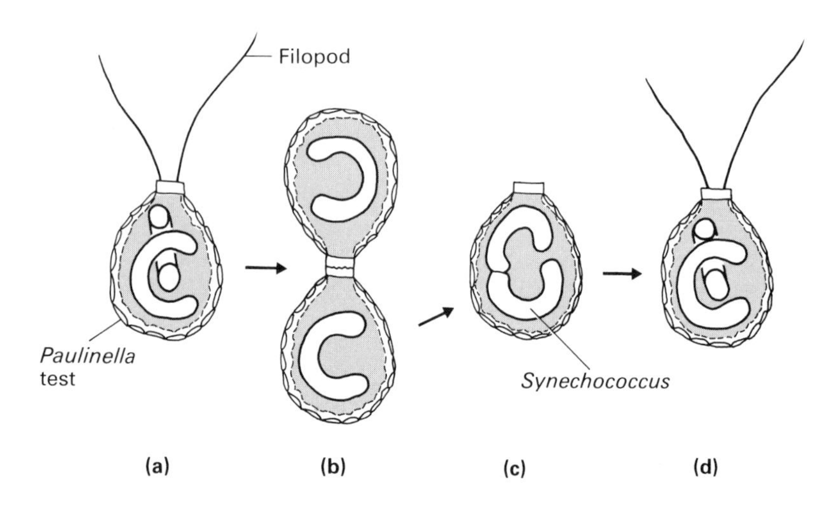

Fig. 2.10 Division cycles of the testate amoeba, *Paulinella*, and its blue-green algal symbiote, *Synechococcus* (see text). Stage (**b**) shows the amoeba dividing, stage (**c**) the alga dividing. (From Kies.[175])

simple plasmalemma, quite different from the membrane plus double-layered cell wall which is typical in the cyanophyceans. The lack of such structures is almost certainly a secondary consequence of an endosymbiotic existence.

In a few instances, algal symbiotes are not sequestered within the cells of the animal macrosymbiote. The so-called 'algal gardens' of tropical clams are organized in this way. Clusters of algal cells, probably dinoflagellates, are found in the mantle edges of these gigantic bivalve molluscs. The mantle edges are protruded out of the shell valves into the sunlight in shallow coastal waters, enabling the symbiotes to photosynthesize. These algae are largely in extra-cellular locations, many lying freely in the blood spaces of the mantle.[320]

Closely analogous to endozoic algal associations is the phenomenon of chloroplasts acting as 'symbiotic' organelles[349] (see Chapter 3, p. 113). Certain saccoglossan marine molluscs feed on green algae with robust chloroplasts (see Fig. 3.21). The herbivorous molluscs can retain the photosynthetic organelles in a functional condition in their gut cells for up to three months. Outer chloroplast membranes survive this indignity and they appear to be in direct contact with the host cell cytoplasm. This most intimate form of 'associate' contact is really equivalent to Smyth's 'cytoplasm-to-cytoplasm' category. The opisthobranch sea slugs appear to be adept at the retention of functional organelles from their food. Another family, the aeolids, have representatives which feed on cnidarians and retain functional cnidocysts from this prickly diet. The organelles are stacked in gut cells in cerata, which are long dorsal projections of the sea slug's body wall. When attacked by a potential predator the aeolid can extrude cnidocysts which explosively evert their poisonous threads.[112]

LICHEN ASSOCIATIONS[157,274,321]

Lichens are highly integrated symbiotic associations between algae and fungi. A discussion of the ways in which photosynthate and organic compounds containing fixed atmospheric nitrogen pass from the algal cells to the fungal hyphae in a lichen thallus can be found in the next chapter (pp. 106–111). Such nutrient translocation is dependent upon intimate physical associations between the algal cells and the fungal hyphae (Fig. 2.11). These associations are concentrated in an algal layer usually near the outer surface of the thallus. The relationships between the mycobionts (fungi) and phycobionts (algae) in different groups of lichens vary considerably. In some forms, mycobiont hyphae are pressed tightly against the algal cells and are called appressoria. In other genera, fungal hyphae penetrate the thin cell walls of the phycobionts and are termed haustoria.[157]

This much was known from early light-microscopical investigations. More recent electron-microscopical studies[274] have revealed much more about the precise nature of the fungal-algal interface. The mycobiont hyphae are usually multinucleate syncytia with a normal complement of eukaryotic organelles including ribosomes, cristate mitochondria and endoplasmic reticulum. The hyphae are covered with fungal cell walls consisting of fine filamentous material probably containing aminoglycans and proteins. These walls can form a thick matrix between hyphae in alga-free regions of the lichen thallus. Near

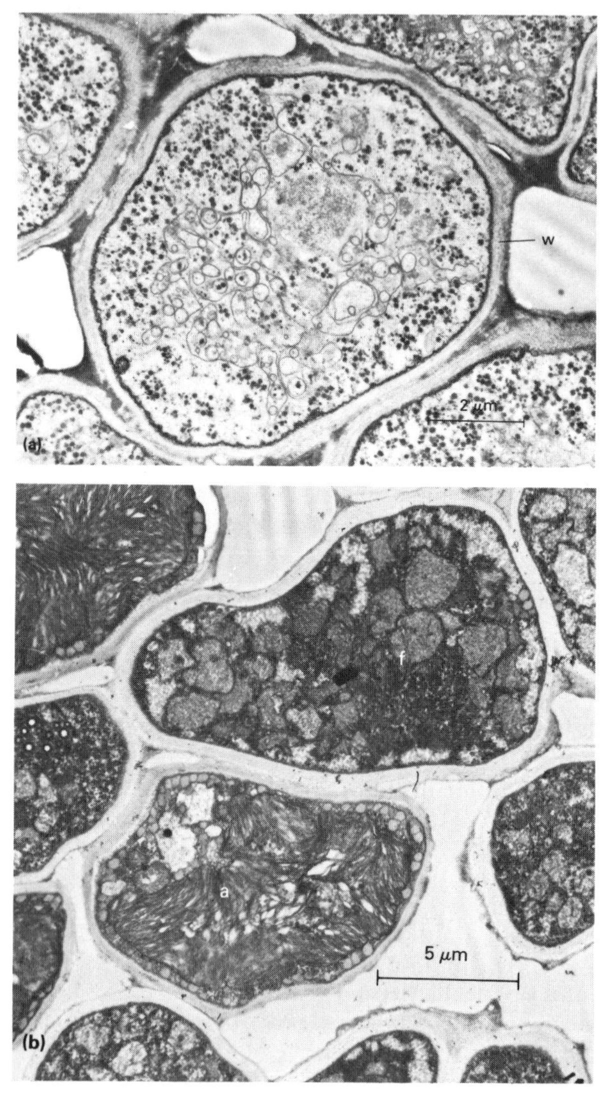

Fig. 2.11 Mycobiont and phycobiont associations in a lichen. Transmission electron micrographs of the lichen, *Dermatocarpon hepaticum*, from the Judean Mountains, Israel. (From Galun, Poran and Ben-Shaul.[123]) (**a**) Section through close-packed fungal hyphae in the lower medulla zone. w, fungal wall. (**b**) Fungal-algal contact in the lower part of the algal layer. Note the wall material round both myco- and phycobionts. f, fungal hypha; a, algal cell (*Myrmecia biatorellae*).

algae, however, the walls become thinner and in many appressoria and haustoria the wall can be only 150 nm thick. Phycobionts are surrounded by cell walls or sheaths consisting of cellulose in the case of green algal associations, and mucilages in the lichens containing blue-green algae. Phycobiont walls are often stratified and can extend from 100 to 100 nm from the phycobiont plasma membrane.

Light microscopists differentiated haustorial invaginations into intracellular types which they imagined penetrated the algal cytoplasm and intramembranous (intrathecal) types which appeared to spare the

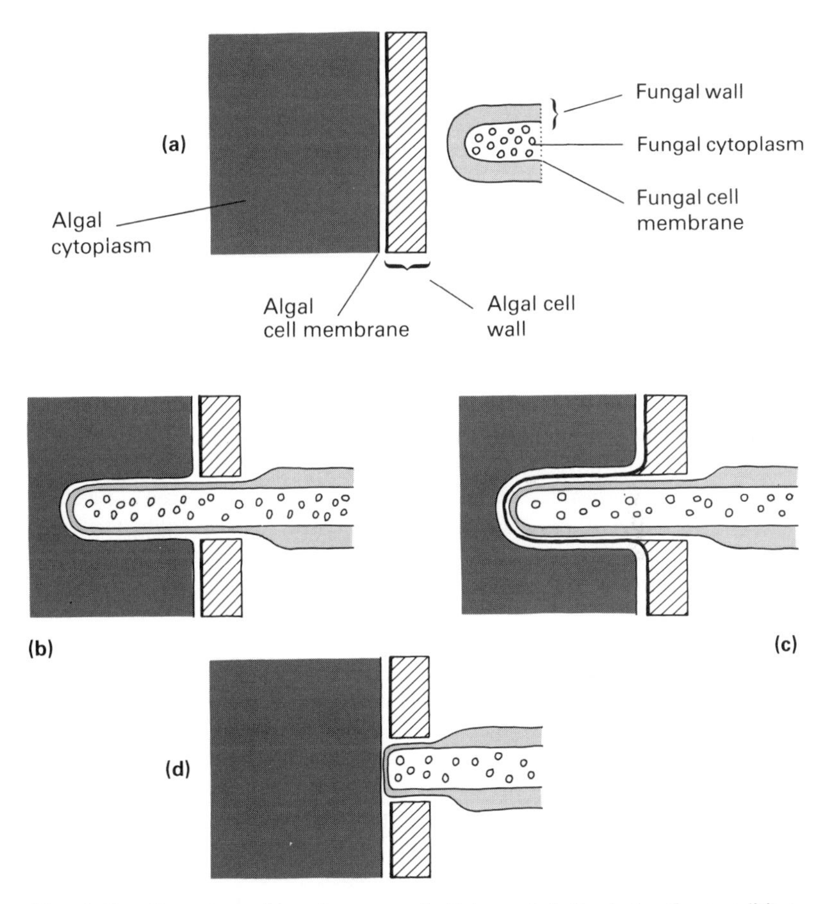

Fig. 2.12 Fungal-algal interface types in lichens. (**a**) Illustrates the condition where no contact between fungus and alga occurs. (**b**), (**c**), and (**d**) Demonstrate a variety of categories of interaction (see text).

algal plasma membrane and to penetrate only the algal cell wall. The electron microscope has shown that these categories are not really tenable ones. No haustoria normally seem to disrupt algal cell membranes, they only invaginate them. Figure 2.12 illustrates a spectrum of fungal-algal interface types which have been demonstrated in different lichens. It ranges from deeply 'penetrating' haustoria that disrupt the entire thickness of the algal cell wall (Fig. 2.12b) to 'intramembranous' types that simply contact the algal protoplast through a hole in the wall (Fig. 2.12d). At these interfaces only fungal wall material separates fungal and algal cell membranes. Some haustoria, however, do not produce full thickness lysis of the plant wall but invaginate inner layers of the latter along with the algal plasma membrane (Fig. 2.12c). Most haustoria contain lysosome-like bodies that are perhaps involved in weakening phycobiont walls to allow membrane invagination. Both haustoria and appressoria show characteristic narrow invaginations of the mycobiont plasma membrane. They are thought to be concerned with nutrient transport from alga to fungus. Many haustoria, especially the intramembranous types, contain an increased complement of mitochondria. This too might be an adaptation for the accomplishment of the active transport of organic nutrients into the hyphae.

The trend from highly penetrative haustoria to surface appressoria is correlated with an increasing integration of the two components of the lichen symbiosis. Primitive lichen associations where lichen form is similar to an unlichenized fungus are only a small advance on fungal mycelia which parasitize algal cells. In these forms disruptive haustoria predominate. At the other end of the spectrum, one finds lichens showing novel morphology which has arisen from the close symbiotic integration of two or more organisms. These almost always possess appressoria.

PROTOZOAN ASSOCIATIONS

Commensal protozoans

Sessile protozoans attach themselves to a wide range of living and inanimate objects in aqueous environments. It seems consistent to retain the description commensal for those species which show some degree of specificity in their choice of substrate animals. Such discriminatory powers usually suggest that adaptation on the protozoan's part has taken place to integrate its biology with that of the substrate organism. Restricting oneself to this subset of sessile protozoans, it becomes obvious that the crucial interface phenomenon that occurs between substrate organism and commensal is one of attachment. If a commensal protozoan has committed itself to an existence with a particular substrate organism, the physical contact with this partner must be vital. Unfortunately we know very little about the physico-chemical nature of these contacts. To take a specific example, the gill lamellae of the common freshwater amphipod, *Gammarus pulex*, frequently carries a characteristic mixed fauna of sessile commensal ciliates.[318] Typically, several specimens of the chonotrich, *Spirochona gemmipara*, the suctorian, *Dendrocometes* and a peritrich of the genus *Epistylis* can be found attached to a single gill. In the cases of at least two of these species, namely *Spirochona* and *Dendrocometes*, the ciliates are found almost exclusively on *Gammarus* surfaces, other substrates being rarely utilized. Individuals of each of these commensals have an unchangeable attachment site on the thin chitin-protein cuticle of the gill. The protozoan cells appear to be cemented onto the cuticle by their proximal ends but that statement almost exhausts the knowledge that we have about these association interfaces. There is certainly a fruitful field for study here concerning the settlement activity of the free-swimming ciliated dispersal stages used by each of these commensals. How are attachment sites chosen? Is the surface of *Gammarus* distinguishable by the dispersal stages from other surfaces? Is there recognition of the presence of conspecific attachment sites as occurs in, for instance, the settlement of barnacle larvae? What is the nature of the interface cement if it is present at all? These and many other related questions remain unanswered for a wide range of commensal protozoans which display substrate animal specificity.

Parasitic protozoans[3,364]

Whereas commensal protozoans do not derive their nutrients from their substrate organisms, parasitic forms usually utilize their host in this way. Because of this, interface activities in host-protozoan parasite associations can relate to trophic as well as attachment functions. Equally, because the host-parasite relationship is one displaying more integrative properties than the commensal one, the surfaces of parasitic protozoa can also become involved in physiological and regulatory interactions with the host.

Extracellular parasitic protozoans

Extracellular parasitic protozoa have developed an impressive series of specialized attachment organelles which adhere to host cells and their secreted cell coats. Some are based on the production of desmosome-like contacts. A desmosome (macula adherens) is a discrete area of close contact between apposed surface membranes of one cell or of different cells. A gap of between 10 and 20 nm exists between the membranes. Dense plaques are found subtending the apposed areas and the inner leaflet of the membranes in these areas is thickened. Fine fibrils extend out from the plaques into the adjacent cytoplasm. A hemidesmosome is an assymetrical attachment zone of this type, where the contact is achieved between a cell surface and a membrane or non-cellular substrate which does not form the reciprocal mirror-image of attachment structures.

The anterior mucron attachment organelle by which some gregarines hold on to host cell surfaces appears to be of desmosomal or hemidesmosomal type.[310] Brooker has made an extensive series of investigations into the formation of desmosomal contacts by parasitic flagellates in the family Trypanosomatidae[49–51]. Trypanosomatids adhere to surfaces with their single anterior flagellum and the adhesion usually occurs in a specific area of the gut of an insect host. Brooker[51] has shown that haptomonad forms of the insect trypanosomatid *Crithidia fasciculata* attach to the cuticle-lined fore- and hind-gut of the mosquito, *Anopheles gambiae*, by hemidesmosomes formed at the surface of a contorted flagellum. At the attachment site the flagellar membrane follows the irregular contours of the surface to which it is apposed. This surface is in fact the chitinous covering of the gut epithelium which is surmounted by a thin fibrous coat (Fig. 2.13). The membrane of the flagellum tip and the cuticle surface approach to within 18–20 nm of each other with the intervening space filled with the fibrous coat material. Hemidesmosomal specializations occur in the zone of contact; increased density of the inner membrane leaflet, attachment plaque and fibrils all being observable. In addition, extra desmosome contacts are formed between the sides of the attachment flagellum and the inner membrane of the pocket (reservoir) from which the flagellum arises. It appears that the terminal hemidesmosome holds the flagellum against the gut wall of the mosquito and the reservoir-flagellum desmosomes strengthen the anchorage of the flagellum to the haptomonad itself.

The attachment strategy adopted by some gregarines and haptomonad flagellates appears to be one of semi-permanent cementing together of membranes. Other protozoan parasites need to be able to be alternately mobile and sessile and for such organisms a more dynamically reversible attachment mechanism is required. The gut dwelling zooflagellates of the genus *Giardia* provide a good example of this latter situation.

Most *Giardia* species are inhabitants of the vertebrate alimentary tract. They are exceptional in possessing bilateral symmetry as a result of their laterally duplicated karyomastigont apparatus. All forms have a prominent concave depression on the ventral surface which, following the discovery of these flagellates by Grassi in 1881, was regarded as an organelle of attachment and termed an attachment or sucking disc. The kite-shaped cells of *Giardia muris* adhere strongly to

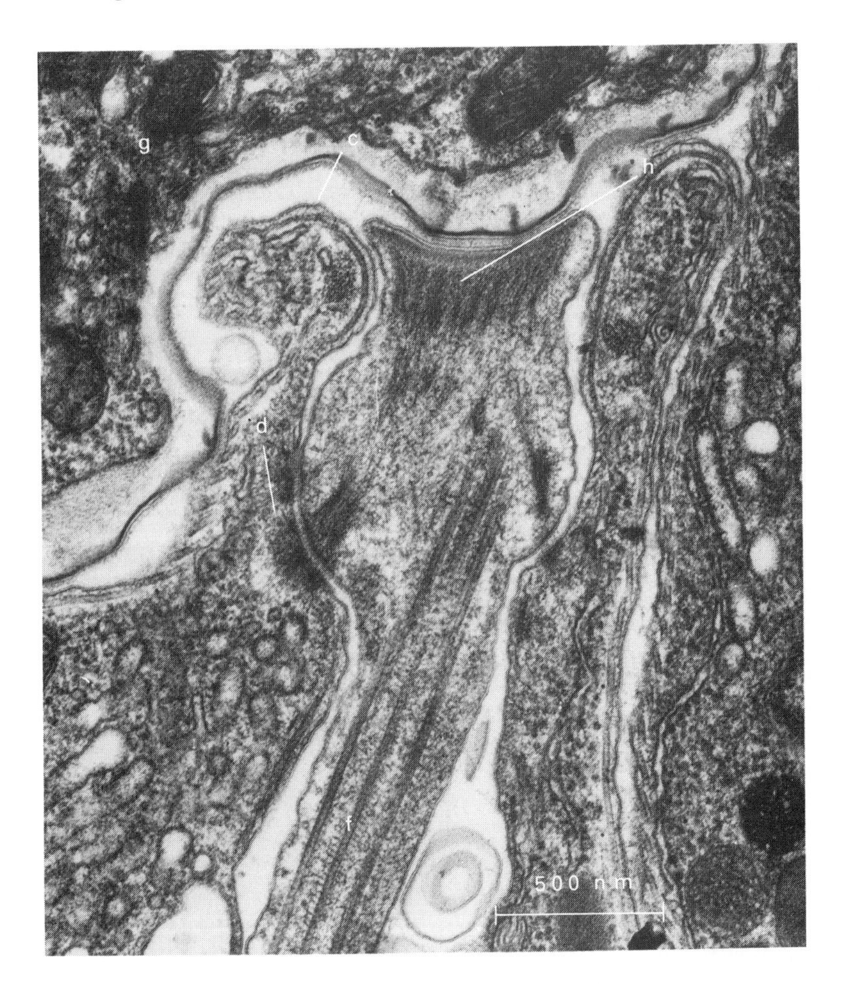

Fig. 2.13 The attachment of the flagellum of *Crithidia fasciculata* to the cuticular lining of the rectum of *Anopheles gambiae*, c, cuticular lining of rectum; d, desmosomal adhesion between flagellum and the wall of the flagellum pocket; f, flagellum; g, rectal epithelial cell; h, hemidesmosomal contact between flagellum and cuticular lining of rectum. (From Brooker.[51])

the brush border surfaces of the gut epithelial cells in the duodenum and jejunum of mice by their ventral surfaces. As a result, for many years the active sucking function of the ventral disc apparatus (the depression) was regarded as being self-evident.

When a careful electron microscopical examination of this parasite was first made by Friend,[122] it appeared as though this long-held view was to be supplanted. Friend found that the concavity of the disc was supported interiorly by an apparently rigid set of 25 nm diameter microtubules which were linked to the ventral membrane by side arms and which carried heavy cross-linked vertical ribbons. Friend regarded this massive complex as a firm 'pontoon' which served to support the structures which he thought were the effective attachment organelles. These were the ventrolateral flanges, mobile flaps of cytoplasm supported by two lateral plates with a periodic banded substructure. The plates were assumed to be composed of contractile paramyosin which could apply the flanges to the gut cell brush border as 'grasping organelles'. This revolutionary reinterpretation only held sway in an unmodified fashion for seven years, because recently, Holberton, in two impressively thorough and elegant papers[145–6] has reinstated the primary adhesive role of the disc apparatus. The new interpretation depended on critical observations of living flagellates, both attached to glass and swimming freely, with high resolution phase-contrast light microscopy. His work re-emphasizes the crucial importance of accurate observations on living cells in interpreting static ultrastructural images.

Holberton noted that when *Giardia* was unattached, the anterior, posterolateral and caudal pairs of flagella, which are not associated with the disc, were beating actively. When the cell attached to a glass surface though, these flagella remained quiescent. By contrast, the two ventral flagella, which arise near the posterior end of the disc on the ventral surface, always beat strongly and in synchrony during attachment with frequencies of up to 18 Hz (Fig. 2.15). When this beating stops, liquid flows into the disc chamber and the cell detaches from the surface. Holberton deduced that strong synchronous beating of the ventral flagella was necessary for attachment. By combining this initial postulate with further ultrastructural findings he has produced a detailed hydrodynamic theory[146] which accounts for the observed structures and behaviour excellently.

The disc is not principally an active suction cup. It cannot be because it is not a complete disc. The ventral flagella pass out through the posterior wall of the disc via a discontinuity in the wall called a throat. It communicates posteriorly with a divergent ventro-caudal groove and anteriorly with the temporary channels, formed laterally

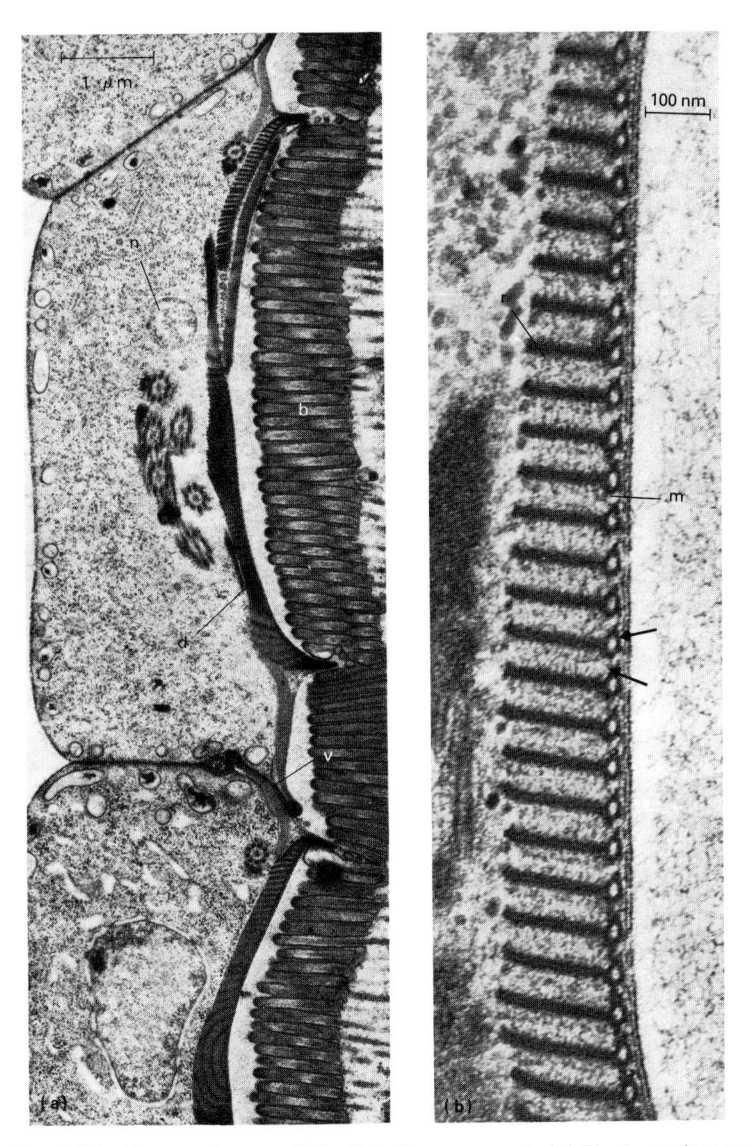

Fig. 2.14 The attachment of *Giardia* to the mouse gut. (**a**) Electron micrograph of three individuals of *Giardia* attached to the mouse gut. In the central individual note how the line of host microvilli is distorted upwards into the cavity of the ventral disc of *Giardia*. b, brush border of mouse gut cell; d, ventral disc; n, nucleus of *Giardia*; v, ventrolateral flange. (**b**) Electron micrograph of the microtubule array beneath the plasma membrane of the ventral disc. m, microtubule; r, dorsal ribbon. Arrows indicate side arms on microtubule. (Both from Holberton.[145])

when the ventrolateral flanges are curved towards the substrate (see Fig. 2.15b). Strong beating of the ventral flagella pulls water from the flagellate's anterior edge along the ventro-lateral channels and out posteriorly through the divergent groove. The constraints to flow, imposed by the disc wall and the regions in contact with the substrate,

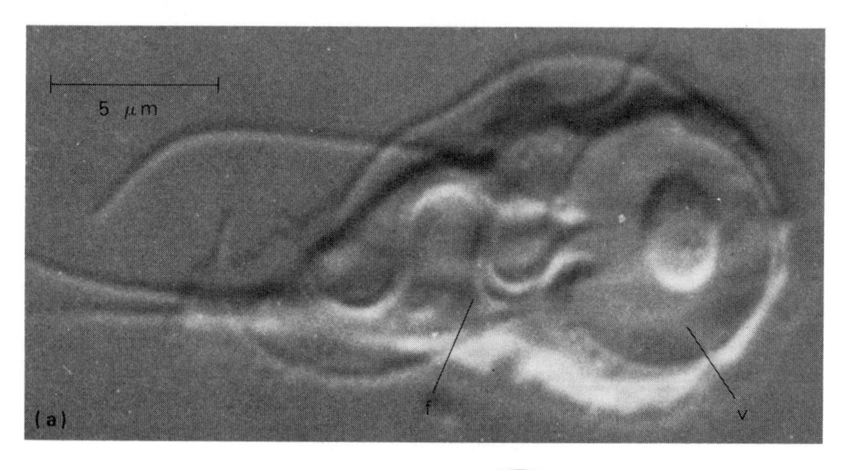

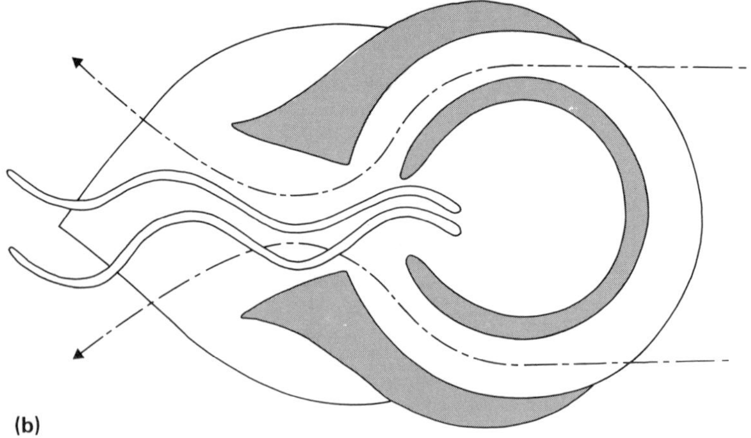

Fig. 2.15 The attachment of *Giardia* to the mouse gut. (**a**) Interference contrast photomicrograph of a living *Giardia*. v, the ventral disc; f, free ventral flagella. (Photograph supplied by Dr. D. Holberton, University of Hull.) (**b**) Diagram illustrating the flow patterns in the water on the ventral side of an attached *Giardia* set up by the beating of the free ventral pair of flagella. The stippled zones indicate areas of contact with the substrate, the two arrows give the general pattern of water flow. (Based on information in Holberton.[145–6])

ensure that water flows through the throat and ventro-caudal groove in a convergent-divergent fashion. This flow pattern produces a suction pressure which is maximal at the disc throat where the flow suffers greatest contraction. The microtubule strengthening of the roof of the disc (m, Fig. 2.14b) which resists downward buckling enables this suction pressure to operate through the disc on the underlying substrate. Modelling this qualitative description by quantitative viscous-flow equations enabled Holberton[146] to show that observed flagellar beat rates could generate suctions of the order of 10^2 dynes cm^{-2}. Suction pressures of this order could account for the upward deformations (see Fig. 2.14a) of host cell brush borders observed under attached discs in electron micrographs. It is also possible that the microtubules associated with the roof of the disc aid in the production of the suction pressure.

Intracellular parasitic protozoans

Intracellular parasitic protoza[3,364] will often develop and grow only in particular types of host cells in a specific host species. The host cells concerned usually remain alive and functioning, albeit in an altered state for most of the parasite's period of residence within them. This latter fact means that the entry of the parasite's invasive form into the host cell and its intracellular activity must be precisely regulated so as to avoid unwanted host cell damage. Inevitably, as we are dealing with close interactions between two cells, cellular interface phenomena are of vital importance in both these aspects of the host-parasite association.

Parasitic protozoa can enter host cells by one of two basic methods. They can punch a hole in the plasmalemma of the target cell which seals behind them as they pass through into the cytoplasm. An alternative tactic involves the parasite cells taking advantage of the intrinsic phagocytic ability of many host cells. In this case, the parasites are taken into the host cytoplasm within an invaginated vacuole of host plasma membrane. It might be thought that the boring method would result in the parasite being in direct contact with host cytoplasm whereas the engulfment mode would imply the presence of a host-produced parasitophorous vesicle around the parasite. As we shall see from a closer study of particular intracellular protozoans, these expectations are not always fulfilled.

The spore stage in the life cycle of microsporidian protozoans is designed to act like a hypodermic syringe. It injects the sporoplasm, or young feeding form, of the parasite into the cytoplasm of a host cell. Most investigations into the functioning of the spores have been carried out on the genus *Nosema*[199,372]. The resistant, inactivated spore has a long, hollow tube, the polar filament, which is invaginated inside it, like the finger of a glove turned inside out. The membranous wall of this filament is continuous with that of the spore. When the spore is ingested by an appropriate host, such as a crab, the tube is evaginated with considerable force.[372] It finally can be more than twenty times the original length of the spore. The filament, which can penetrate host cell membranes during extrusion, forms a channel along which the nucleated sporoplasm passes from the spore lumen to the host cell cytoplasm. Initially, within the host cell, the sporoplasm is surrounded by two membranes, one the bounding membrane of the sporoplasm itself and the outer one apparently a continuation of filament wall.[372] One envelope, presumably the latter, is rapidly lost so that the intracellular parasite ultimately has only a single membrane at its surface and is in direct contact with the host cytoplasm.

The sporozoites of various coccidians also actively penetrate host cell membranes. In their case, however, the whole cell passes directly through the membrane. Roberts *et al.*[291] gained useful information at a light microscopical level by observing that sporozoites of *Eimeria larimerensis* were able to enter host cells within a few seconds of first touching them. With very little deceleration a gliding sporozoite insinuates itself through an extremely fine hole in the plasma membrane. This rapid penetration, anterior end foremost, is thought to be correlated in the eimerians with the presence of a complex anterior collection of organelles, including a conoid, which together

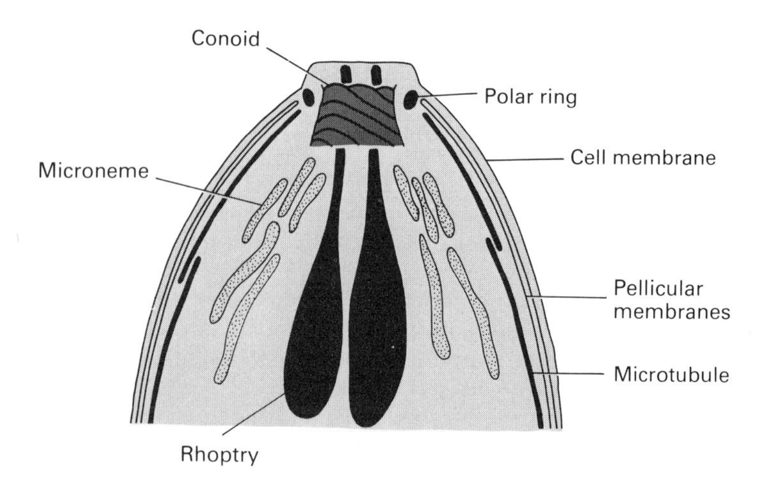

Fig. 2.16 Diagram of the ultrastructural organization of the anterior end of a typical eimerian.

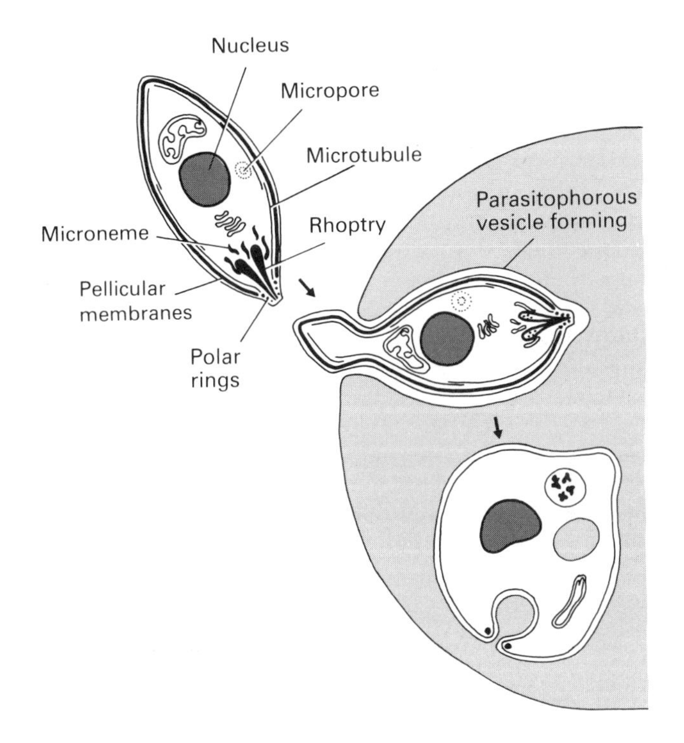

Fig. 2.17 Schematic diagram illustrating some of the changes that occur in a *Plasmodium* merozoite during the invasion of a red blood cell. On entering the red blood cell and transforming into a trophozoite, the merozoite loses polar rings rhoptries, sub-pellicular microtubules and pellicular membranes.

might have lytic and mechanically penetrative properties (see Fig. 2.16). Despite their ability to penetrate cells directly, most eimerians within cells are surrounded by a parasitophorous vesicle which presumably forms after penetration has occurred.

It was at first assumed that all parasitic protozoans with a conoid apparatus at their anterior ends would penetrate host cells by membrane rupture. Recently, however, it has been shown that the conoid-bearing parasites, *Toxoplasma gondii*[164] and *Lankasteria culicis*,[313] can enter cells inside a phagocytic vesicle.

An invasive phase in the life cycles of the malarial parasites in the genus *Plasmodium* penetrates host erythrocytes. The merozoite stages, as they are called, have several similarities to the penetrative form of the eimerians. They lack a conoid but possess polar rings, rhoptries, micronemes and subpellicular microtubules. Under the outer plasmalemma of the merozoite is a closely apposed pair of extra membranes, the inner membrane complex (or pellicular membranes). Merozoites enter red blood cells by invaginating the host cell membrane to form a parasitophorous vesicle of host origin. Once in this intracellular location the merozoite undergoes a profound transformation to the trophozoite form which eats its way through the haemoglobin-filled cytoplasm of the host cell. The trophozoite loses its polar rings, rhoptries and microtubules and has only a single plasma membrane[3] (Fig. 2.17). A much fuller account of this complex penetration process has been provided recently by Bannister.[26]

The piroplasms are in a separate class of the subphylum Sporozoa from the malarial parasites, but they too exist in intraerythrocytic sites. In contrast to the malarial trophozoites, though, piroplasms in the genus *Babesia* have only a simple bounding membrane separating them from the host cytoplasm (Fig. 2.18). This is surprising because the penetrative merozoites of this genus have rhoptries, micronemes and an inner membrane complex like that of the malarial merozoites.

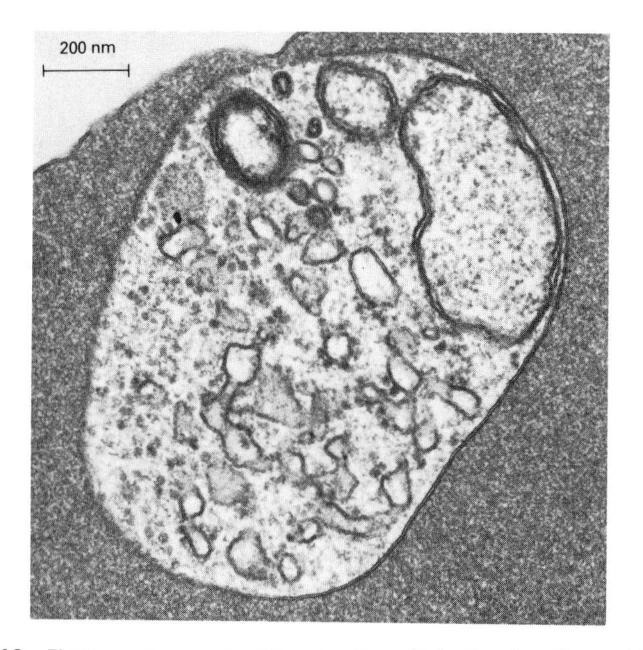

Fig. 2.18 Electron micrograph of the piroplasm *Babesia microti* in a red blood cell. Note the absence of a parasitophorous vesicle. (Electron micrograph supplied by E. J. Wills and J. E. Richmond, Clinical Research Centre, Harrow.)

Unfortunately, electron-microscopical observations have yet to be made on piroplasm merozoites in the process of entering host cells. Such images might explain the reasons for the absence of a parasitophorous vacuole.

THE SURFACES OF PARASITIC FLATWORMS

In the phylum Platyhelminthes there are three main groups in which all the members are parasitic when sexually mature adult worms. The monogeneans show most correspondence with the free living turbellarian flatworms. Most are ectoparasites of cold-blooded vertebrates and have direct life cycles. Digenean flukes are usually endoparasitic and are to be found in a variety of internal locations within poikilothermic and homeothermic vertebrates. The vast majority have indirect life cycles involving asexual larval reproduction in the molluscan intermediate host. The gutless cestodes or tapeworms are among those platyhelminths showing the most extreme modifications for an endoparasitic way of life. As adults they inhabit an internal space of a vertebrate host which, in most instances, is the gut lumen. Life cycles, except for a few odd exceptions, are indirect.

As well as these groups there are also other platyhelminth taxa in which some members show the development of parasitic modes of existence markedly in contrast to the free living habits of their close taxonomic relatives. One of the best researched examples is *Kronborgia*,[71–2] a rhabdocoele turbellarian which is as specialized for endoparasitism as a tapeworm, but always kills its host (see p. 12).

Until the early 1960s what might be called the 'armour-plated parasite' concept was the accepted attitude among parasitologists where the nature of the parasitic flatworm's surface was concerned. Especially in the case of endoparasitic digeneans and tapeworms it was assumed that the interface between parasite and host was a resistant cuticle secreted by the surface cells of the worm. This interpretation was held to explain the self-evident ability of these parasites to exist in the lumen of the vertebrate gut. The cuticle armour was supposed to protect the parasites against vicissitudes such as pH changes, surfactants like bile salts and the action of digestive lytic enzymes. Adherence to this conceptual framework, of course, posed additional questions. Not the least of these was the need to explain how a tapeworm, without a gut, could absorb nutrients through a supposedly impenetrable cuticle.

This and other puzzles were rendered irrelevant by the discovery that the outer surface of parasitic flatworms was a living cytoplasmic layer. In monogeneans, digeneans and cestodes[189,192] the outer layer is syncytial; in forms like *Kronborgia*,[46] the covering is usually cellular in organization. In all cases, though, the ultimate boundary between the parasite and host cells or fluids is a living plasma membrane. This new knowledge has necessarily placed many of the relationships between hosts and parasitic platyhelminths on a new conceptual footing. If the parasitic surface is alive, it can respond dynamically and sensitively to changes in external conditions. Its protection from potentially damaging external influences is more likely to be subtle and biochemical than a total physical exclusion of the influences. Equally there need be few barriers to the transport of substances into the parasite from the host or in the contrary direction.

The organization of the syncytial outer covering of monogeneans, digeneans and cestodes shows a common topological plan (Fig. 2.19). The outer surface of an entire parasitic worm might well be a single unbroken plasma membrane, penetrated only by openings for sensory endings (see Chapter 5, p. 159 *et seq.*), mouth, genital apertures, excretory opening and gland ducts. It is always overlain by a glycocalyx, a sheet of polysaccharide-containing material (see Ito[156] for a review of the structure and function of the glycocalyx). The outer plasma membrane encloses a continuous, non-nucleated layer of cytoplasm (the distal cytoplasm or external tegument) which is what

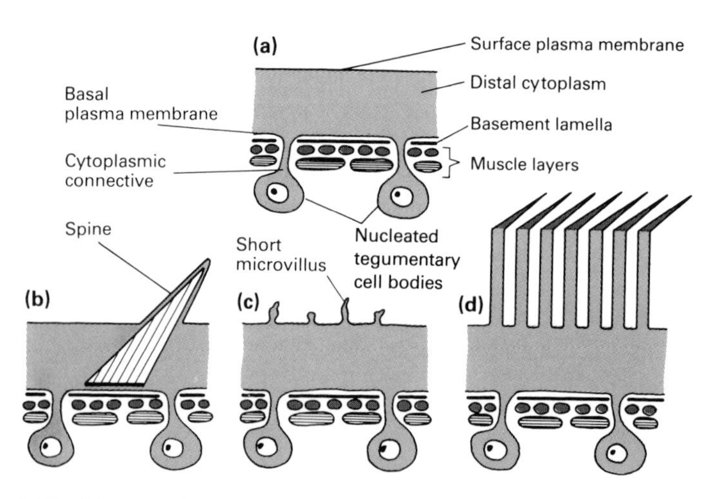

Fig. 2.19 The organization of the teguments of parasitic platyhelminths. (**a**) A generalized plan for all parasitic platyhelminths. (**b**) Typical digenean organization, with an intracytoplasmic spine in the distal cytoplasm. (**c**) Typical monogenean organization, with short microvilli apically. (**d**) Typical cestode organization, with a dense, regular array of apical microtriches.

earlier workers thought to be a cuticle. This living layer contains mitochondria and several different types of inclusions.

The distal cytoplasm is delimited from deeper tissues by a basal plasma membrane which lies on top of a basement lamella of collagen fibres. The nucleated portions of the tegument are found below the basement lamella and usually below the underlying body wall muscle cells. Cytoplasmic connectives penetrate the lamella and muscle layers and join the distal cytoplasm to the nucleated regions. Each inward projection of the tegument containing a single nucleus is called a tegumentary cell body, although of course it is not strictly a cell but a portion of a multinucleate syncytium. The cell bodies are actively secretory. They possess granular endoplasmic reticulum, Golgi complex and polyribosomes with which they elaborate the inclusions found in the distal cytoplasm and the precursors of the glycocalyx, transporting material up to the cytoplasmic connectives to do so. The complex comprising distal cytoplasm plus cell bodies has been called integument and epidermis as well as tegument, the word used here. Its basic organization is modified in ways appropriate to its special functions in the different parasitic groups, as well as undergoing profound modifications during the life cycle of any single species.

Monogeneans[209]

The teguments of the adults of a number of different monogenean species have been examined ultrastructurally. Those investigated include typical skin- and gill-inhabiting forms from fish, *Polystoma* from the frog bladder and *Amphibdella* which, in an unorthodox fashion, lives in the blood stream of its elasmobranch host, the electric ray, *Torpedo*. In most of these species the outer membrane of the external tegument is complicated by a sparse scattering of microvilli which are thin, finger-shaped projections. In *Polystoma* and *Rajonchocotyle*, a gill-dwelling form, microvilli are absent and multiple folds are present instead. The distal cytoplasm of all monogenean teguments contains membrane-bound inclusions. Two main kinds have been described, electron-dense granules which are often rod-shaped and vesicles with electron-lucent cores. Their functions are not well understood, but by analogy with teguments which have been probed experimentally in greater depth, it seems likely that they contribute to the production of the glycocalyx covering.

This polysaccharide-containing cell coat is always present externally above a simple trilaminate plasma membrane. Lyons[209] has speculated on its function in monogeneans. She has suggested that it could protect against mechanically or bacterially induced damage or

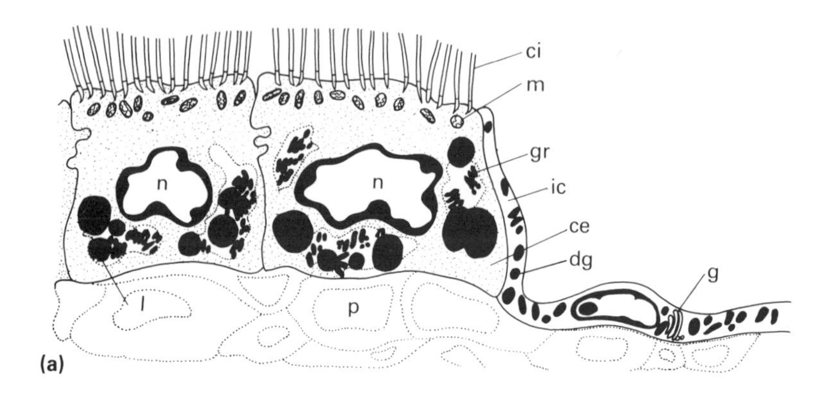
ci
m
gr
ic
ce
dg
g
n
n
l
p
(a)

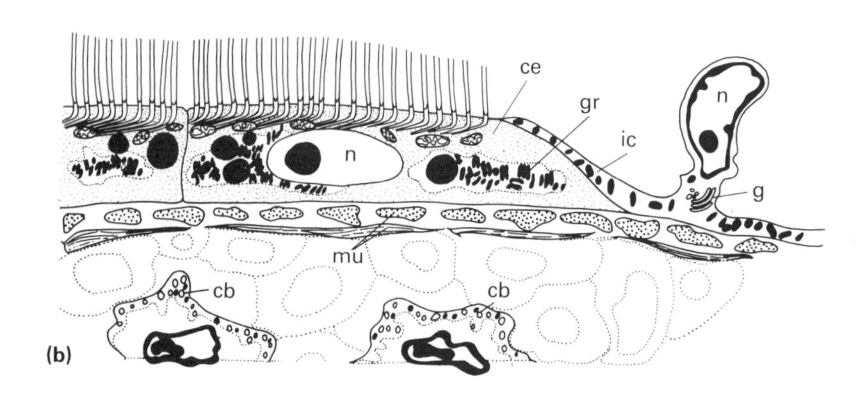
ce
gr
ic
n
n
g
mu
cb
cb
(b)

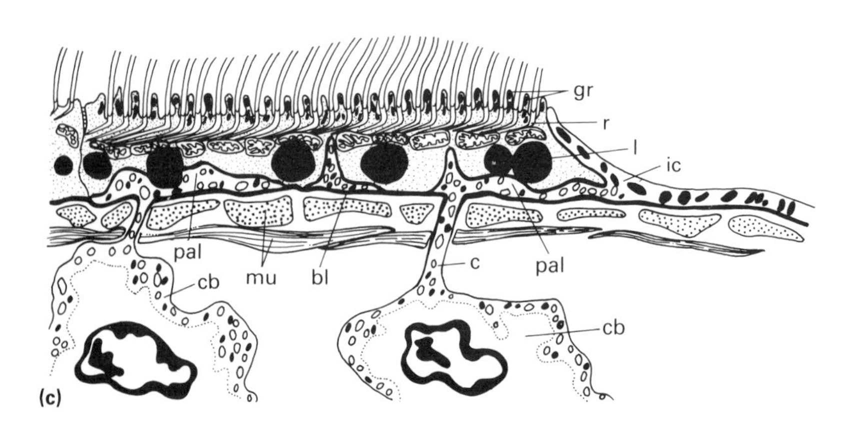
gr
r
l
ic
pal
mu
bl
c
pal
cb
cb
(c)

be implicated in ionic regulation in freshwater ectoparasitic forms, perhaps binding Ca^{++} ions prior to uptake. Nutrient transfer across the tegument has yet to be demonstrated in monogeneans. This is not surprising as they have well developed guts and oral structures for bulk feeding (see Chapter 3, p. 82).

Lyons[210] has followed the embryological development of the tegument of the skin-dwelling monogenean of the sole, *Entobdella soleae* (see Fig. 2.20). The distal cytoplasm of the adult is constructed under ciliated surface cells of the free-swimming oncomiracidium larva. It is composed partly of cytoplasm from interciliary cells that lose their nuclei. More important contributions are made by cells originating in the parenchyma that migrate centrifugally. They send out cytoplasmic connectives through differentiating muscle layers to make a presumptive adult distal cytoplasm under the ciliated cells. At the same time they fuse laterally with the interciliary cytoplasm. The nucleated regions of the tegument forming cells remain beneath the muscle layers as tegumentary cell bodies. When the oncomiracidium attaches to its fish final host the ciliated cells are shed and the uncovered external tegument becomes the outer surface of the juvenile worm. The centrifugal movement of central cells that occurs during the ontogeny of the tegument mirrors similar outward cell translocations that occur in free-living turbellarian flatworms for the replacement of epidermal cells.[317]

Cestodes[327–8]

Llewellyn[198] has convincingly demonstrated that greater phylogenetic affinity exists between monogeneans and cestodes than exists between either of these groups and the digeneans. In this context it is interesting that the adult cestode tegument can be understood as an extreme development of the microvillous surface found on many monogeneans.

Fig 2.20 Developmental stages in the formation of the larval epidermis of *Entobdella soleae.* (**a**) Early stage. Junction of surface ciliated cells with the apparently syncytial interciliary cytoplasm. (**b**) Middle stage. Ciliated cells have flattened. A nucleus in the interciliary cytoplasm projects from the surface. Cell bodies of the 'presumptive adult' cytoplasm are differentiating in the parenchyma. (**c**) Hatched oncomiracidium. Discontinuous layer of 'presumptive adult' cytoplasm underneath ciliated cells. The layer is continuous with the parenchymal cell bodies and shows connections with the interciliary cytoplasm. The latter and the ciliated cells have lost their nuclei. bl, basement lamella; c, cytoplasmic connection; cb, cell body of 'presumptive adult' layer; ci, cilia; ce, ciliated epidermal cell; dg, dense granule; g, Golgi apparatus; gr, granules; ic, interciliary cytoplasm; l, lipid; m, mitochondrion; mu, muscle; n, nucleus; p, parenchyma; pal, presumptive adult layer; r, ciliary rootlet. (From Lyons.[210])

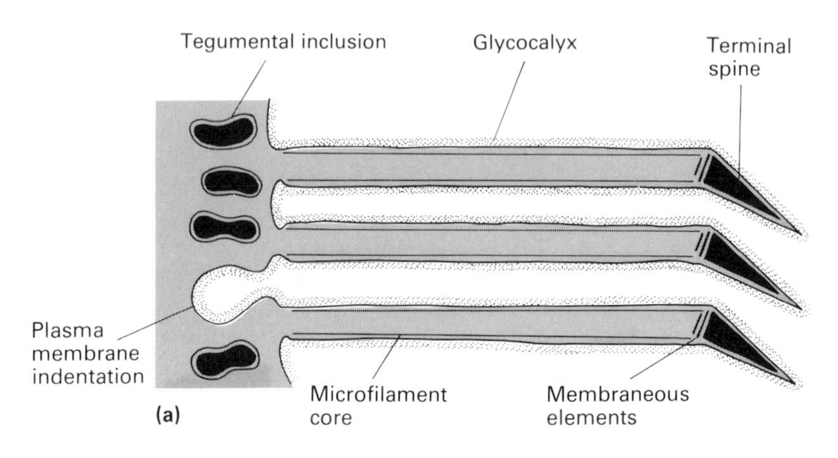
Tegumental inclusion
Glycocalyx
Terminal spine
Plasma membrane indentation
(a)
Microfilament core
Membraneous elements

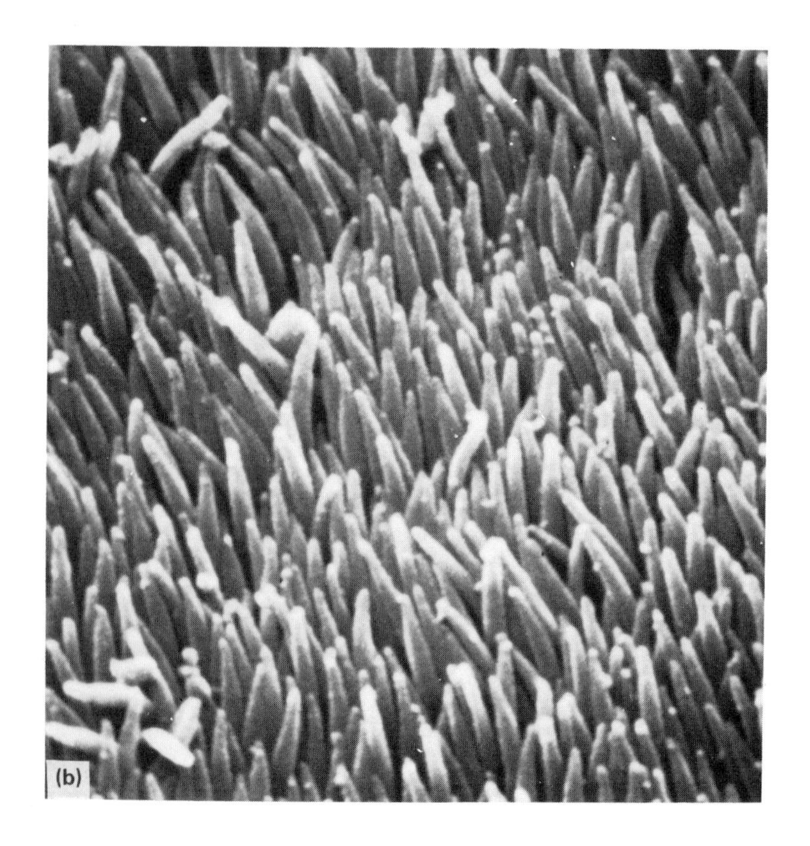
(b)

All adult tapeworms yet examined are covered with a dense array of specialized microvilli that in scanning electron micrographs looks like the pile on a carpet (Fig. 2.21). Most of these microvilli have an angled, tapering electron-dense spine at their tip within the enveloping plasma membranes and are termed microtriches. Each microthrix is braced internally by a core of longitudinal microfilaments that connect with a terminal web of similar filaments in the distal cytoplasm. The expansion of the external plasma membrane area which the microtriches represent is partly an adaptation for maximizing nutrient absorption. In the absence of a gut this external surface is the only one at the parasite's disposal for the uptake of organic nutrients (see Chapter 3, p. 92). The electron-dense spines on the microtriches have usually been regarded as an attachment modification, with the implication that they could interlock with the brush border microvilli on adjacent gut epithelial cells of the host. This seductively simple and plausible idea has yet to be substantiated. The only study which has been made on the actual brush border-microtriches interface[225] merely showed a close approximation of the two sets of finger-like projections; no real interdigitation was observed. Equally, cestodes in guts are in constant motion and it is difficult to imagine how interdigitations could be made and broken rapidly enough to allow such mobility. Perhaps by a ratchet mechanism, the spines, which all point posteriorly, merely make backward worm movement more difficult than anteriad locomotion. In such circumstances almost any muscular activity of the tapeworm would help to maintain its position against backward gut peristalsis.

The external plasma membrane of the microtriches can be of simple trilaminate form[200] or more complex. In *Echinococcus granulosus*[160] adults the shafts of the microtriches have a surface that consists of a continuous trilaminate membrane covered discontinuously by another trilaminate membrane structure. The cestode surface membrane contains complex carrier systems for the active transport of nutrient molecules as well as intrinsic digestive enzymes (see Chapter 3, p. 102). Because of this it would not be surprising if this functional complexity often resulted in ultrastructurally observable departures from normal trilaminate membrane morphology.

The nature and development of the glycocalyx that covers the outer tegumentary membrane of tapeworms has been extensively stud-

Fig. 2.21 (**a**) Diagram illustrating the typical ultrastructural organization of the microtriches on a cestode tegument. (**b**) Scanning electron micrograph of the microtriches on a proglottid of *Hymenolepis diminuta*. All the tips of the microtriches point posteriorly along the worm (upwards in the micrograph). (Micrograph supplied by Ms. H. M. Paterson, Imperial College, London.)

ied.[201–3,259,350] If evidence from different species is amalgamated to produce a general description it seems that a distinct glycocalyx is ubiquitous among tapeworms. It is definitely produced by the tegumental complex of the worm itself and does not receive significant contributions from the mucopolysaccharides of the host's gut. The coat is polyanionic in charge topography and is rich in neutral and acidic carbohydrates. Some of the components of the glycocalyx can be labelled by supplying the tegumental cell bodies with a 'pulse' of a tritiated sugar, ^{3}H-galactose. The label appears first in the Golgi complexes of the cell bodies, then moves up the cytoplasmic connectives to the distal cytoplasm. It appears to be a constituent of the carbohydrate-containing macromolecules in some of the membrane-bound vesicles in the distal cytoplasm. These vesicles discharge their contents to the exterior and add molecules externally to the glycocalyx. There is a considerable loss of radioactivity from the glycocalyx within six hours of the synthesis of labelled constituents in the cell bodies. This loss suggests that there is a considerable and continuous replacement of glycocalyx material. It is likely that this delicate and dynamic process is part of the cestode's defences against surface attack by host digestive enzymes. In addition the glycocalyx is an absorptive zone for host-produced enzymes that are utilized by the tapeworm (see Chapter 3, p. 103) and probably also plays a part in the direct inhibition of potentially harmful host enzymes such as trypsin.[265]

Lumsden *et al*[204] have recently provided a detailed analysis of the ontogeny of the microthrix-covered surface of the cestode of the cat, *Spirometra*. The process shows intriguing parallels with the analogous monogenean developmental story described by Lyons.[210] *Spirometra* belongs to the primitive pseudophyllidean group of tapeworms which possess a ciliated, free-swimming larva, the coracidium. Within the ciliated outer coat of this larval form is the precursor of the next larval stage, the procercoid. In *Spirometra*, this larva develops in a copepod intermediate host and possesses microtriches. At the coracidium stage the procercoid precursor below the ciliated cells is covered by non-merging cytoplasmic projections of cells whose nucleated regions lies beneath the surface musculature. Differentiation of the tegument begins with the shedding of the ciliated coat. The cytoplasmic projections fuse laterally with one another, forming a thin external tegument which initially bears no microvillous projections. Later a series of 'waves' of membrane-bound granules move from the nucleated regions of the tegument to the young distal cytoplasm. The first wave is associated with the formation of simple microvilli and a later population of granules contributes to the elaboration of the spined procercoid microtriches.

Digeneans[115,143,189,192]

It was Threadgold[346] in 1963 who first demonstrated the living, tegumental surface organization of a digenean in his work on adults of the liver fluke, *Fasciola*. Subsequent work has shown that this sort of ultrastructural architecture exists or is developing in the surface layers of all the larval stages (i.e. miracidia, sporocysts, rediae and cercariae) as well as adult worms. Each of these developmental stages shows characteristic modifications of the tegumentary structures. These specializations demonstrate well the developmental plasticity of the digenean surface and also the way in which regional differentiation of a continuous syncytial tegument can occur.

In the free-swimming, cilia-bearing miracidium larva of *Fasciola* a layer of nucleated, ciliated cells overlays a basement lamella and muscle layers. Linear cytoplasmic projections of sub-lamellar cell bodies extend between the ciliated cells.[331,380] The miracidium is a non-feeding larva and has a short life span circumscribed by the finite nutrient reserves that it contains. Death is inevitable on the exhaustion of these reserves, unless in its swimming it finds a suitable snail host in which to continue its development utilizing snail nutrients. On penetrating a snail (Fig. 2.22), the ciliated cells are shed. With the now familiar pattern of centrifugal cytoplasmic extension and lateral fusion, the linear cytoplasmic extensions ('ridges' of Southgate[331]) fan

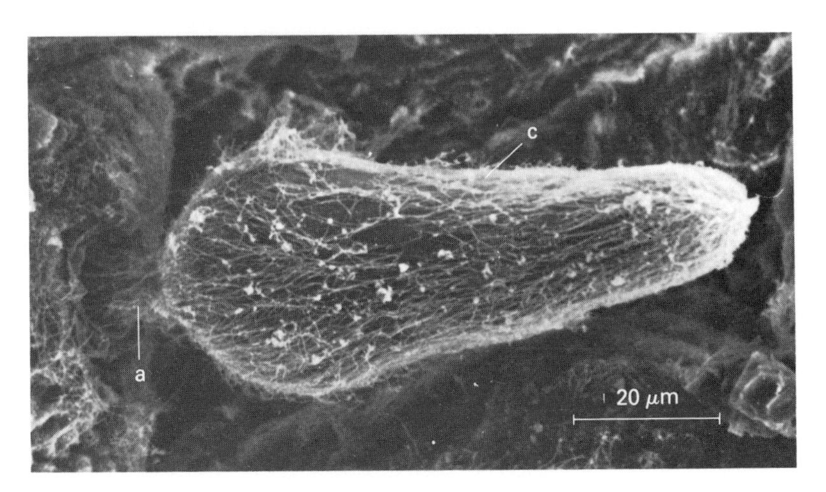

Fig. 2.22 Scanning electron micrograph of the miracidium of *Trichobilharzia ocellata* penetrating the epithelial cells of the freshwater snail *Lymnaea stagnalis*. a, apical papilla; c, cilia of miracidial epithelial cells. (Micrograph supplied by Dr. H. D. Blankespoor, Museum of Zoology, University of Michigan, U.S.A.)

out to form the early tegument of the next larval stage, the sporocyst.

Within five hours of penetrating the molluscan host the *Fasciola* larva, now a young sporocyst, has adapted its previously smooth external tegument by the construction of many plasma membrane folds. It is a characteristic of sporocysts that they have no mouth nor gut. They must support the production of daughter sporocysts or rediae, which is their asexual reproductive functión, by the uptake of nutrients from the snail across their body walls. In this respect the dense array of surface folds[331] or microvilli[177] on sporocyst teguments must be regarded as analogous to the similar surface amplification that occurs at the tapeworm outer surface.

Most radial larval stages have a mouth, muscular pharynx and an unlobed, caecal gut (Fig. 2.23). Ultrastructural studies show, however, that the external tegument of these larvae often retains microvilli[288] or intense surface folding.[178] This ultrastructural evidence suggests that small molecular weight organic nutrients can be absorbed directly via the body wall (see also Fig. 3.17).

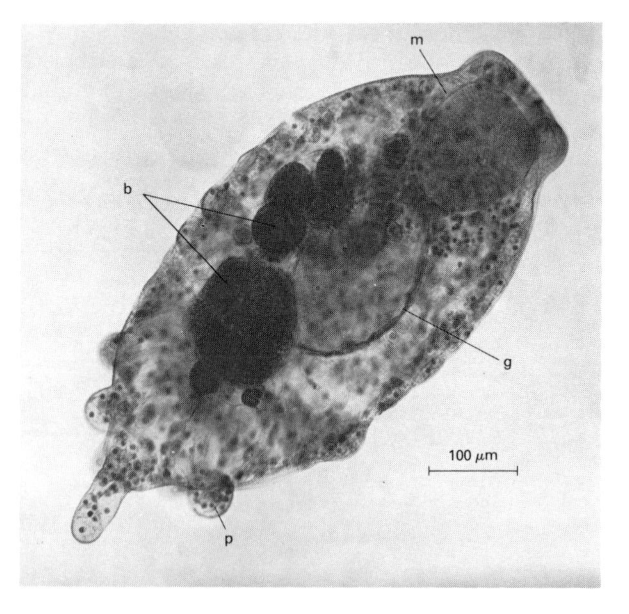

Fig. 2.23 Photomicrograph of the redia of the digenean *Transversotrema patialense* from the digestive gland of the snail *Melanoides tuberculata*. b, developing germinal balls; g, unlobed, simple gut; m, anterior muscular portion of gut; p, projection of body wall.

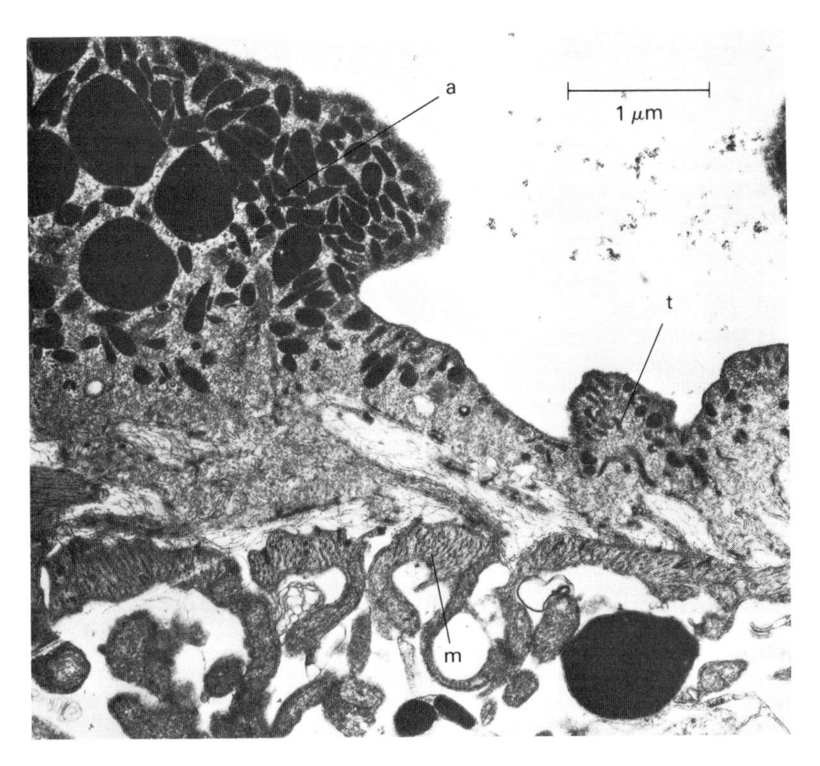

Fig. 2.24 Regional differentiation of the digenean tegument. An electron micrograph of the junctional zone between normal arm process tegument and adhesive pad tegument on the tail of the cercaria of *Transversotrema patialense*. a, thickened tegumental distal cytoplasm of the adhesive pad, containing adhesive granules; m, subtegumental muscle; t, thin, normal distal cytoplasm of the arm process. (From Whitfield *et al.*[375])

Rediae or daughter sporocysts produce cercariae. These are another motile, non-feeding developmental phase which must usually swim using a muscular posterior tail in order to reach the next host in the life cycle. The cercarial tegument is the direct developmental precursor of the adult's tegumentary layer and is often very complex. In particular, from point to point over the syncytial surface of the cercaria, regional differentiations of the tegument can exist. At a gross level the tegument of the head often contains spines whereas that of the tail does not. At a more topographically precise level one may cite the tegumentary specializations of the tail in the cercaria of the ectoparasitic digenean, *Transversotrema patialense*. Here, anterior projections of the tail

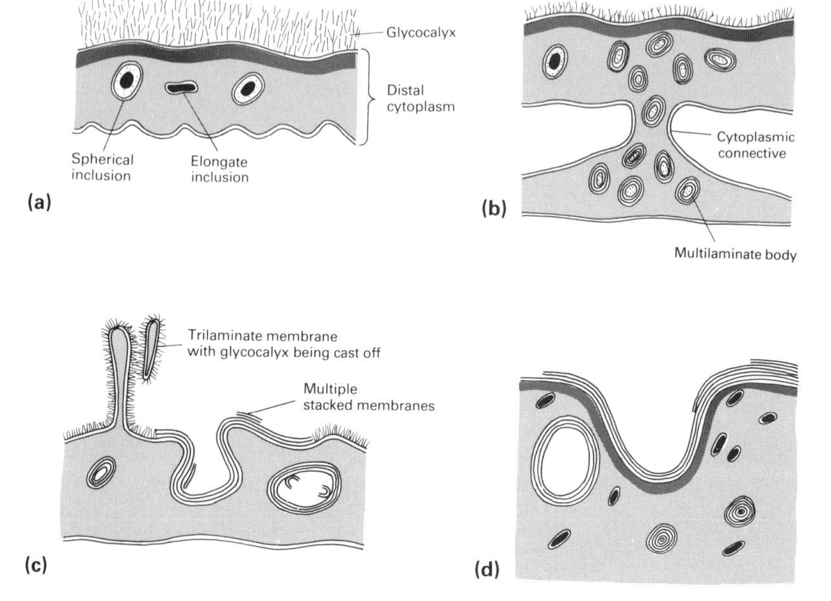

Fig. 2.25 Tegumental changes during the cercaria-schistosomulum transformation of *Schistosoma mansoni.* See text for details. (Based on Hockley and McLaren.[144])

possess adhesive pads (see Fig. 5.14b). These are used to attach the cercaria to its final host, a fish. Electron-microscopical studies[375] have shown that the pads are specialized regions of the general tail tegument. In the pad zone the external tegument becomes greatly thickened and packed with membrane-bound adhesive inclusions. They release their contents at the tegumentary surface on contact with the fish host and hold the cercaria onto the fish's scales (Fig. 2.24). The cytoplasmic connectives which fuse with the pad zone contain developing adhesive inclusions but a few μm away similar connectives, linking to normal distal cytoplasm, have none. It appears a general phenomenon that localized groups of tegumentary cell bodies can show cytodifferentiation and influence the composition of a similarly localized region of external tegument.

The immediate environment of the cercaria of *T. patialense* is the same as that of the adult worm, namely freshwater. For most digeneans this correspondence does not apply. The cercaria of *Schistosoma mansoni*, for instance, after swimming in freshwater penetrates the skin of man. Tegumental changes linked with this transition from a free-swimming freshwater existence to one in man's tissues are diverse and instructive.[143–4] The changes are illustrated in Fig. 2.25.

In the free-swimming cercaria the external tegument is bounded by a trilaminate plasma membrane surmounted by a thick fibrous glycocalyx. Immediately beneath the membrane is a prominent dense layer. Both glycocalyx and dense layer might be concerned with ionic regulation in a hypo-osmotic environment. Certainly both these features are lost quickly after penetration of the vertebrate skin. The distal cytoplasm of the cercaria contains elongate and spherical membrane-bound inclusions as well as large intracytoplasmic spines with a regular packing of protein subunits. Cytoplasmic connectives between the external tegument and tegumentary cell bodies are difficult to find in the cercaria and some workers[144] have even suggested that the connectives might only have a temporary existence.

When the schistosome cercaria penetrates the skin it is termed a schistosomulum. In contrast to the antecedent cercarial phase this juvenile adult worm is adapted to hyper-osmotic environments and high salinity conditions. It is freshwater intolerant. These profound physiological alterations are paralleled by changes in tegumentary ultrastructure. Thirty minutes after skin penetration the schistosomulum distal cytoplasm still has a surface coat and a trilaminate membrane, but multilaminate bodies comprising concentric whorls of trilaminate membranes have appeared in the external tegument. These have moved up cytoplasmic connectives from tegumentary cell bodies which Hockley and McLaren[144] suggest only connect with the distal cytoplasm after skin penetration. One hour after entry into the host the tegument presents a mosaic appearance. In some areas the bounding membrane is still trilaminate and carries a glycocalyx, but such regions appear to be forming elongate external projections of membrane and coat which are cast off from the schistosomulum. At the same time, in other areas, the multilaminate bodies open out at the surface and add multiple stacked membrane sets to the pre-existing external membrane. As a result of these processes, within three hours most of the outer membrane is heptalaminate (i.e. three stacked membranes), parts are pentalaminate and there are membrane fragments 'fraying' off the surface.

The schistosomula develop into the dioecious adult schistosomes. These live in permanently copulating, linked pairs in host blood vessels. Their surfaces are pentalaminate like those of the three hour old schistosomulum or possess even more layers of membrane. It is very likely that the production of multiple membranes is somehow correlated with the parasite's evasion of the host's immune responses (see Chapter 4, p. 130 *et seq.*). As a schistosomulum migrates, grows and develops into an adult worm, the tegumentary surface increases greatly in area. Within four days of entry, while the young worms are still migrating in the lungs, the outer membrane becomes invaginated into complex superficial infoldings. There is some evidence for the pinocytosis of host-derived materials via these depressions.[323] As the external tegument becomes thicker these infoldings become tortuous, branching channels which almost certainly have a role in the uptake of soluble organic nutrients. (See p. 86).

Work on the tegumental plasticity of *S. mansoni* has demonstrated the probable importance of the addition of new subtegumental cells to a pre-existing tegument at different times in a developmental sequence.[144] A similar conclusion has been reached by Bennett and Threadgold[32] in studies on the ontogeny of the adult tegument in *Fasciola hepatica*. These workers looked particularly at changes that occurred from the time when juvenile worms excysted in the host gut until their arrival in the bile duct after a trans-peritoneal migration. During this linked migration and tegumentary development two different types of morphogenetic event occurred. Firstly, tegumentary cell bodies that were already connected to the distal cytoplasm changed their secretory behaviour and began to produce new types of inclusions (type 0 cell to type 1 cell transformation). Secondly, a new set of cells (type 2 cells) which differentiated from embryonic cells in the parenchyma, made cytoplasmic connections with the distal cytoplasm. These newcomer cell bodies produced yet another type of cytoplasmic inclusion that also passed into the external tegument (Fig. 2.26).

Bennett and Threadgold[32] also analysed the possible origins of the prominent tegumental spines of the adult *Fasciola* surface. These are used to hold the worm in position in the host's bile duct. Spines are very common among adult digeneans (see Fig. 2.27) and a description of their elaboration in *Fasciola* can probably serve, in many respects, as a general model for the group. In *Fasciola* the spines are only synthesized after arrival in the final host. Their intracytoplasmic location means that they can only be synthesized in the tegumental cell bodies or the distal cytoplasm. There is evidence that spines are proteinaceous so their assembly must be linked either with the

Fig. 2.26 The transformation of the tegument of *Fasciola hepatica* during its migration from the gut of its final host to the bile duct of the host's liver. Stage (**a**) represents the organization of the tegument of a recently excysted juvenile worm in the gut, stage (**f**) its configuration in a young worm recently arrived in the bile duct. Intervening stages demonstrate the tegumental organization on the migration through the gut wall, across the peritoneal cavity and through the liver parenchyma cells (see text for details). (From Bennet and Threadgold.[32])

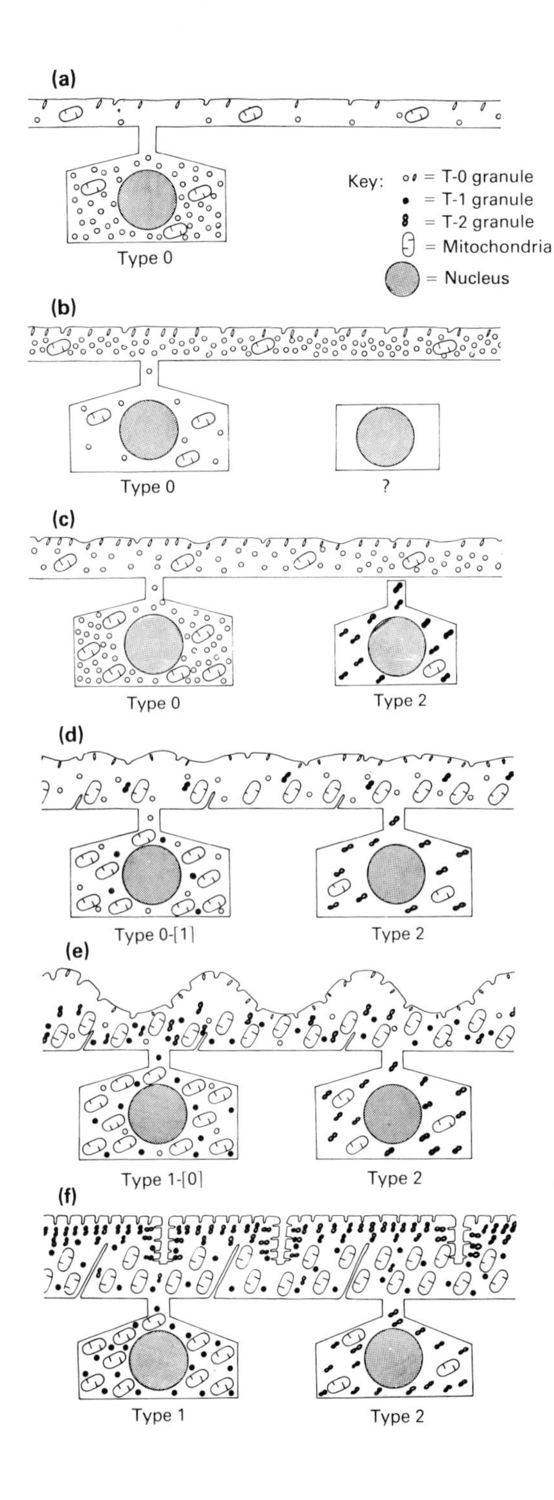
(a)
Key:
= T-0 granule
= T-1 granule
= T-2 granule
= Mitochondria
= Nucleus
Type 0
(b)
Type 0
?
(c)
Type 0
Type 2
(d)
Type 0-[1]
Type 2
(e)
Type 1-[0]
Type 2
(f)
Type 1
Type 2

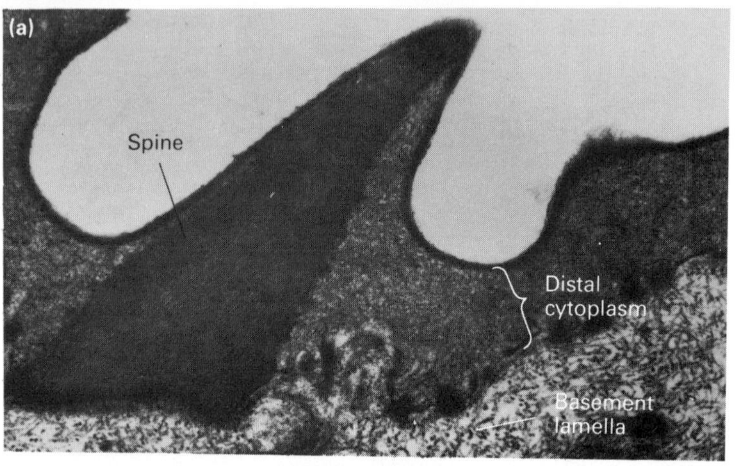
(a)
Spine
Distal cytoplasm
Basement lamella

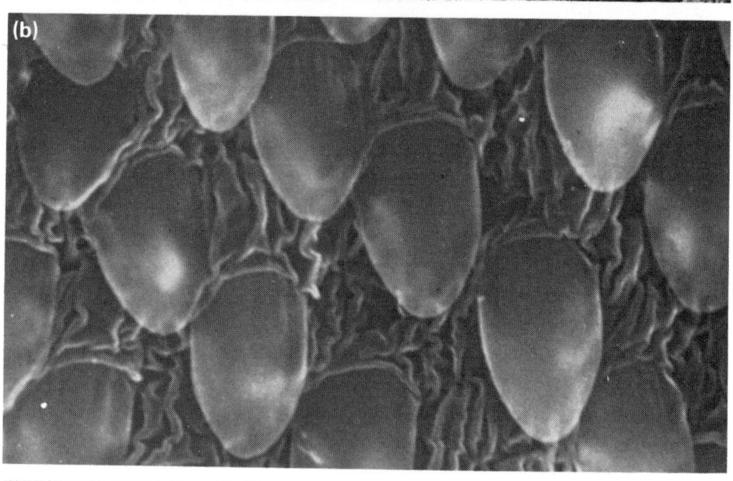
(b)

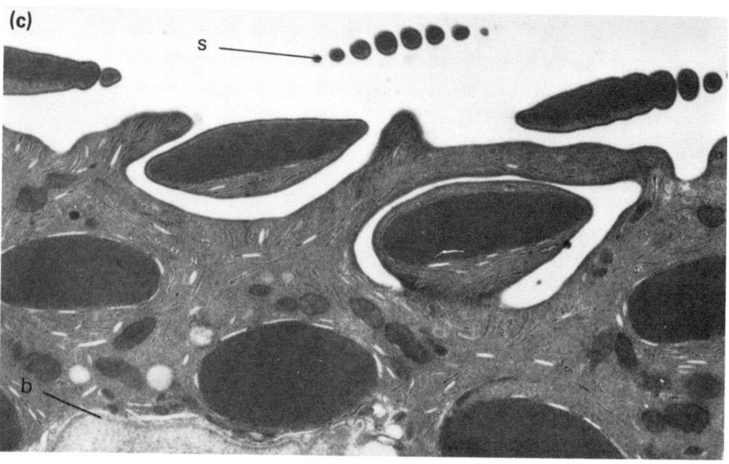
(c)
s
b

granular endoplasmic reticulum-Golgi complex system or the polyribosomal system or both. Neither of these types of protein-synthesizing machinery are present in the distal cytoplasm so synthesis must occur in the cell bodies. It seems that spine protein monomers are produced on polyribosomes in these nucleated areas and passed up the cytoplasmic connectives to the distal cytoplasm. Here they self-assemble the crystalline spine, probably using regions of the basal membrane of the distal cytoplasm as a foundation plate organizing centre for monomer aggregation. It is intriguing to ponder how such a synthetic and assembly system produces on the one hand neatly patterned arrays of spines and on the other complex, non-pyramidical spine conformations like the flattened, serrated spines of the cercaria of *Neophasis lageniformis*[179] and the adults of *Microphallus sp.*[96] (Fig. 2.27).

Rhabdocoele parasites

Rhabdocoele turbellarians are usually small, ciliated free-living flatworms with a simple sac-like gut. Within the same taxon however exist commensal[159] and endoparasitic forms.[72] The most remarkable surface specializations occur in members of the parasitic family Fecampidae which inhabit crustaceans and myzostomids. Within the family only one species has been subjected to rigorous modern investigations, namely *Kronborgia amphipodicola*, which as its specific name suggests is an endoparasite of amphipods. The separate male and female worms of this parasite develop in the haemocoele of the marine, tube-building amphipod, *Ampelisca macrocephala* (see Fig. 1.5). Neither sex possesses eyes, mouth, pharynx nor intestine. Central parenchymous spaces are filled with reproductive organs. Nutrients must be taken up across the body wall from the haemolymph of the host, and the epidermal ultrastructures of the adult female worms seems well adapted for such a function.[46] In most areas the surface is cellular with adjacent epidermal cells linked laterally by apical demosomes. The outer membrane of these cells covers many cilia and an extensive mat of thin, branched microvilli (Fig. 2.28).

Fig. 2.27 Electron micrographs of digenean tegumental spines. (**a**) Transverse section through a spine in the dorsal tegument of the cercarial head of *Transversotrema patialense*. (**b**) Scanning electron micrograph of part of the dorsal spine array of an adult *Transversotrema patialense*. Note the regular pattern of spines and their simple outline. (**c**) Transverse section through serrated spines in the tegument of an adult *Microphallus* sp. b, basement lamella; s, serrated spine tip. (Micrograph (**c**) from Davies.[96])

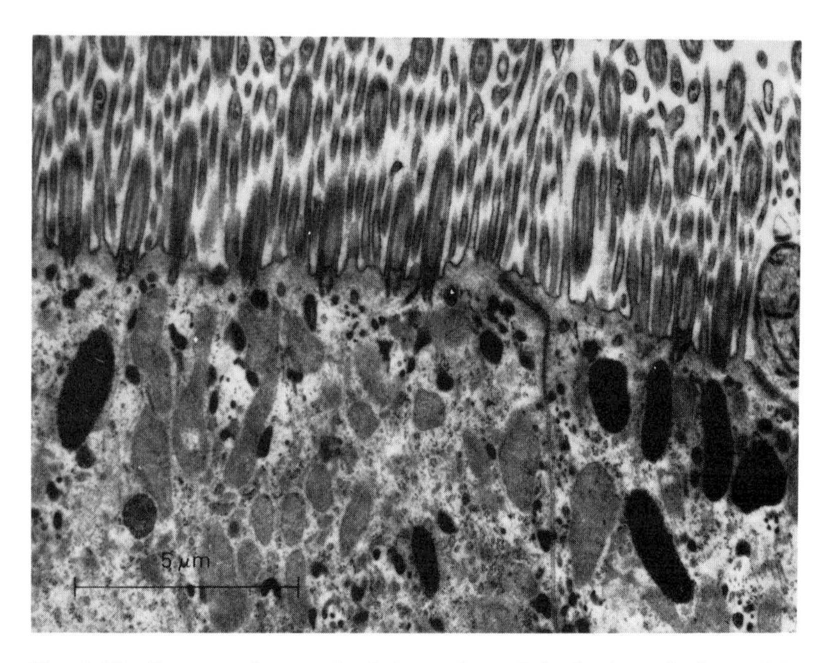

Fig. 2.28 Electron micrograph of the surface of the body wall of an adult female *Kronborgia amphipodicola* from the haemocoele of the amphipod *Ampelisca*. Note the dense array of cilia and thin microvilli at the surface. (From Bresciani and Køie.[46])

ACANTHOCEPHALANS[87,189,192]

Endoparasitic platyhelminths that can absorb nutrients transtegumentally, such as sporocysts, adult *Schistosoma* and tapeworms, have experimented with a wide range of methods of expanding their external surface areas. Most platyhelminth answers to this morphological problem involve external projections (folds, microvilli or microtriches). Only the schistosome response includes extensive inward membrane invaginations.

The pseudocoelomate acanthocephalans are the aschelminth analogue of the tapeworms. They have no gut and inhabit the intestines of vertebrates. Their body wall is highly specialized for nutrient uptake and ultrastructurally appears to be the ultimate development of the surface invagination strategy. In all forms yet investigated the epidermis is syncytial and its outer surface is expanded by high densities of narrow invaginations variously called pores,[89] tegumentary pore-canals[383] or crypts[57] (see Fig. 2.29). The

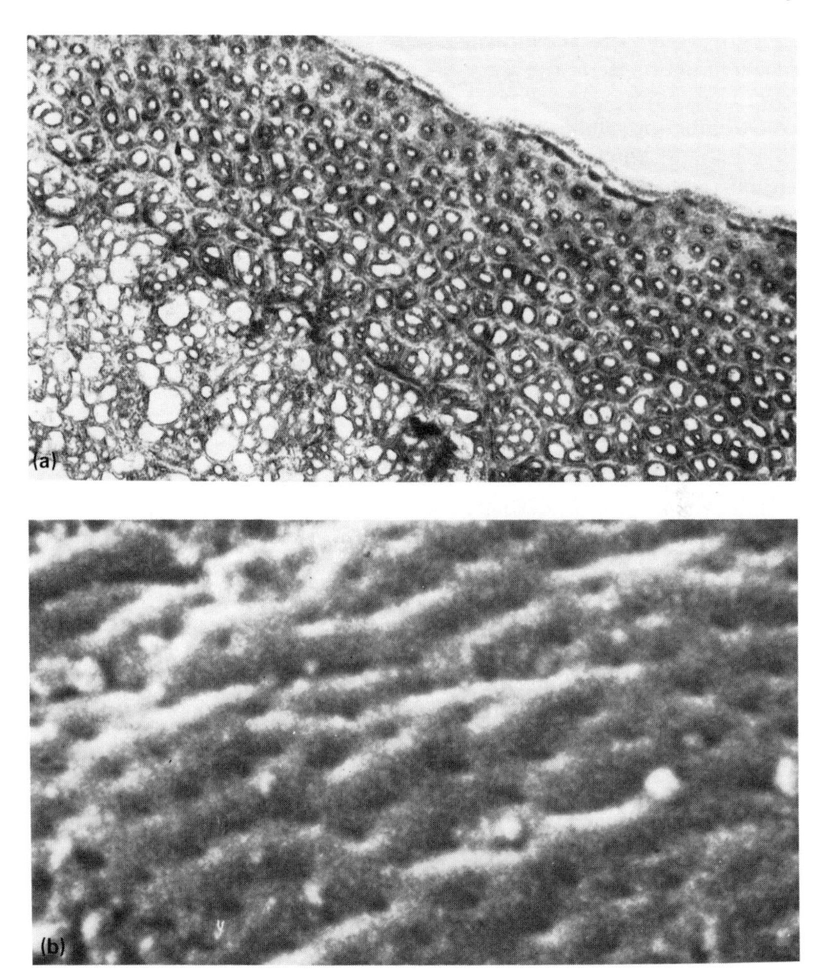

Fig. 2.29 Electron micrographs of acanthocephalan tegumentary pore canals. (**a**) Transmission electron micrograph of an oblique section through the body wall surface of the trunk of the palaeacanthocephalan *Polymorphus minutus*. The pore canals are cut in oblique transverse section and their profiles show that the canals branch as they descend deeper into the body wall. (**b**) Scanning electron micrograph of the trunk surface of the archiacanthocephalan of the pig, *Macracanthorhynchus hirudinaceus*. Note the surface dimples marking the opening of each pore canal.

invaginations represent inward depressions of the common limiting plasma membrane of the syncytial epidermis. Each invagination has a minute opening at its outer end (12–42 nm across in *Moniliformis*[57]) but dilates internally to form branching channels which can attain diameters of almost 0.5 μm. The outer parts of the invaginations are supported in a dense terminal web of filamentous material. External to the general outer surface is a thick glycocalyx. Taken together the additional membrane area represented by the invagination linings constitute a huge increase in external surface area which is certainly implicated in nutrient uptake (see Chapter 3, p. 103). Byram and Fisher[57] have calculated that for young male worms of *Moniliformis dubius*, which have a surface density of 15 invaginations per μm^2, the invaginations increase the available surface area 44 times.

NEMATODES AND CRUSTACEANS[189,192]

The nematodes do actually have a cuticle. A resistant, relatively impermeable exoskeleton of structural proteins like collagen is secreted by a syncytial epidermis, the hypodermis. In parasitic forms the cuticle is permeable to water and some ions but rarely to organic molecules. Perhaps the nematodes are the nearest real approach to the 'armour plated' parasite which is now a discredited model for platyhelminths and acanthocephalans.

Even this apparently absolute constraint, however, has not stopped a few endoparasitic nematodes from developing an absorptive outer surface. All the known examples involve larval or adult worms that live in the haemocoeles of insects (see also Chapter 3, p. 104). Transcuticular uptake of organic food has been demonstrated conclusively in such forms,[301] but the ultrastructural basis for this ability is a matter of some controversy. In one study,[290] adult female nematodes in the genus *Bradynema* were described as having a naked hypodermis bearing branched microvilli. No cuticle was supposed to be present. In another investigation,[255] the surface of parasitic heterosexual female specimens of *Heterotylenchus* was interpreted as a cuticle traversed by hollow cytoplasmic canals arising from the outer hypodermal surface and possibly opening at the cuticular surface. Similar canals have been described in the cuticle of the adult females of the nematode *Mermis nigrescens*[190] which are free living.

Crustacean endoparasites without guts which possess a chitin-protein exoskeletal cuticle have had to solve the same basic nutrient uptake problem as *Bradynema* and *Heterotylenchus*. The best known example is the endoparasitic cirripede of crabs called *Sacculina*. This form loses all external resemblance to a free-living barnacle and consists of a brood sac in which eggs are stored and a series of rootlets which ramify through the crab's tissues. The rootlets are absorptive, but are covered by cuticle probably composed of a chitin-protein mixture. The cuticle is extremely thin and is presumably permeable to organic nutrients.

MECHANICAL ANCHORAGE

One large-scale aspect of the interface between parasite and host or between symbiotic partners which has received much attention is that of the biomechanics of attachment or anchorage. Hooks, spines, suckers, clamps, pulleys, snap fasteners, distension bulbs, levers and glue all play their part in a fascinating range of both straightforward and bizarre engineering solutions to the problem of maintaining contact between two dissimilar organisms. Recently, Nachtigall[253] has made an extensive comparative and categorizing review of attachment methods and there seems little point in trying to list host-parasite and symbiotic examples from his 50 sub-categories of linkage modes. Two examples, however, one parasitic and one symbiotic, might serve to map out the range of complexity encountered in this interesting field.

Acanthocephalus ranae

Hammond[134–5] has analysed the method whereby the acanthocephalan parasite, *A. ranae*, attaches itself to the gut of its anuran host (see Fig. 2.30). The hooked and evertible proboscis is pushed into the gut wall of the host so that the backward-pointing and recurved hooks on the proboscis are embedded in host tissue. Then the contraction of neck retractor muscles pulls the body hard against the gut wall. At the same time this muscular activity squeezes fluid from paired internal organs, called lemnisci, into spaces in the proboscis wall. As fluid is forced into this wall, it expands the wall thickness and locks host gut tissue between the recurved ends of the hooks and the proboscis wall. As long as the muscles stay contracted this hydraulic mechanism provides extremely firm anchorage.

Remoras[253]

Teleost fish in the family Echeneidae (the remoras) show extreme morphological specialization for strong but rapidly reversible attachment to large, carnivorous 'host' fish like sharks and marlins. All remoras have oval suction devices on the dorsum of their heads apparently developed from the dorsal fin. With these devices they remain clamped for long periods to the surface of their larger partners. Some such relationships may be commensal in character but others are

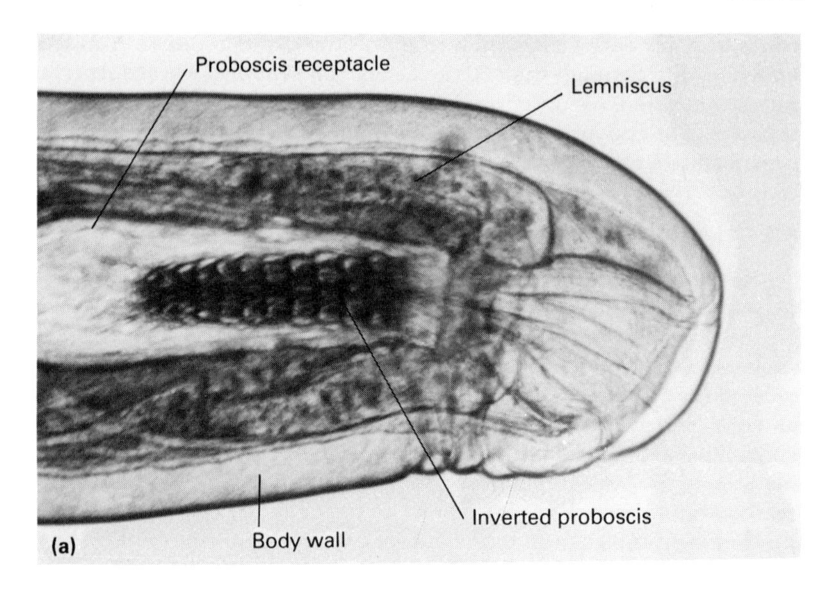

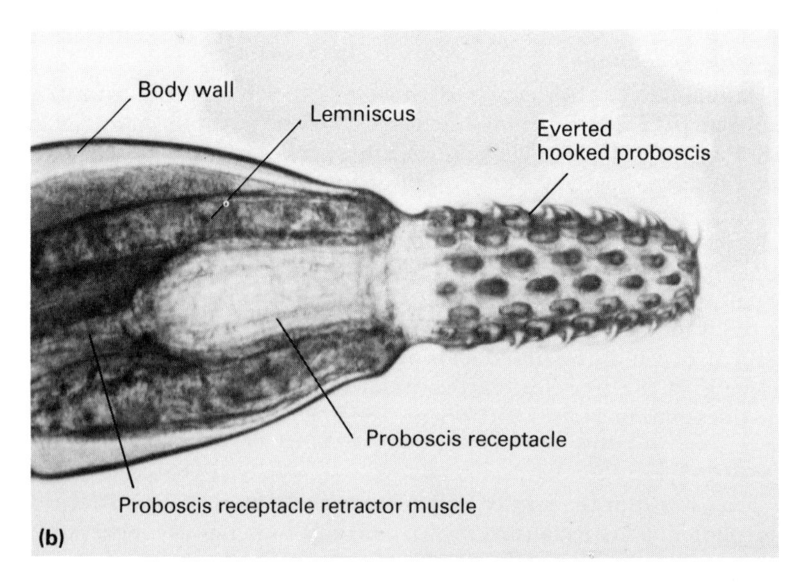

Fig. 2.30 The proboscis apparatus of the palaeacanthocephalan *Acanthocephalus ranae*. (**a**) Inverted. (**b**) Everted. (Photomicrographs of living *A. ranae* supplied by Dr. R. A. Hammond, University College, Cardiff.)

probably symbiotic as the remoras feed on crustacean ectoparasites of the larger fish as well as food scraps from the 'host's' prey. The sucker is an oval depression, the floor of which is divided up into compartments by two lateral rows of toothed lamellae, each supported by an internal bony plate. In the non-attached condition the plates have their free edges pointing forwards and lie flat. During attachment the remora brings the sucker up against the larger fish and then pushes forwards. The toothed plates engage with the 'host's' surface and are pushed upright. While changing configuration like this they increase the volume of the sucker depression and apply a suction force as well as propping the sucker cavity in this 'suck' configuration. For quick detachment the remora moves backwards, the plates fold back again and the suction pressure is released. In Polynesia fishermen use remoras to catch turtles. They tie the remora to a line, let it swim and attach to a turtle and then pull both fish and turtle back into the boat!

3

Nutrient Exchanges in Associations

INTRODUCTION

An investigation of the movement of nutrients between the partners in a symbiotic or parasitic relationship leads one close to the central aspects of an understanding of such associations. Most of the more useful attempts to define parasitism and symbiosis have had to concern themselves to some extent with this phenomenon of nutrient transfer. Specializations for the passage of food substances between partners often have influences that can be observed in other compartments of association biology. There is also every reason to believe that in an evolutionary sense the acquisition of food resources has been a potent factor in maintaining otherwise unstable relationships. Indeed, it is a selection pressure that has probably resulted in the initiation of associations. Structural, behavioural and physiological adaptations of partners can often be related to the all-important trophic requirements. Only in a few symbioses and host-parasite relationships is nutrient exchange of peripheral significance.

The investigation of the passage of nutrients between associates has been a productive and rapidly widening field of research in recent years. Modern methods of analysis, especially biochemical ones, have enabled the research worker to interpose himself experimentally between the associating organisms and to monitor exchanges between them. Investigations of this sort, although of very great importance, are vitiated by intrinsic problems of interpretation. The observer must interfere with the fluxes under study if he is to identify and measure them. One has constantly to ask to what extent the investigator's perturbation has altered the nature of the system under study. This basic problem has prompted, on the one hand, the use of increasingly indirect and subtle means of monitoring exchanges, and on the other a realization that conclusions about exchange phenomena must be based on zones of coinciding evidence gained from independent investigatory techniques.

PARASITE NUTRITION

Bacterial nutrition

Parasitic bacteria utilize almost all types of eukaryotic organisms as hosts. The bacterial parasites whose nutritional requirements and nutrient uptake mechanisms have been best studied are those that cause overt diseases in man and other vertebrates. They are all heterotrophs.

Microorganisms that live and feed inside the cells of their hosts must have defence mechanisms that enable them to withstand the lysosomal attack which eukaryotic cells usually mount against ingested material. Recent studies have provided clues about the way in which pathogenic mycobacteria achieve this protection. These bacteria are able to feed and reproduce inside macrophages which have ingested them. Electron microscopy reveals that *Mycobacterium tuberculosis* enclosed within phagosomal vacuoles within macrophages does not elicit lysosomal discharge into the phagosomes. Assay work has shown that macrophages containing live mycobacteria (*M. microti*) have abnormally high intracellular levels of cyclic AMP and suggests that this substance is of bacterial origin (Fig. 3.1). It is known that in experimental conditions high intracellular concentrations of cyclic AMP mediate pharmacological inhibition of lysosomal discharge in leucocytes, so it seems that cyclic AMP synthesized and released by the bacteria protects them from the bacteriocidal lysosome contents. The cyclic AMP appears to make the fusion of phagosomal and lysosomal membranes more difficult to accomplish. Macrophages which have ingested cells of another mycobacterium, *M. lepraemurium*, show no cyclic AMP increase and here it is assumed that the survival of the bacteria is a function of the specialized lipid coat which surrounds each parasite and appears to offer protection against lysosomal enzymes. Another intracellular microorganism that can inhibit lysosome-phagosome fusion is *Toxoplasma gondii*; it would be interesting to know whether host cells containing this parasite also have enhanced cyclic AMP levels.

At least one parasitic bacterium uses other prokaryotic bacterial cells as its hosts.[1,336–7] Vibrio bacteria in the genus *Bdellovibrio* utilize

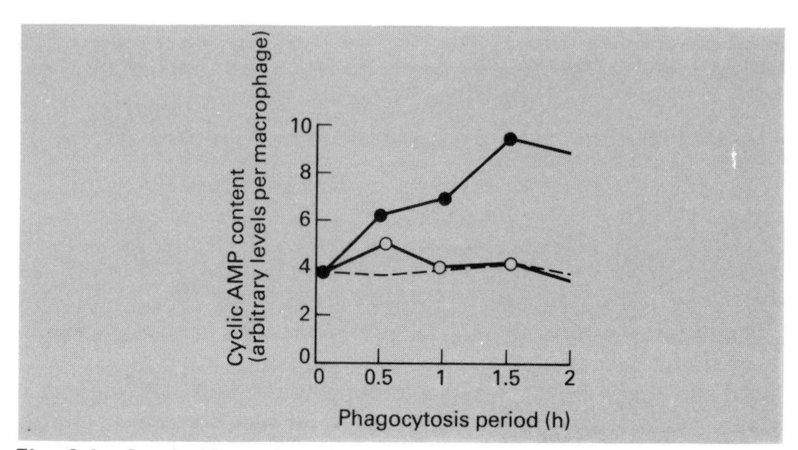

Fig. 3.1 Graph illustrating the probable synthesis of cyclic AMP by *Mycobacterium microti* inside macrophages. Monolayers of macrophages ($8{\cdot}4 \times 10^5$ cells) were exposed to levels of about 1000 bacteria per macrophage. Cyclic AMP levels in the cultures were then assayed through two hours of phagocytosis. (— — — —) control, i.e. no bacteria; (●———●) live bacteria; (○———○) heat killed bacteria. (Redrawn from Lowrie, D. B., Jackett, P. S. and Ratcliffe, N. A. (1975). *Nature, Lond.*, **254**, 600–2.

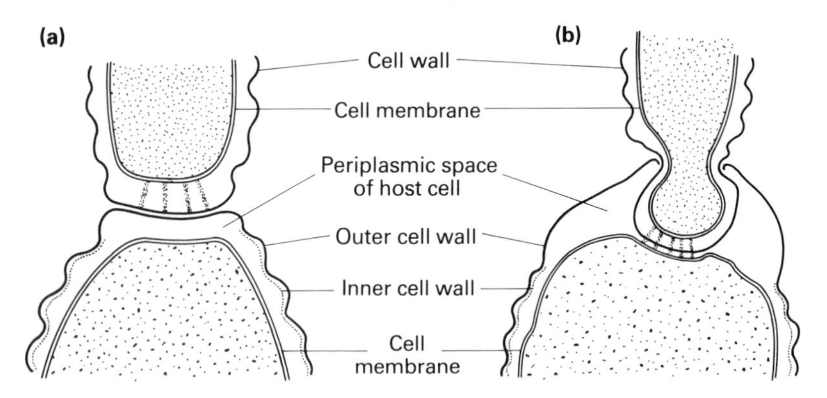

Fig. 3.2 The early phases of the interaction of *Bdellovibrio* with an *Escherichia coli* host cell. (**a**) *Bdellovibrio*, the upper cell, has made contact with the host cell. The cell walls flatten against one another and strands of electron dense material extend from the protoplast to the cell wall of *Bdellovibrio* adjacent to the area of contact. Beneath this zone the inner cell wall of the host cell is disrupted. (**b**) *Bdellovibrio* has penetrated the cell wall of the host cell but the cell membrane of the latter remains intact. Around the zone of penetration the host periplasmic space has enlarged and *Bdellovibrio* expands into it. The strands of dense material appear to be continuous with similar strands attached to the host cell membrane. (Based on electron micrographs in Abram *et al.*[1])

Gram-negative bacteria like *E. coli* in a parasitic manner (Fig. 3.2). The cell wall of the host is breached in one place by the secretion of proteases, lipases and neuraminidases. The cell membrane, however, is not digested and *Bdellovibrio* takes up residence in the periplasmic space between the cell wall and the cell membrane of the host bacterium. In this location it forms a specialized 'infection cushion' attachment to the host's cell membrane and proceeds to extract nutrients from the host's cytoplasm in an efficient and meticulously controlled fashion. The food source is progressively degraded, but even when 30% of the cytoplasm has been removed, 40 minutes after attachment, the host cell still retains the capacity to recover if the parasitic vibrio is removed. This can be done experimentally by the use of sodium desoxycholate to which the parasites are extremely susceptible. The maintenance of the host's viability for much of the period during which nutrient removal takes place appears obligatory for the transfer itself. By the time nutrient extraction is concluded, however, the host cell is dead and the original *Bdellovibrio* has produced a chain of daughter cells at the nutrient expense of the host. Different strains of *Bdellovibrio* demonstrate an interestingly wide range of dependence on interactions with host cells. Some are facultative parasites, others obligate, and there are even strains which are entirely free-living. Some forms are entirely specific to a particular host species and strain, whereas contrasting types show a far more catholic appetite for a range of different bacterial cells. The short absolute life spans of individual bacterial cells make it difficult to discuss relationship permanence in bacterial associations using the same criteria that are used for more long-lived metazoans. The *Bdellovibrio*-host bacterium association, for instance, although terminated by the host's inevitable death after a few hours, in fact represents a large proportion of the generation times of both organisms. The host's final demise means that the relationship shows one of the important attributes of the parasitoid type of interaction.

Protozoan nutrition

Protozoan parasites, like the bacteria discussed above, are often able to absorb nutrient molecules across their general cell surface. Monosaccharides, amino acids and other small organic moieties may be removed from the host's free pools of such compounds by diffusion or active transport mechanisms. Experiments to establish, *in vitro*, the nutritional culture requirements of parasitic protozoans have told us much about these nutrient exchanges. They also provide information concerning the changes in nutrient needs that occur during the different phases of those protozoans with complex life cycles. In this latter regard, the trypanosome flagellates are, by far, the most heavily investigated group. Honigberg[151] has pointed out that there are only a few groups of parasitic organisms which have been the subject of more biochemical and physiological study than the family Trypanosomatidae.

Trypanosomatids and other flagellates

The bloodstream-inhabiting forms of the pathogenic species, *Trypanosoma brucei*, are obligate extracellular parasites. They live in human plasma which has glucose levels of about 1 mg/ml. In this homeostatically maintained superabundance of a convenient energy source, the slender trypomastigote phases of this parasite display a relatively huge carbohydrate consumption. They can take up glucose equal to their own dry weight in 1–2 hours. They show no signs of storing polysaccharide reserves and are utterly dependent on the uptake of exogenous glucose from the plasma, with transport apparently taking place directly across the general cell surface. The absorbed glucose is metabolized purely as an energy source, its carbons contributing nothing to the synthetic activities of the parasite. Glucose is converted almost stoichiometrically by glycolysis to pyruvate in a metabolic pathway which is aerobic but contains none of the normal components of a Kreb's cycle nor phosphorylating cytochrome chain (Fig. 3.3). Pyruvate produced in this way passes out of the flagellates into the blood.[116,151,368]

The reduced NAD formed during synthesis of the pyruvate waste product is reoxidized by a coupled oxidation-reduction reaction in which dihydroxyacetone phosphate is reduced to α-glycerophosphate. Reoxidation of the latter is achieved with the oxygen which is plentifully available in the plasma, capillary partial pressures in man being about 95 mm Hg. In suboptimal concentrations of oxygen, not all of the glycerophosphate can be reoxidized and some is cleaved into inorganic phosphate and glycerol. Overall, the respiration of the slender trypomastigotes is extremely wasteful of both glucose and oxygen in terms of the efficiency of ATP production. In fact, the parasites hand back to the host the vast majority of the utilizable bond energy in the carbohydrate that they absorb. The excreted pyruvate can be used aerobically by the host's mitochondria in cytochrome-linked phosphorylation. Of the 38 ATP molecules that most eukaryotic cells can synthesize from one glucose molecule in aerobic conditions, the slender trypomastigotes can only form two. It is the lack of a cytochrome system which explains most of this poor performance. It also explains the insensitivity of ATP production in these flagellates to inhibitors like cyanide, azide and antimycin. Cyanide and azide inhibit

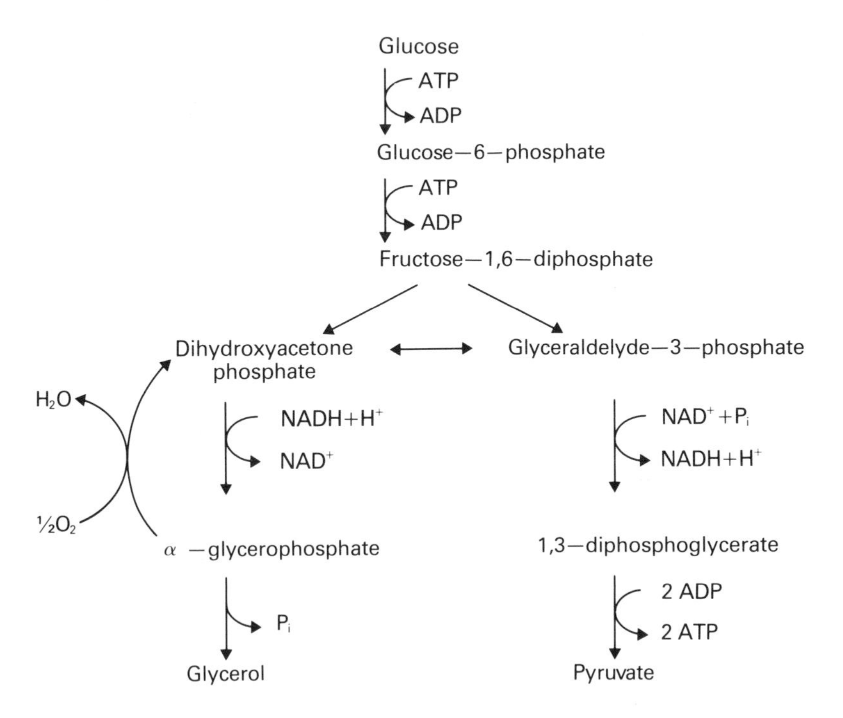

Fig. 3.3 Glycolysis in bloodstream trypanosomes. Certain reversible reactions are shown as though they were unidirectional for simplicity. P_i, inorganic phosphate. (Based on metabolic pathways described by Honigberg.[151])

the cytochrome a_3 to oxygen link and antimycin the cytochrome b to cytochrome c_1 link in the cytochrome respiratory chain. They might be expected to be without effect in an animal cell with no such system. Inhibitors of the iodoacetate type, however, which have profound repressive effects on dehydrogenase enzymes, do adversely affect trypanosome ATP production. They probably inhibit the important oxidation-reduction reactions linked with NAD reoxidation. In all eukaryotic cells the functional subunits of Kreb's cycle and the cytochrome chain are located on the inner mitochondrial membrane. Trypanosomes have a single extensive mitochondrion and Vickerman[363,365] has shown that in slender trypomastigotes it is thin and has sparse, short inner membrane cristae. This finding provides elegant ultrastructural confirmation of the biochemical indications that Kreb's cycle and cytochrome-based phosphorylation are lacking in these parasites.

It might be thought that these gaps in the metabolic repertoire of some bloodstream trypanosomes represented a form of genetic erosion. In a situation where there is no negative selection pressure on a low efficiency of glucose utilization this would seem to be a reasonable hypothesis, especially as in vertebrate blood plasma glucose and oxygen could not be expected to be limiting resources. The hypothesis is, however, untenable because of the crucial fact that the forms of *T. brucei* that live in the gut of the insect host have a functional Kreb's cycle and cytochrome system correlatable structurally with an enlarged mitochondrion containing numerous cristae. Unfortunately, little direct information is available concerning these interesting endoparasites of invertebrates but most workers consider that the laboratory culture forms of the parsite provide realistic physiological models for the phases that live in the gut of the tsetse fly. Such model flagellates, like the insect forms, can achieve far more efficient yields of ATP from glucose than can the profligate bloodstream types. It appears that in the relatively low glucose and oxygen concentrations of the insect gut increased respiratory efficiency is at a premium. In fact, the glucose and oxygen uptake by culture forms occurs at rates representing only about 10% of that demonstrated by the slender trypomastigotes and amino acids provide an important additional energy source.

It seems very likely that the new environment provided by the insect requires a new pattern of gene expression that results in the synthesis of the components of Kreb's cycle and the cytochrome chain. In some polymorphic forms of *T. brucei* this metabolic switch occurs in the stumpy trypomastigotes while they are still in the vertebrate blood rather than being directly induced by conditions in the insect. Thus the metabolic deficits demonstrated by blood-inhabiting trypanosomes are no more examples of real genetic loss than is the fact that human fibroblasts do not synthesize haemoglobin or insulin although they must have the genetic information to do so. In both cases we are dealing with instances of cytodifferentiation and gene expression rather than gene loss.

Several trypanosomatids and even examples of the free-living flagellate genus *Bodo*, thought to represent an antecedent group to the Trypanosomatidae, possess intracytoplasmic symbiotes which are probably bacteria as they contain 70S ribosomes and their protein synthesis can be stopped with chloramphenicol, an antibiotic. The symbiotes appear to have nutritional significance. Two species of *Crithidia* isolated from bugs, that is *C. deani* and *C. oncopelti*, have unexpectedly simple nutritional requirements in culture when compared with other members of this trypanosomatid genus.

Table 3.1 The organic constituents of the defined culture medium in which *Crithidia deanei* can be grown *in vitro*. (From Mundim, Roitman, Hermans and Kitajima.)[246]

Compound	Concentration ($g\ l^{-1}$)
Sucrose	20
β-glycerophosphate	20
Potassium citrate	10
Succinic acid	1·0
Citric acid	0·5
L-tyrosine	0·4
L-methionine	0·2
Malic acid	0·2
Nicotinamide	0·002
Folic acid	0·002
Thiamine HCl	0·006
Biotin	0·00008

Specifically, in respect of amino acids the former has an obligate requirement for only methionine and tyrosine[246] (see Table 3.1), the latter a requirement for methionine.[254] Both species contain symbiotes (Fig. 3.4) and it is assumed that these prokaryotes are synthesizing some of the extra nutrients that the flagellates would need if aposymbiotic (without symbiotes). Recently, Chang and Trager[66] have removed the bacterial symbiotes (called bipolar bodies) from culture lines of *C. oncopelti* by prolonged treatment in a concentration of 800 μg/ml of chloramphenicol. The aposymbiotic flagellates showed new requirements for exogenous haemin and other nutritional factors present in liver extract.

A few trypanosomatids can obtain nutrients, grow and reproduce mitotically in intracellular sites. Perhaps the best example is *Leishmania donovani*, the causative agent of human visceral leishmaniasis. These flagellates can be said to live dangerously! Not only are they capable of living within macrophages, they do so within active phagolysosomes.[64] The survival of the parasites in this most hazardous of locations is apparently based upon resistance to macrophage lysosomal enzymes.

Bulk ingestion of host macromolecules

Although most parasitic protozoans can probably absorb small organic molecules across their cell membranes, several types can augment this uptake by the bulk ingestion of host macromolecules. The ingestion takes the form of entry of macromolecules in solution as well as in solid aggregates of host material. It is probably unwise in

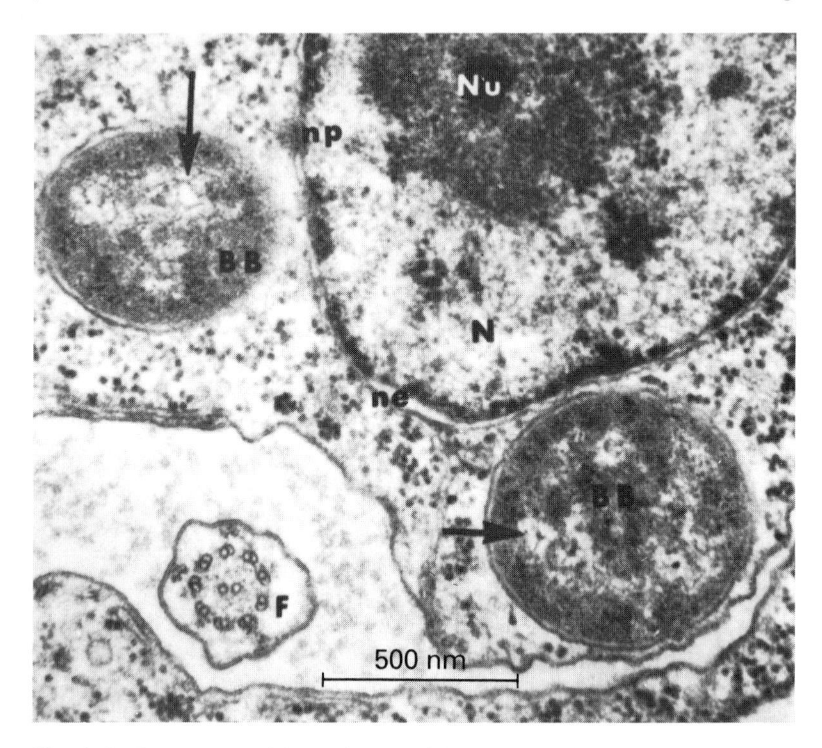

Fig. 3.4 Bacterial symbiotes (termed bipolar bodies) in the cytoplasm of *Crithidia deanei.* BB, bipolar body; F, flagellum of *Crithidia*; N, nucleus; ne, nuclear envelope; np nuclear pore; Nu, nucleolus. Arrows show fine fibrillar material in the bipolar bodies. (From Mundim *et al.*[246])

parasitic situations to describe such classes of uptake as pinocytosis and phagocytosis respectively. This is because in most real cases it must be assumed that solid and dissolved macromolecules are taken together into infolding vesicles of parasite plasma membrane. The evidence for bulk ingestion is almost entirely static and morphological, in most reports involving only the ultrastructural demonstration of cytostomal regions of the parasitic plasma membrane. These regions can range from temporary plasmalemmal indentations to organelles of permanent position in the cell and of some internal complexity. In most cases the cytostome itself is subtended by a cytostomal cavity into which the nutrients pass.

The mechanisms by which the parasitic cytostomal areas operate are entirely undetermined. By analogy with the process used in macromolecule uptake by other cells like rat macrophages[185] it is likely that the cytostomal plasma membrane is modified to interact specifically with different types of nutrient of host origin. Perhaps such interactions induce new membrane elaboration at the site of membrane-nutrient interplay, enlarging the size of the vesicles that form inwards from the cytostome. Nothing is known directly about the manner in which components of the ingested nutrients gain access to the cytoplasm proper. Analogies with non-parasitic uptake systems[260] suggest that the closed vesicles formed beneath the cytostome are equivalent to prelysosomes (heterophagosomes). These could then be expected to interact by fusion of membranes with primary lysosomes containing digestive enzymes. Such fusion produces a secondary lysosome (phagolysosome) which is the site of activity of lysosomal enzymes on the ingested nutrients. Products of digestion could then pass out of the secondary lysosome into the cytoplasm. Postlysosomes (residual bodies) containing undigested remnants of lysosomal activity might be expected in the late phases of the processing of macromolecular nutrients (Fig. 3.5).

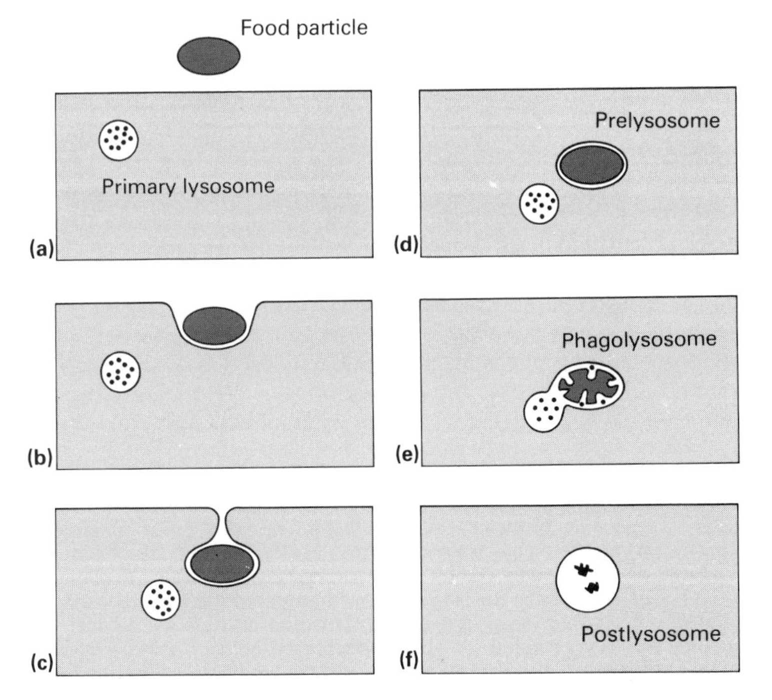

Fig. 3.5 (a–f) Diagrammatic representation of the involvement of lysosomes in the bulk uptake of macromolecules into a cell (see text for details).

This scheme for bulk nutrient uptake and assimilation by parasitic protozoans has the virtue of avoiding imprecise terms like food vacuole. It also achieves terminological comparability with the more intensively researched non-parasitic examples of such processes. Does it also tally with the actual electron-microscopical evidence relating to protozoan parasites? Although all of the components of the suggested sequence have yet to be definitely demonstrated in a particular host-parasite interaction, each of them has been observed.

A considerable amount of work has been carried out on the feeding behaviour of *Entamoeba histolytica* which can cause amoebic dysentery in man. Early ultrastructural work[109–10] pointed to the conclusion that the lysis of mammalian host cells by the amoebae depended on contact with surface-active lysosomes at the periphery of the parasites. These studies suggested that the lysosomes were equipped with tubular 'triggers' which initiated lysosomal discharge on contact with host cells. This plausible explanation now seems unlikely to be correct. McCaul[213] has shown, in elegant investigations using mammalian cell line monolayers in culture, that surface lysosomes play no part in initiating host cell damage. His alternative interpretation of the evidence is that the cells are first injured by an amoebic toxin, perhaps one associated with the plasma membrane of *Entamoeba*, and that the cells are then ingested along a large phagocytic channel in the amoebic cytoplasm. Electron micrographs of this process (Fig. 3.6) give a graphic impression of host cells being bodily sucked into the amoeba.

Later stages in the processing of bulk food by a parasitic amoeba have been identified in the amoebo-flagellate *Histomonas* which causes black spot disease in the turkey gut.[193] Here, in an invasive amoeboid phase in extracellular spaces, the parasites have large, blunt pseudopods. Each of these has several cytostomal invaginations which ingest host cell cytoplasm, usually still enclosed by a host plasma membrane. Deeper in the amoebic cytoplasm are found secondary lysosomes (food vacuoles) in which digestion of host material occurs.

Fig. 3.6 *Entamoeba histolytica* ingesting cells from mammalian cultured cell lines. (All micrographs from McCaul.[213]) (**a**) Transmission electron micrograph of a BD-V1 cell (rat epithelial cell line) being ingested by an *Entamoeba* trophozoite. (**b**) Scanning electron micrograph of an *Entamoeba* trophozoite on a monolayer of RK 13 cells (rabbit kidney epithelial cell line). Long mucoid threads are seen attached to the amoeba. (**c**) Photomicrograph of a late stage in the interaction between an *Entamoeba* trophozoite and an RK 13 cell. The preparation has been stained for acid phosphatase activity using β-glycerophosphate as substrate. Note the reaction product in the phagocytic channel and cytoplasm of the trophozoite. b, phagocytic bulb; c, phagocytic channel; e, *Entamoeba* cytoplasm; H, mammalian cell (being sucked into phagocytic channel in (**a**)); the arrow indicates the phagocytic channel.

Malarial trophozoites inhabit the matrix of red blood cells. They are bounded by their own plasma membrane and a second, outer membrane sometimes termed the parasitophorous vacuole. Portions of the erythrocytic matrix, composed mainly of haemoglobin and

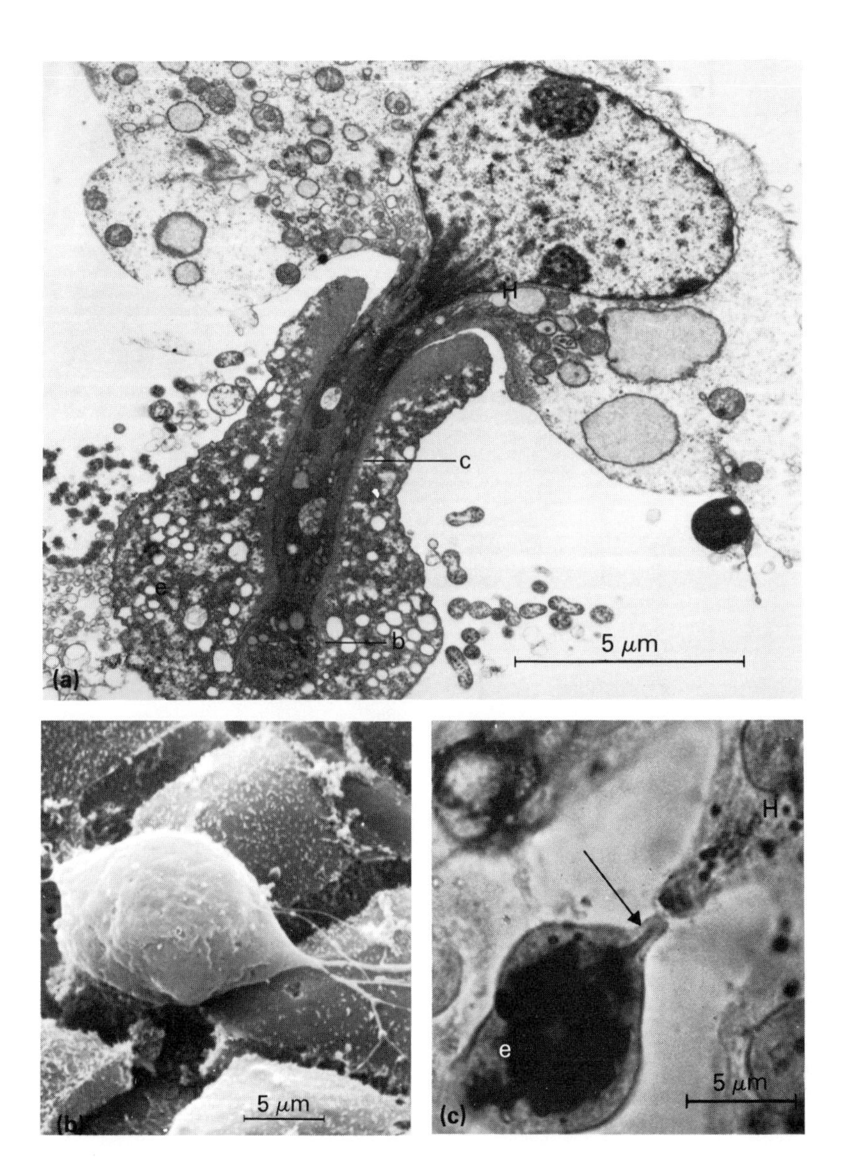

bounded at first by the outer membrane, are ingested by the parasite. Such phagotrophy, as it is called, often takes place through a differentiated cytostome. From the cytostomal cavity, vacuoles of matrix are budded off internally and these are ultimately the sites of haemoglobin digestion. In *Plasmodium elongatum*[2] the pinched-off

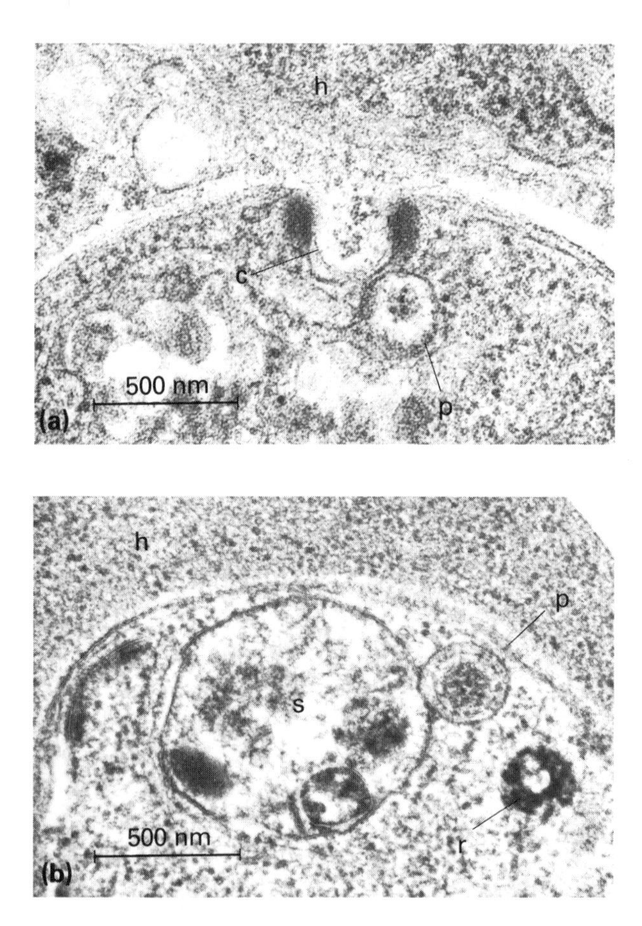

Fig. 3.7 Food ingestion by the avian malarial parasite, *Plasmodium elongatum*. (**a**) The cytostomal region of a parasite cell. Host cytoplasm is present in the cytostomal cavity and a phagosome (prelysosome) has probably just passed into the cytoplasm beneath the cytostome. (**b**) Large digestive vacuole (secondary lysosome, phagolysosome) with a phagosome close to it. c, cytostome; h, host cell cytoplasm; p, prelysosome (phagosome); r, post-lysosome?; s, digestive vacuole. (Transmission electron micrograph from Aikawa *et al.*[2])

vesicles fuse with a larger digestive vacuole before lysis of food occurs. In this case the conjoined vacuoles must be considered a secondary lysosome and the digestive vacuole perhaps a primary lysosome (Fig. 3.7). In *P. berghei* and *P. gallinaceum* electron-microscopical histochemical techniques have demonstrated[4] the typical acid phosphatase activity of lysosomes in the vacuoles containing haemoglobin matrix. Malarial trophozoites even provide examples of what appear to be postlysosomes. Some forms show the presence of haem pigment crystals, representing the indigestible portion of the haemoglobin molecule, at the periphery of the parasitic cell.

Hopefully this general understanding of macromolecular nutrient uptake in malarial parasites can be used as a general conceptual framework for the assessment of related phenomena in other parasitic protozoans for which the nutrients are not so circumscribed and identifiable as is haemoglobin. It is encouraging that, at least at the ultrastructural level, uptake by an extracellular amoebo-flagellate, *Histomonas*, is similar to ingestion by *Plasmodium* within an erythrocyte. Perhaps there really is a basic series of cytoplasmic mechanisms at work here which can be adapted, truncated or refined by different parasitic protozoans ingesting different foods in disparate hosts.

Metazoan nutrition

Members of almost every metazoan phylum have become adapted for parasitic modes of existence. Were each group considered separately, the methods by which these parasites obtain their host-derived food would constitute an unmanageably extensive list. It would contain, because of duplication, a vast amount of redundant information. In order to rationalize the treatment of this spectrum of trophic organizations, it is convenient to consider relatively few basic feeding methods, one or some of which are employed by most metazoan parasites. Ordered in this way the nutritional eccentricities of particular parasites can be understood as modifications of a basic adaptive mode. The specializations of a single parasite's food-gathering techniques can be seen as adjustments of niche parameters to fit particular host conditions and particular patterns of competition.

Most metazoan parasites utilize feeding methods encompassed by the following scheme:

(a) surface browsing
(b) blood feeding
(c) bulk tissue feeding
(d) uptake across the parasitic body wall.

Surface browsing

Ectoparasites: Host epithelia, whether external or internal, represent renewable and croppable fields of nutrients for many parasites. Almost all epithelial tissues are subject to relatively rapid replacement in the healthy host and renewal rates can often be increased in conditions of unusually high cell wastage. A parasite is unlikely to inflict serious pathological damage on its host if it crops the epithelial resource at rates which lie within the attainable rates of cell renewal.

The few digenean species that have adopted ectoparasitic ways of life all seem to be browsers of external host epithelia. *Transversotrema patialense*, an ectoparasite of fish, that inhabits the recesses under its host's scales, appears to feed on mucus and skin epithelial cells whilst holding onto the skin surface with its spined ventral sucker (Fig. 5.5d). *Philophthalmus*, which lives under the optic nictitating membrane of birds, can absorb nutrients across its body wall to a certain extent[257] but probably also ingests the mucopolysaccharide secretions of the orbit as well as sloughed epithelial cells.

Most monoposthocotylinean monogeneans, characterized by their relatively simple posterior suckers, live attached to the skin of fishes. Detailed experimental studies by Kearn[169] have revealed that these skin-dwelling platyhelminths feed almost exclusively on the epidermis of their hosts (Fig. 3.8). Most of the detailed work on this aspect of monogenean biology has been carried out on *Entobdella soleae*, a monopisthocotylinean that lives on the skin of soles. In this marine fish, as in most teleosts, the epidermis is non-keratinized and consists of six or seven layers of epidermal cells plus superficial mucus cells which produce a slime layer that covers the body of the fish. Beneath the epidermis is a collagenous dermis containing coloured chromatophore cells. An adult *E. soleae* attaches to such a skin surface by its posterior sucker. The worm's mouth is situated ventrally in the head region. It opens internally into a voluminous cavity that contains an invaginated pharynx. Distal portions of the pharynx can be evaginated through the mouth to produce a bell-shaped cup that is pressed against the host's epidermis. Within the wall of the bell is a ring of large gland cells. During feeding contacts with the fish's surface, each of which lasts about five minutes, these glands produce a powerful proteolytic secretion which rapidly lyses the epidermal and mucus cells enclosed by the everted pharynx. Peristalsis by the wall of the proximal part of the pharynx pushes the slurry of partly digested food back into a circular intestine. After a feeding contact, fluid food can be seen in all the intestinal caecae being moved by the musculature of the intestinal region.

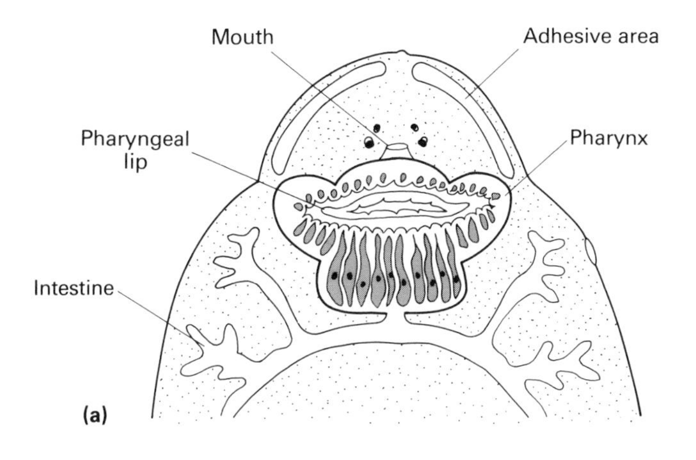

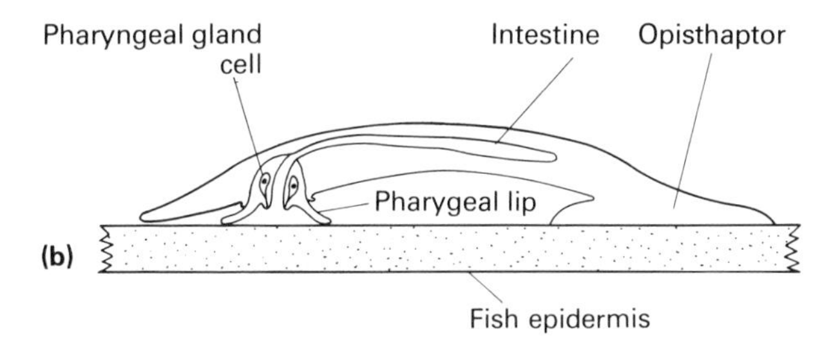

Fig. 3.8 Feeding structures in the monogenean *Entobdella soleae.* (**a**) Organization of the pharyngeal area in the non-everted condition. The tinted cells in the pharynx are the pharyngeal glands. (**b**) A diagram of a sagittal section through a feeding parasite on the epidermis of a fish. (Both redrawn from Kearn.[169])

A few monopisthocotylideans have moved into endoparasitic sites on their fish hosts. *Calicotyle krögeri*, for one, lives in the cloaca of the thornback ray, *Raia clavata.* Despite its changed location it has retained the surface-browsing habits of its presumably ectoparasitic ancestors. Once more the removal of host tissue is carefully regulated. It seems that only mucus and shed cloacal epithelial cells are ingested. Attached, living cells are untouched.

The non-keratinized epidermis of a fish is soft and relatively easily digested. The keratinized outer layers of mammals and birds, however, present far more serious problems to a would-be parasite, being relatively hard and very resistant to enzymatic lysis. The disulphide and hydrogen bonds that lock the fibrillar and matrical components of keratin together can resist the activities of most normal protein-cleaving enzymes. Some parasites that feed by surface browsing have, however, been able to utilize this apparently intractable resource which, like the fish epidermis, is constantly renewed by the host.

The feather lice, or mallophagans, are a group of wingless insects, closely related to the free-living psocids, that have become highly specialized for keratin feeding. They eat bird feathers and epidermis. All of the mallophagan life cycle takes place in physical contact with this food source. Eggs are stuck to feathers by a cement; they are usually found in the grooves between the feather barbs, a position which protects them from the preening activities of the host bird. Nymphs hatch from the eggs and pass through three moults before gaining adulthood. Both larval and mature lice feed on keratin, supplemented by host skin secretions and debris found between the feathers. The psocid-like ancestors of the feather lice probably fed like present-day psocids on organic debris, fungi and lichens. A change from this diet to one of feathers would not have entailed any large-scale alterations in the mouthparts. Mallophagans grasp feather barbs with their rear two pairs of legs and use the anterior legs and labrum to push individual feather barbules into the mouth and past the mandibles. These are positioned either at the front edge of, or within, the box-like, dorso-ventrally flattened head skeleton. The mandibles chop off small lengths of barbule which are passed into the crop. Here strong peristalsis of the crop and comb-like projections of the crop wall together help to triturate the tough feather keratin. Digestion of the ground-up structural protein is made possible by the production of strong reducing conditions in the crop. In these conditions the covalent disulphide bonds (– S – S –) that cross-link polypeptide chains in keratin are split by a reduction reaction to a pair of sulphydryl groups (—SH HS—). With these covalent links ruptured, the remaining hydrogen bonds cannot hinder the subsequent lysis of the keratin fragments by proteolytic enzymes that are secreted into the crop.

Endoparasites : Internal host epithelia are utilized extensively as food sources by many metazoan parasites. Most well documented examples in this category are digeneans. The adult of *Fasciola hepatica*, the liver fluke, inhabits the bile duct of sheep. Juvenile worms reach this site subsequent to a migration that takes them, after metacercarial activation in the gut, through the gut wall, across the peritoneal cavity and into the liver parenchyma. During the wanderings of the juveniles they are bulk feeders, ingesting, for instance, smooth muscle cells in the gut wall and parenchyma cells in the liver. When they reach the lumina of the bile ducts, though, their nutritional strategy changes. The feeding habits of these adult worms have been the subject of much investigation and a fair amount of healthy controversy. The differences of interpretation have been useful, for they have made obvious the need for experimental rigour in nutritional experiments on parasites. An impartial synthesis of the results achieved to date would indicate that the adult flukes were feeders with a catholic taste. They certainly browse the epithelium of the bile duct, but they also seem to be able to feed to an undetermined extent on blood and other hepatic tissues as well as absorbing small organic molecules across their tegument. Dawes and Hughes, in an extensive series of histological investigations,[97] have provided apparently irrefutable morphological evidence that the oral sucker of the adult worm can be used to ingest bile duct epithelium in addition to glandular and collagenous tissues beneath the epithelium. Interestingly, these cell layers become hyperplastic during the infection of a host with *Fasciola*. The hyperplasia is initiated before the flukes actually enter the bile duct, during their liver migration phase. Dawes' results provide a clear-cut and important instance of a host-parasite interaction in which the presence of the parasite triggers an increase in the production of the tissues upon which the parasite feeds.

Fasciola is a large, powerful digenean with sharp tegumental spines. It would be surprising if such an animal did not cause some traumatic injury to the host tissues with which it was in contact. Such trauma as well as the abrading activities of the oral sucker would almost certainly give adult *Fasciola* the opportunity of ingesting other hepatic tissues including blood. Certainly erythrocytes and leucocytes have been observed in the fluke's intestinal caeca and in smears of the gut contents squeezed out through the mouth by compression.[338] Attempts have been made to quantify the possible blood uptake of *Fasciola* by isotopic assay methods, but to date they have not been well controlled enough to provide definitive results. In one study,[271] a proportion of the red blood cells of sheep hosts were labelled with ^{51}Cr. Two hours after the labelling, adult *Fasciola* were recovered from the sheeps' bile ducts and their levels of radioactivity assessed. The flukes were found to contain ^{51}Cr and the quantitative results were interpreted as meaning that adult flukes ingested about 0.03 ml of blood hour^{-1}. Unfortunately, the experimental design employed did not enable the form in which the label had been taken up to be determined. This poses analytical problems as the same study also showed that within two hours of the administration of labelled erythrocytes, high levels of activity were present in bile itself. If ^{51}Cr was present in some soluble, inorganic or organic form in the bile it could be assimilated by the flukes in ways which have no connection with blood feeding. *In vitro* experiments by Mansour[227] on adult *Fasciola* have shown that the worms can absorb glucose across the body wall and use it in glycolysis. In performing such experiments on digeneans it is usually difficult to exclude the possibility of gross nutrient uptake via the mouth. *Fasciola*, however, is large and this fact enabled Mansour to perform the manipulation of ligating the anterior ends of worms by tying a cotton ligature between the oral and ventral suckers. Worms treated in this way could not take up nutrients via the alimentary canal. Despite this restriction, in anaerobic conditions, in a culture system containing 0.011 M glucose, ligated and non-ligated worms took up and metabolized glucose at remarkably similar rates. The mean uptake rate in experiments with unligated *Fasciola* was 19.5 μmoles glucose g wet weight^{-1} hour^{-1}, with ligated worms 18.7 μ moles g wet weight^{-1} hour^{-1}. Mansour's experiments do not allow any deductions to be made concerning the mechanism of this movement of glucose across the tegument as the kinetics of uptake were not studied. The transfer could represent diffusion, active transport or a combination of both. A typical free glucose concentration in vertebrate bile is 5 mM which compares with the 11 mM level used in Mansour's culture fluids. The similarity of these molarities suggests that, whatever the exact mechanism of glucose ingress, the transport probably has a nutritional significance *in vivo*.

More recent work on adult *Schistosoma mansoni*[16] has shown that the tegument in this digenean is the predominant, if not exclusive, site for the uptake of glycine and proline from the serum of the host's blood. If such findings can be extrapolated to *Fasciola* it seems likely that many different classes of small organic molecules may be absorbed trans-tegumentally by the adult in the bile ducts. A wide range of such compounds including amino acids, monosaccharides and fatty acids certainly exist in solution in bile.

Of the digenean browsers on internal host epithelia, the gut dwelling strigeoids are the group which shows the most marked specializations for this nutritional habit.[115] Most inhabit the alimentary canals of birds and certainly utilize the epithelial lining of the gut as food. Like *Fasciola*, however, some probably have a mixed nutrition and can obtain nutrients in ways other than surface browsing. The most elaborately specialized strigeoids are forms like *Apatemon gracilis*. In this digenean the anterior portion of the body, termed the forebody, is a deep cup-shaped structure which enfolds host villi. Within the cup are an oral and a ventral sucker, a lappet on either side of the mouth and an adhesive organ consisting of two large lobes which come into very close contact with villi. The villi within the forebody cup are denuded of epithelium and it is likely that these cells represent an important component of the diet of *Apatemon*. Both the lappets and adhesive organ can secrete extracorporeal digestive enzymes and these serve to initiate maceration and lysis of epidermal cell sheets. The enzymic secretions of the lappets are produced by discrete gland cells whose cytoplasmic necks penetrate the tegumentary distal cytoplasm to reach the exterior. Secretions of the adhesive organ are elaborated in differentiated secretory zones of the syncytial tegumentary cytoplasm. Partly digested host cell debris within the forebody cup is ingested through the mouth. In addition, though, there is strong ultrastructural evidence that the adhesive organ is an absorptive as well as a digestive structure. The epithelium-lacking villi of the host are found in the most intimate contact with specialized regions of the tegument on the adhesive organ lobes. The tegument on the medial faces of the lobes is thrown into a profusion of first order plasma membrane folds and second order microvilli which arise from the edges of the folds. In thin sections of *Apatemon* fixed *in situ* in a duck's gut, the distance between this folded outer surface of the parasite and the lumen of capillaries within villi can be as little as 390 nm. This remarkably short diffusion path and the known 'leakiness' of capillaries to small organic molecules must be strong circumstantial evidence for the possibility of direct nutrient uptake by the attachment organ.

Blood feeding

Although the important vitamins of the B complex are absent, in most other respects vertebrate blood can be regarded as a liquid food containing a wide range of nutrients. Using human blood as a resonably representative example (Table 3.2), mineral salts, vitamins A and C, monosaccharides, amino acids, fatty acids, cholesterol, triglycerides and proteins are all present in the plasma. The cellular and cytoplasmic components, that is erythrocytes, leucocytes and platelets, provide a suspension of solid particulate food. In a given volume of total blood they provide the bulk of the blood protein, mainly in the form of haemoglobin.

For the vertebrate, and especially the homeothermic mammal or bird, the blood has a central importance in an intermeshing array of translocatory and metabolic pathways. Its vital role is correlated with an intricate system of regulatory control mechanisms which maintain the blood's composition in the face of perturbations in its molecular and cellular content. It is this subtle feedback regulation of constitution which makes blood the ultimate renewable resource when

Table 3.2 Blood as a food for parasites. Some constituents of nutritional significance in mammalian (human) blood: expressed as constituents per 100 mls of total blood.

		Haemoglobin	16 g
		Vitamin A	40–60 μg
Na^{+}	310–340 mg	Vitamin C	0·4–1·5 mg
K^{+}	14–20 mg	Cholesterol	150–250 mg
Ca^{++}	9–11 mg	Glucose	60–100 mg
Mg^{++}	1–3 mg	Total lipids	450–850 mg
Erythrocytes	5×10^{11}	Total fatty acids	190–420 mg
Leucocytes	5×10^{8}	Neutral lipid	0–150 mg
Platelets	$1{\cdot}5 \times 10^{10}$	Serum protein	6–8 g
		Pyruvate	1–2 mg
		Amino acid nitrogen	3–5·5 mg

used as a food by parasites. It has been utilized in this way by many metazoan parasites. Ectoparasites excavate openings into blood vessels from the exterior, insert tubular feeding devices of various types or lacerate the dermal capillary bed to make a feeding pool of blood. Endoparasites can be found either residing within blood vessels in a luminal location or removing blood from vessels whilst actually located outside the vascular system in, for instance, the gut.

Blood contains a potent immunological defence system, it clots and it lacks certain important nutrients like B vitamins. Each of these three facts can cause potential problems for blood-feeding parasites. The lack of B vitamins affects both ecto- and endoparasitic animals. If blood is the principal source of nutrients for such parasites, they often contain symbiotic microorganisms which supply the missing substances. These endosymbiotes are rickettsial or bacterial and in arthropods are found in specialised organs called mycetomes. Mycetomes have been described in blood-feeding suckling lice (Anoplura), the bed bug (*Cimex*), fleas (Syphonaptera), ticks and certain mites (Acarina). In the human body louse there is a mycetome which lies mid-ventrally in the wall of the midgut. In embryonic lice the symbiotes are found in a group of cells situated in the yolk of the gut lumen. These cells approach the mycetomal area and transfer the symbiotes to the mycetomal pouch. Symbiotes remain in this position in larval forms and in male lice. In the female, however, most move to the oviduct at the time of the last moult. From this location they enter each egg, ensuring that all lice obtain their quota of symbiotes. If the mycetome is removed experimentally from females before symbiote migration occurs, the louse rapidly dies and eggs which are laid before death are not viable. Equally, if, by centrifugation of eggs and embryos, the position of the mycetomal pouch is altered before the symbiotes enter it, the larval forms do not survive.[56] Results such as these suggest that substances produced by the symbiotes are crucial for normal growth and development as well as egg production.

Blood clotting can interfere with the feeding of blood-consuming parasites. To combat this interference, substances that inhibit the clotting process are often produced by the parasites at the feeding site.

The immunological properties of blood are potentially most damaging to those few parasites that are long-term inhabitants of the lumina of blood vessels. Forms which have been carefully studied, notably the digenean schistosomes, show elaborate antigenic subterfuges which ensure survival in an immunologically hostile environment (see page 130 *et seq.*, Chapter 4).

Apart from helminths like schistosomes, all other blood-feeding metazoans have to penetrate blood vessels from the exterior in order to feed. Convergent evolution has produced rather similar piercing and sucking mouthparts with associated supporting and sensory structures in a wide variety of parasitic insects. In the different taxa, components which perform parallel functions in the feeding process have often been drived from non-homologous mouth parts. To take only one contrasting pair, *Pediculus humanis*, the human body louse, and *Cimex*, the bed bug, both feed on human blood by inserting a thin, hollow feeding tube into a venule or small vein in the skin, but the origins of their analogous feeding structures are quite disparate (Fig. 3.9).

The polyopisthocotylidean monogeneans feed mainly on blood.[197] They have a terminal, rather than a ventral, mouth and the typical gill-inhabiting forms on fish use this to penetrate the thin epidermal and dermal tissues that separate the external environment from the branchial capillaries of their hosts. Host red blood cells are quickly haemolysed in the intestines of these flukes. This fact probably explains the difficulties encountered by workers who have tried to identify blood in the guts purely on morphological grounds. Erythrocytes are only seen in the gut of monogeneans that were feeding within a few minutes of examination.

Hookworms are strongyle nematodes that live in the vertebrate small intestine and feed both by browsing host mucosal epithelium and by removing capillary blood from gut villi. *Ancylostoma duodenale* and *Necator americanus* are hookworms that cause serious disease in man. The loss of blood caused by the worms is an important aspect of their pathogenesis. Such hookworms possess a buccal cavity which contains cutting plates and teeth composed of hard, sclerotized protein. They feed by drawing a plug of villus tissue into the buccal cavity and rasping it with the cutting plates. Anticoagulants seem to be

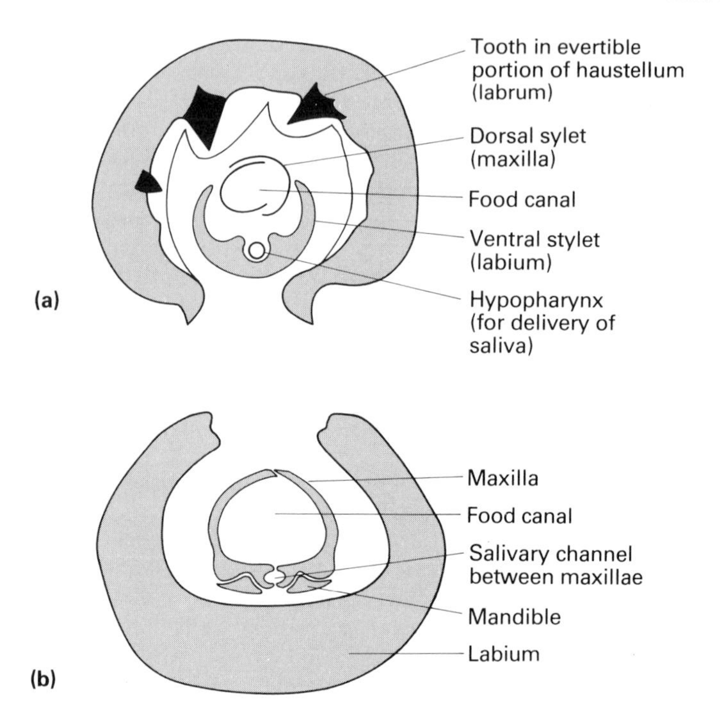

Fig. 3.9 Transverse sections through mouthparts modified for blood sucking from a human host in (**a**) *Pediculus humanus* (body louse) and (**b**) *Cimex lectularius* (bed bug). ((**a**) redrawn from Buxton,[56] (**b**) from Askew.[18])

produced by glands in the pharyngeal region. In the dog hookworm, *Ancylostoma caninum*, the muscular pharynx itself has been shown to pump host blood into the nematode's intestine at a high rate. A female worm can ingest between 46 and 60 mm^3 of blood per day. The transit time of the blood through the worm's gut is only about 10 minutes and about 50% of the ingested red blood cells remain intact after passing through the worm. The fates of the serum and cellular components of the blood meal are not fully understood, but it seems likely that small organic molecules in the serum must be assimilated. Detailed work on the feeding of hookworms has only become possible since the development by Roche and Torres[292] of an *in vitro* chamber designed specifically for such studies. Worms are held by and penetrate a thin rubber membrane which divides a chamber of saline into two sub-chambers. In one is placed host blood with its erythrocytes labelled with ^{51}Cr; the other sub-chamber contains physiological unlabelled saline. With the nematodes head end in the blood, analysis of the saline can provide data on the transit time of blood through the nematode, rates of nutrient uptake and the amount of digestion which has occurred.

Schistosomes are among the very few metazoan parasites that live within vertebrate blood vessels and also feed there. With their anteriorly positioned mouths they begin to ingest red blood cells about 14 days after they have entered their final host. There is considerable indirect evidence that the worms utilize haemoglobin as a food with digestion being entirely extracellular. The black pigment within the schistosomes gut caeca that makes the worms so easy to locate during the dissection of infected hosts is a haem-derived product. Homogenates of worms certainly contain a proteolytic enzyme which degrades haemoglobin rather specifically,[347] and it has been shown, using haemoglobin with its leucines isotopically labelled, that these labelled amino acids become distributed in schistosome proteins after feeding.[385] More recently Lawrence,[188] using the extremely useful ^{51}Cr labelled erythrocytes, has estimated the rates of ingestion and turnover times of red blood cells engulfed by *Schistosoma mansoni in vivo*. In mice, egg-producing female worms ingest red blood cells at the rate of about 330000 cells per hour whereas males with their lower anabolic requirements ingested cells at the lower rate of about 39000 cells per hour. The worms completely exchanged their caecal contents in three to four hours, and the total caecal capacity of an *in copulo* pair of worms was found to be about 1330000 red blood cells. The differential cell uptake rates of male and female worms have considerable significance. They suggest that red blood cell acquisition really does have trophic importance. If erythrocyte processing was carried out by the worms entirely for antigenic camouflaging (see page 130, Chapter 4) one would predict that the male with its larger surface areas would ingest more cells than the female. The fact that the converse is true points forcibly at a nutritional interpretation for the results.

Bulk feeding

The solid tissues of host organisms provide a ready supply of small molecular weight and macromolecular nutrients for those parasites that have the means of entering the tissues and ingesting them. The feeding methods of such parasites, termed here bulk feeders, do not in general show unusual association-specific adaptations. The food tissues involved do not pose extreme problems of digestion and the methods of food ingestion characteristic of each group can be employed in a relatively unmodified manner.

Ectoparasitic molluscs like the pyramidellids use a feeding proboscis similar to that of their non-parasitic gastropod relatives to abrade food from the solid mantle tissues of their bivalve hosts. Juvenile worms of *Fasciola* eat tunnels for themselves through liver parenchyma and digenean redia larvae ingest snail digestive gland tissues in which they live. In all these cases a parasite embedded in its food source appears to do nothing more complicated than eat its way through it.

Nutrient uptake across the parasitic body wall

Scattered throughout the spectrum of metazoan taxa that possess endoparasitic members are individual parasitic species, genera and even phyla that lack a gut completely or are without a normally functioning alimentary canal.

Cestodes Historically it was the tapeworms that first prompted experimental investigation of the nutritional biology of gutless endoparsites. No single example of an adult or larval tapeworm possesses a gut or any embryological remnant of one. As in the similar case of the free-living pogonophorans, this stark negative fact has stimulated a continuing enquiry into the manner in which medium-sized animals can obtain nutrients without an alimentary canal. In the absence of a gut, and given the clue that all tapeworms, both mature and larval, live in host locations which are rich in small organic molecules, attention turned to the possibility that such nutrients could be absorbed across the general body surface of cestodes. Before any direct evidence was forthcoming, this interface between parasite and host fluids was tacitly assumed to be the site of the uptake of food. For once, this, the obvious assumption, was well-founded. A succession of recent publications, at least fifty in number in the past decade, have provided particular confirmations of the central assumption. With increasingly sophisticated biochemical procedures, investigators have succeeded in monitoring the uptake of amino and imino acids, fatty acids, glycerol, monoglycerides, acetate, butyrate, purines, pyrimidines and nucleosides as well as the monosaccharides like glucose which seems always to be the first nutrient that experimentalists instinctively take from their laboratory shelves.

Most nutrients appear to pass into tapeworms from their immediate external environment by diffusion, active transport mechanisms or a combination of both processes.[13] In diffusive movement, net transfer of a particular nutrient substance is dependent on the concentration difference in that substance that exists between the external environment and the internal tissues of the tapeworm. The direction of net flow is always from the high concentration to the low – that is, down the concentration gradient. When the process occurs through living, biological interfaces and between complex mixtures of organic solutes it may show anomalies compared with the theoretical, simple solution situation. Over a wide range of concentrations, however, the diffusive rate will be a linear function of solute concentration difference (Fig. 3.10a). In diffusive movement no energy expenditure is involved apart from the kinetic energy of the solute. Stereospecific effects are generally absent.

Active transport, in contrast to diffusion, is a transfer process that occurs only across biological membranes. It is characterized by the obviously adaptive feature of solutes moving across membranes against a concentration gradient. This movement cannot be attributed to the net kinetic energy differences between the solutions on either side of the membrane. The movement can occur only in this 'uphill' direction, by the intervention of an additional energy source. This source is a respiratory, biological one provided by the cell itself. Some of the membrane-associated macromolecules that are essential for the mediation of this activity have enzyme-like behaviour which manifests itself in three interrelated characteristics of active transport. Firstly, such transport shows stereospecificity with respect to D– and L– isomers of organic molecules. Secondly, the uptake kinetics of active transport, that is the behaviour of the initial rate of nutrient uptake as a function of external nutrient concentration, show the same type of saturation kinetics that enzyme-substrate interactions do (Fig. 3.10b). Lastly, some component of the uptake sequence, probably a receptor phase, can display competitive inhibition behaviour paralleling the similar phenomena that occur at the active sites of enzymes (Fig. 3.10c). The passage of an organic nutrient into a tapeworm by active transport can also be non-competitively inhibited by substances that act as general inhibitors of ATP production. The action of these inhibitors is presumably to interfere with the membrane-associated energy source that drives the uptake against a concentration gradient.

Most useful investigations into the mechanisms of tapeworm nutrition have resolved themselves into studies of the transport kinetics of the uptake of particular nutrients with and without potential competitive inhibiting substances and non-competitive inhibitors of energy metabolism. These investigations usually make it possible to decide how the nutrient is being absorbed.

The basic investigatory technique in the analysis of tapeworm nutrition is superficially simple. One places living parasites in a physiologically adequate medium containing a single nutrient substance that is isotopically labelled, usually with ^{14}C or ^{3}H. At the

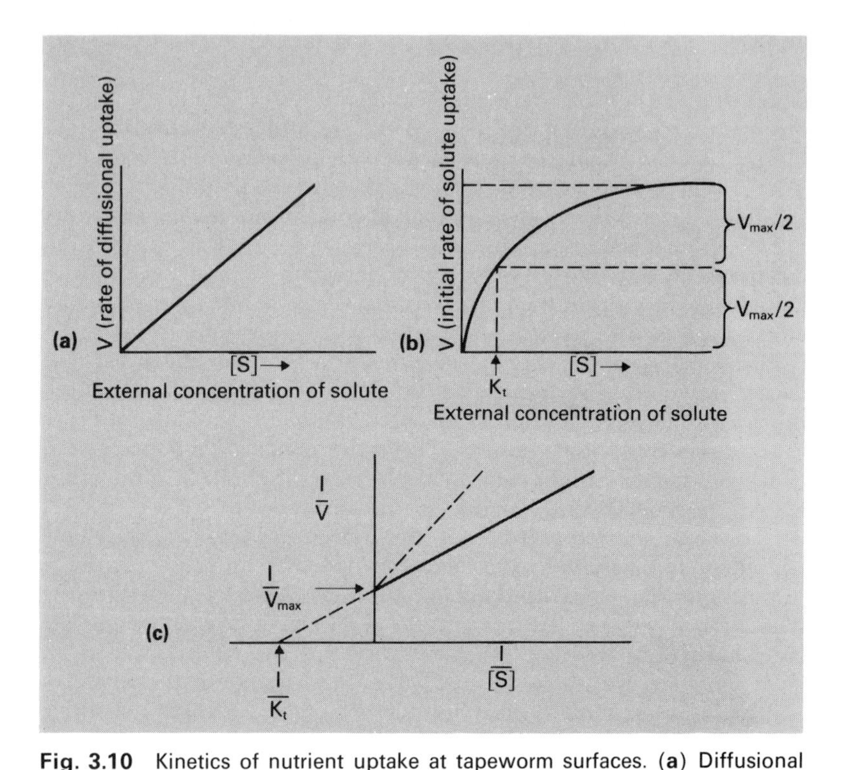

Fig. 3.10 Kinetics of nutrient uptake at tapeworm surfaces. (**a**) Diffusional uptake – the rate of diffusional uptake, V, is directly proportional to the external concentration of the solute nutrient [S]. (**b**) Active (mediated) transport – the curve demonstrates saturation kinetics with the initial rate of solute nutrient uptake V reaching a maximum rate V_{max} at a limiting value of [S]. K_t, the transport constant, equivalent to the K_m or Michaelis constant of enzyme kinetics, defines the external concentration of solute nutrient which will support an uptake rate of $K_{max}/2$. (**c**) Saturation kinetics curve plotted in the double reciprocal form, $\frac{1}{V}$ against $\frac{1}{[S]}$ which produces a linear relationship enabling V_{max} and K_t to be directly calculated as intercepts. (———) represents relationship in conditions with no inhibition, (-—-—-—) the relationship when competitive inhibition occurs. In the latter situation the intercept on the V axis is the same, i.e. V_{max} is unchanged, but $\frac{1}{K_t}$ is reduced, i.e. K_t is increased. Such inhibited curves can be used to calculate the inhibition constant K_i which is derived from the equation:

$$\text{Slope of inhibited curve} = \frac{K_t}{V_{max}}\left(1 + \frac{[I]}{K_i}\right)$$

where K_t is uninhibited transport constant and [I] is the concentration of inhibitor used. Inhibitions are usually expressed in terms of $\frac{1}{K_i}$.

end of an incubation period at a typical host temperature the worms are removed, their surfaces are washed and then the isotopic label inside them is assayed by a variety of quantitative techniques such as gas flow and liquid scintillation counting. Unfortunately, any simple-minded interpretation of findings obtained in this way is liable to result in invalid generalizations or real factual misconceptions. A number of factors, which make for non-repeatable results or artefactual tendencies, have to be constantly taken into account:

(a) Tapeworms of the same species but of different ages (Fig. 3.11), from different hosts (Table 3.3) or from hosts fed on different diets, can exhibit different absolute nutrient uptake characteristics. Standardization of the history of experimental cestodes is therefore essential if repeatable results are sought.

(b) The isotopic label that is taken up by a tapeworm as part of a nutrient molecule will not necessarily stay covalently bound to that molecule for more than a few minutes if the nutrient is rapidly metabolized after uptake. This dynamic problem has been overcome experimentally in two distinct ways. Some workers have resorted to high activity tracers and very short-term incubations (30 seconds to 2 minutes). Others have used close molecular analogues of the nutrients under study which are not metabolized by the helminth but which seem to be taken up in a fashion similar to that experienced by the actual nutrient.

(c) The demonstration of saturation kinetics *per se* does not necessarily imply that active transport processes are operative. Saturation kinetics merely show that there is a limiting step in the uptake mechanism with adsorption-type characteristics. Proof of uptake against a concentration gradient and the demonstration of the efficiency of respiratory inhibitors in diminishing uptake are far better indicators of active transport processes.

The differential importance of diffusion and active transport in the capture of a particular nutrient molecules by a particular tapeworm cannot, at present, be predicted from first principles. None-the-less, it may be imagined that the range of external concentrations of a substance that a tapeworm species is likely to encounter in its normal host location will have some bearing on this matter, as will the complex competitive interactions for uptake sites that will occur at the surfaces of cestodes living in natural, very complicated mixtures of nutrients.

A few specific examples will demonstrate the wide variety of modes of nutrient uptake in cestodes. Read, Rothman and Simmons,[286] in a pioneering paper which laid the ground rules for such investigations, examined the uptake of amino acids by that work-horse of the

Table 3.3 Relative $1/K_i$ values of amino acids as competitive inhibitors of L-methionine uptake by 10 day old *Hymenolepis diminuta* (methionine taken as unity), when the tapeworms were grown in hamsters and rats respectively. (From data in Read, Rothman and Simmons.[286])

Amino acid	Hamster origin	Rat origin
Alanine	0·30	0·26
Aspartic acid	0·22	0.13*
Glycine	0·25	0·06*
Hydroxyproline	0·19	0·06*
Histidine	0·08	0·14*
Isoleucine	0·37	0·26
Leucine	0·41	0·40
Methionine	1·00	1·00
Phenylalanine	0·09	0·02*
Proline	0·18	0·11*
Threonine	0·67	0·43
Valine	0·22	0·14

* indicates a significant difference in the degree of inhibition in respect of tapeworms from the two different hosts.

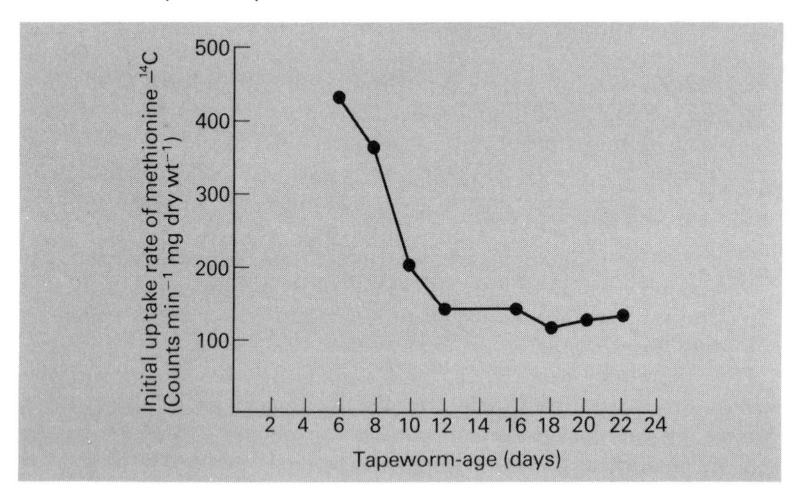

Fig. 3.11 Age-dependent amino acid uptake in *Hymenolepis diminuta.* The initial uptake rate of methionine was assessed in short (2 minute) incubations for tapeworms of different ages. (Based on data in Read *et al.*[286])

cestodology field, adult *Hymenolepis diminuta*. This cyclophyllidean tapeworm is a convenient parasite of laboratory rats. With this cestode it was shown that most amino acids were taken up by saturatable processes which could occur against concentration gradients and

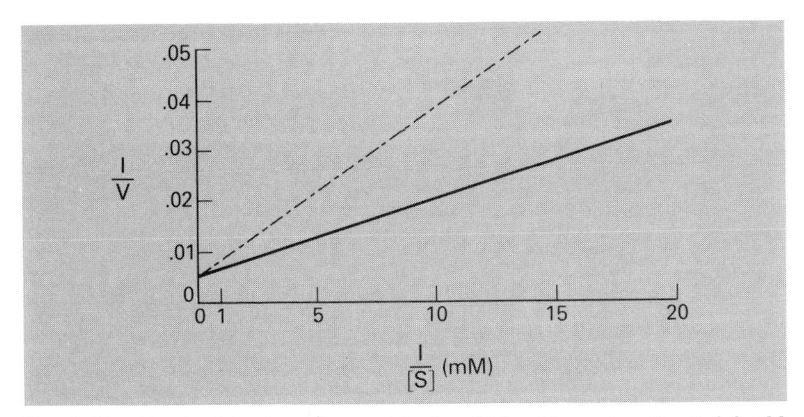

Fig. 3.12 Uptake kinetics of L-methionine-^{14}C with and without 1.0 mM leucine as a competitive inhibitor. (———) methionine alone, (– - — - -) methionine plus leucine. [S] = mM and V = μM h^{-1} per gram ethanol extracted dry weight of tapeworm, based on 2 minute incubations using 10 day old worms from rats. (Data from Read *et al.*[286])

which showed stereospecificity. They were thus active transport processes. Fig. 3.12 illustrates the uptake kinetics of L-methionine with and without leucine acting as a competitive inhibitor, and Table 3.4 the stereospecific differences between the uptake parameters for L-methionine and D-methionine. A consequence of these parameter values is that at all external concentrations L-methionine is always taken up faster than D-methionine. By an examination of many of the possible competitive interactions between different pairs of amino acids presented to the worms together, it proved possible to delimit groups of amino acids that had high reciprocal inhibitory activities. Such groupings of amino acids presumably describe the contours of specificity of particular uptake sites or loci as they are often called. In adult *H. diminuta* there might be four or five such sets of loci for amino

Table 3.4 Transport characteristics of the methionine uptake locus on 10 days old *Hymenolepis diminuta* from rats. (Data from Read, Rothman and Simmons.[286])

	V_{max} (μM h^{-1} g dry wt^{-1})	K_t (mM)
D-methionine	111	0·368
L-methionine	183	0·307

acids,[13] one for dicarboxylic amino acids like aspartic acid, one for diamino amino acids like arginine and two or three separate types of locus for neutral amino acids. The uptake of L-methionine has been shown to be pH dependent,[286] with a maximum velocity between pH 8 and pH 9, suggesting that the ionic species $R-CH(NH_2)-COO^-$ is the form adsorbed onto the uptake locus. The transport can be inhibited non-competitively by iodoacetate (Fig. 3.13) which inhibits several enzymes of respiratory metabolism by combining with sulphydryl groups.

More recently, and using the same adult tapeworm system, the uptake of water soluble B group vitamins has been looked at.[264] It is known that *H. diminuta* requires such vitamins for normal growth and development. The tapeworms are markedly and detrimentally affected[277] by living in a rat both fed on a vitamin B-deficient diet and prevented from eating its own voided faeces, the latter being an important source of B vitamins for rats. The vitamins pyridoxine and riboflavin are absorbed by *H. diminuta* in instructively distinct ways. The rate of ^{3}H-pyridoxine uptake by the worms is linear over the

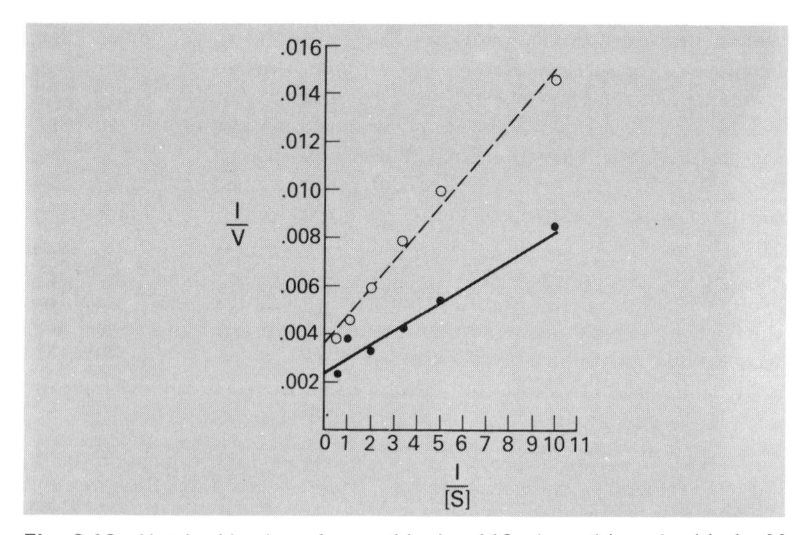

Fig. 3.13 Uptake kinetics of -methionine-^{14}C alone (•) and with 1 mM iodoacetate (o) plotted from the experimental data of Read, Rothman and Simmons.[286] The data has been plotted in the double reciprocal form where [S] = mM and V = μM h^{-1} per g dry weight. The iodoacetate inhibition is non-competitive, with the least squares linear fits to the data points giving non-coincident intercepts on the 1/V axis. The best estimates for the different V_{max} values are 418.38 μM methionine h^{-1} per g dry weight with no inhibitor and 273.45 μM h^{-1} per g dry weight with 1 mM iodoacetate present.

concentration range 0.005–10 mM (Fig. 3.14). Uptake of 0.05 mM ^{3}H-pyridoxine is not inhibited by high relative concentrations of unlabelled pyridoxine, pyridoxal, pyridoxal-5′-phosphate and several other pyridoxine derivatives that might be expected to act as competitive inhibitors if active transport were occurring. Equally, the uptake of labelled pyridoxine is linear with respect to time for the whole of a thirty minute incubation period. These uptake characteristics represent a classically uncomplicated instance of simple diffusive uptake.

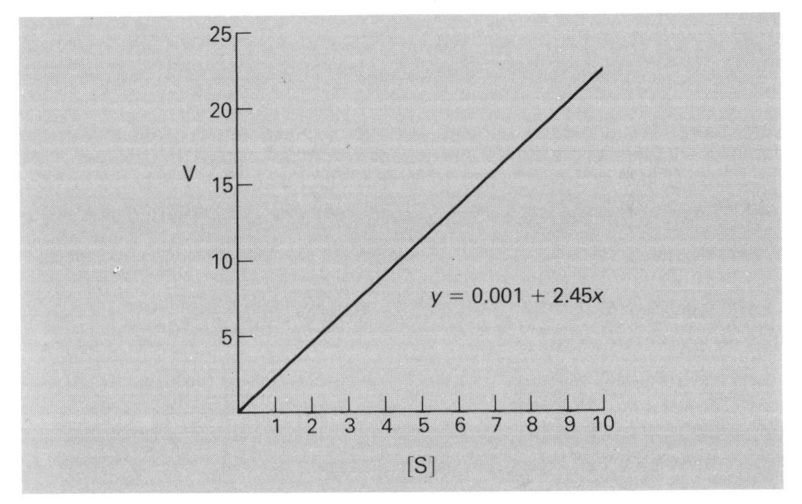

Fig. 3.14 Linear relationship between uptake rate V (μM h^{-1} per g ethanol extracted dry weight) of ^{3}H-pyridoxine by *Hymenolepis diminuta* and external concentration of ^{3}H-pyridoxine [S] (mM). (Redrawn from the data of Pappas and Read.[264])

^{14}C-riboflavin passes into *H. diminuta* with more complex kinetics (Fig. 3.15). The absorption is not linear over the concentration range of 3 to 250 nM and can be competitively inhibited by unlabelled riboflavin, riboflavin-5′-phosphate and flavin adenine dinucleotide. The complex uptake curve which still has a positive gradient at high substrate concentrations, coupled with inhibitor findings, together suggest that diffusive and active transport (mediated) processes are playing a cooperative part in the uptake of riboflavin.

In experimental rats it can be shown[277] that the deleterious effects of the absence of pyridoxine on the tapeworms can be reversed by an intraperitoneal injection of 10 μg of pyridoxine HCl daily. This vitamin flow can be demonstrated biochemically[264] by injecting rats

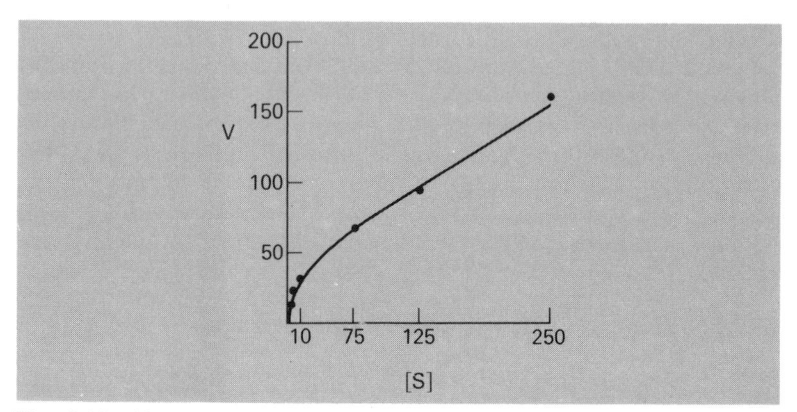

Fig. 3.15 Non-linear relationship between uptake rate V ($nM\,h^{-1}$ per g ethanol extracted dry weight) and external concentration [S] (nM) of ^{14}C-riboflavin. (Redrawn from the data of Pappas and Read.[264])

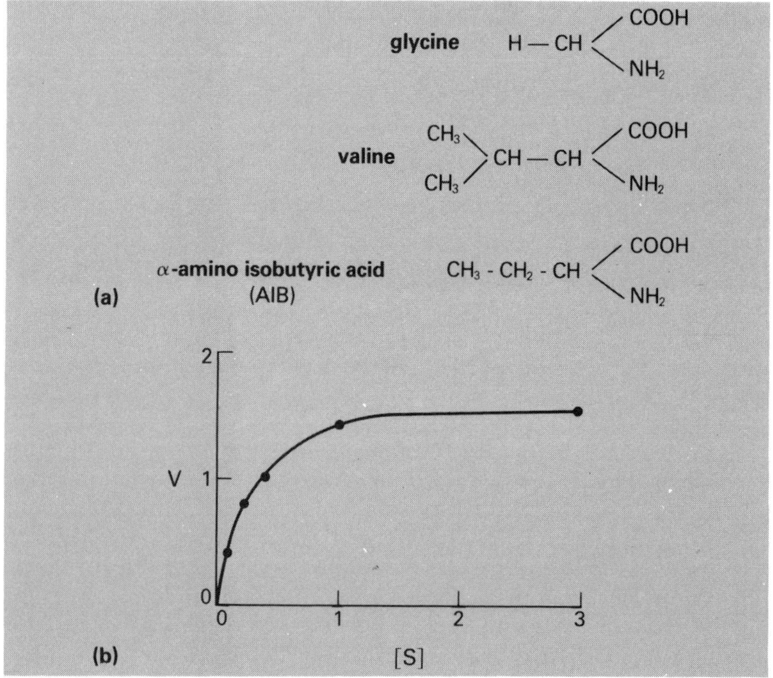

Fig. 3.16 (**a**) Structural formulae of AIB and some other neutral amino acids. (**b**) α-amino isobutyric acid (AIB) uptake by cysticercoids of *Hymenolepis diminuta.* V, uptake (nM per 2 min per 100 cysticercoids); [S] external concentration of AIB (mM). (Redrawn from Arme and Coates.[14])

intraperitoneally with ^{3}H-pyridoxine from day 2 to day 11 of an *H. diminuta* infection. On day 11 the worms contain ^{3}H-pyridoxine This presumed passage of the vitamin to the worms from the peritoneum is thought to be a consequence of an exocrino-enteric circulation process whereby substances, by diffusion or active transport, pass into the intestinal lumen and are subsequently resorbed.

Although less extensive work has been performed on larval tapeworms, they too appear to absorb nutrients from their hosts by a combination of diffusive and mediated methods. Arme and Coates[14] used α-amino isobutyric acid (AIB) to probe the amino-acid uptake characteristics of the cystercercoid larva of *H. diminuta*. This larval phase can develop in the haemolymph-filled haemocoeles of a wide variety of insects including larval and adult moths, beetles, dermapterans, flea larvae, myriapods and orthopterans. AIB is a

Table 3.5 Proportional inhibition of the uninhibited uptake of 0·4 mM AIB by cysticercoids of *Hymenolepis diminuta* caused by potentially inhibitory compounds at a concentration of 8 mM. (Based on data in Arme and Coates.[42])

Compound	Proportional inhibition	Compound	Proportional inhibition
Glucose	0·00	L-alanine	0·93
Galactose	0·00	L-aspartic acid	0·28
Fructose	0·00	Cycloleucine	0·88
Lactose	0·00	L-histidine	0·39
Adenine	0·00	L-leucine	0·69
Uracil	0·00	L-lysine	0·00
Sodium acetate	0·00	L-arginine	0·00
Sodium stearate	0·00	L-methionine	0·91
Betaine	0·00	L-tryptophan	0·42
Sarcosine	0·00	L-valine	0·49
		D-alanine	0·86
		D-methionine	0·81
		D-valine	0·61

synthetic neutral amino acid that appears to be taken up like other neutral amino acids by tapeworms but is metabolically inert once absorbed. Cystercercoid larvae take up AIB with typical saturation kinetics (Fig. 3.16). Arme and Coates[14] tested a wide range of small organic molecules as inhibitors of AIB ingress (Table 3.5). The pattern of results gives interesting information on the specificity of the locus (or loci) at which AIB is being absorbed. Sugars, bases, fatty acids and basic amino acids have no inhibitory effect and therefore cannot interact significantly with the relevant locus. Among the other amino acids, largest inhibitions are produced by other neutral acids ranging from 49% inhibition (L-valine) to 93% (L-alanine). An acidic amino acid-like L-aspartic produces a low but significant amount of inhibition (28%). If these inhibitions are assumed to be of a competitive nature, the locus involved seems to be the alanine-preferring neutral amino-acid type with some overlapping affinity for acidic amino acids.

Most early investigators of tapeworm nutrition assumed that the surface of the helminth was simply an absorptive one, or at least concentrated experimentally on that aspect of its functioning. More recent findings, however, suggest that the surface of tapeworms must be regarded as a digestive-absorptive interface. This understanding parallels modern views on the nature of the brush border region of the mammalian intestinal cell. Indeed, millions of years of competition for the same nutrients in the same physiological niche have produced a remarkable convergence of biochemical and ultrastructural organization between the gut brush border and the tapeworm's microtriche-covered tegument.

The outer surface of adult *H. diminuta* has intrinsic phosphohydrolase enzymes. One such system splits fructose diphosphate into fructose and inorganic phosphate. The enzyme concerned demonstrates substrate inhibition and its action on fructose diphosphate can be competitively inhibited by glucose-6-phosphate.[15] As *H. diminuta* is virtually impermeable to fructose, the activity of the enzyme can be monitored by immersing tapeworms in fructose diphosphate and assaying the external medium for the produced fructose. The principle significance of this enzyme's activity cannot be nutritional as the fructose liberated cannot be absorbed by the worm. The demonstration of its activity is, however, conceptually important as it proves that this intrinsic surface enzyme must be superficial *in situ* because the substrate, fructose-1,6-diphosphate cannot penetrate the tapeworm's surface. In other instances the surface phosphohydrolase activity of *H. diminuta* must have distinct nutritional importance. Glucose-6-phosphate hydrolysis provides glucose which is spatially available for uptake at monosaccharide uptake loci,[102] and nucleotide phosphohydrolysis produces nucleosides which can also interact with their own specific uptake sites.[266]

Intrinsic hydrolytic enzymes do not constitute the entire digestive repertoire of the adult *H. diminuta* surface. Read,[285] elaborating upon earlier findings by Taylor and Thomas,[343] has shown that extrinsic, host-produced digestive enzymes can interact with the glycocalyx of the tapeworm's surface and thus have their activity enhanced. The tapeworms cannot themselves digest starch at their surface. They have no intrinsic amylolytic activity. Read showed, however, that pancreatic α-amylase, introduced into an incubation medium, has its activity augmented by the presence of living tapeworms. Augmentation is not produced by worms killed in 70% alcohol nor by saline in which live worms have been kept for thirty minutes. Maximum relative enhancement is attained at very low enzyme concentrations and the enhancement effect is easily reversed by washing the worms. Further, the enhancement is partially blocked by incubating the worms with high molecular weight polycations such as heparin and poly-L-glutamic acid but not by polyanions like poly-L-lysine. All these findings suggest that the amylase becomes reversibly and superficially bound as a polycation to the polyanionic mucopolysaccharide glycocalyx of the tapeworm. Some aspect of this binding appears to favour the increased activity of the enzyme's active site, perhaps by an allosteric effect.

The modern concept of the tapeworm surface has thus become rather multifactorial in terms of nutrient uptake. Two basic transfer processes exist, diffusion and active transport, the latter occurring at a number of different loci with distinct substrate specificities. Intrinsic phosphohydrolases have superficial sites in the membrane of the outer surface itself and appear to have close spatial relationships with active transport loci appropriate for their hydrolytic products. This steric organization means that the products are more likely to be absorbed than to float off into the external medium. In addition, host-produced digestive enzymes can bind to the glycocalyx immediately outside the worm's outer plasma membrane. Due to the localization of this enzyme appropriation, the products of their catalytic activity, when formed, will be closer to the tapeworm's surface than that of the host's gut cells.

Acanthocephalan nutrition The acanthocephalans, or spiny-headed worms, are the pseudocoelomate equivalent to the acoelomate tapeworms. The adults are gutless parasites that live attached by an anterior proboscis to the gut wall of the vertebrate intestine. Although less studied than the cestodes, acanthocephalans appear to display a set of nutrient uptake modes remarkably similar to that possessed by their cestode cousins. They have, however, been shown to possess one digestive-absorptive ability that has not yet been demonstrated in tapeworms. Uglem *et al*,[355] working on the rat archiacanthocephalan, *Moniliformis dubius*, have provided good evidence that intrinsic aminopeptidases exist in the outer plasmalemma of the tegument. Due to the oligopeptide-splitting activity of these enzymes, peptides such as leucylleucine and alanylalanine exert an apparent competitive inhibition on the active transport of leucine into the worms. In fact, the peptides cannot themselves interact with the leucine-preferring neutral amino-acid locus, but such loci are situated very close to the relevant aminopeptidase locations. This relationship means that leucine arising from aminopeptidase activity is readily available for uptake by the transport locus and can compete with exogenous leucine. In fact, it has been shown[355] that over 90% of the leucine produced by the hydrolysis of leucylleucine at the worm's surface does not diffuse into the external medium but is absorbed by the worm. Cystacanth larvae of *M. dubius* from the haemocoeles of cockroaches show no aminopeptidase activity. If, however, the larvae are pretreated with a variety of surface active agents like pancreatic lipase or sodium taurocholate, aminopeptidase activity becomes apparent, presumably because of some unmasking effect in the plasma membrane. *In vivo*, agents like bile salts in the gut of the rat final host must induce this enzyme activation.

Other endoparasites without guts Extreme adaptation for endoparasitism in host locations with easily available small organic nutrients has led to a secondary loss of a functional gut in a number of independent evolutionary progressions.

Parasitic molluscs in the genera *Exteroxenos* and *Thyonicola*, which occur in the body cavities of holothurian echinoderms, have entirely lost gut tissues.

Among the nematodes, work in this area has concentrated on entomophilic forms, which live in the haemocoelic spaces of insects. Larval female nematodes of the genus *Bradynema* (Order: Tylenchida) enter the haemocoele of the phorid dipteran *Megaselia halterata*, moult, grow and become sexually mature there. In this development the feeding apparatus degenerates and the mouth and anus disappear. When examined ultrastructurally,[290] these female worms appear to have lost any cuticular layer external to the hypodermis. The outer plasma membrane of the latter is thrown into large numbers of thin microvilli, each about 200 nm long. This expanded surface area may be assumed to be the location of nutrient transport mechanisms. In another tylenchid nematode, *Sphaerularia bombi*,[278] nutrients are thought to be taken up from the host's haemocoele across the parasite's uterine wall which is prolapsed and everted. The short parasitic larval phase in the life cycle of the mermithid nematode, *Mermis nigrescens*,[301] takes place in the haemocoele of the locust and other insects. The parasite's gut is atypical and consists of a non-muscular oesophagus and a cuticular intestine. ^{14}C-glucose is taken up *in vitro* across the nematode's cuticle. This transcuticular transport can be inhibited by 2,4-dinitrophenol, which prevents synthesis of ATP. It can also be reduced by the polyphenol phloretin, which is a non-competitive inhibitor of the active transport of glucose. Such pieces of experimental evidence indicate that larval *Mermis* can transport glucose across its cuticle by a process which is, at least in part, a mediated one.

Finally, at least two types of endoparasitic platyhelminths apart from tapeworms lack guts. In both cases, the distal cytoplasm of the parasite's tegument possesses a dense surface array of microvilli or surface folds.

The sporocysts of digenean life cycles develop from miracidia which have penetrated either the blood system or digestive gland of the molluscan intermediate host. These early larval stages have no gut. The mother sporocyst of *Schistosoma mansoni*, obtained from the snail *Biomphalaria glabrata*, is bounded by a halo of branching, anastomosing cytoplasmic processes between 1 and 2.5 μm long. Both microvilli and microplicae (folds) are present. Histochemical studies on

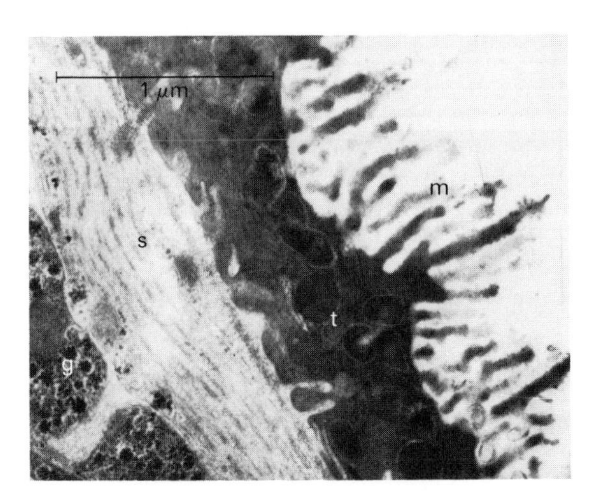

Fig. 3.17 The microvillous outer surface of the sporocyst tegument of an unidentified digenean which can produce micracidia asexually within the sporocyst (Mohandas, A. (1975). *J. Helminth.*, **49**, 167–71.) The species, which is found in the freshwater snail *Melanoides tuberculata*, is probably similar to *Cercaria multiplicata*. The whole tegumental plasma membrane is thrown into branched microvilli. g, glycogen in subtegumental cells; m, microvilli; s, subtegumental muscles; t, tegument. (Whitfield and Mohandas, unpublished observations.)

schistosome sporocysts[180] show alkaline phosphatases and non-specific phosphohydrolases to be localized near the microvilli. Thus, both ultrastructural and cytochemical indications point to the microvillous surface of sporocysts having a digestive-absorptive function (see Fig. 3.17).

Less well known than digenean sporocysts are the remarkable dioecious adult worms of the endoparasitic turbellarian *Kronborgia amphipodicola*.[71–2] These rhabdocoeles, belonging to the family Fecampidae, develop in the haemocoele of marine amphipod crustaceans. They are almost unique among adult turbellarians in their complete lack of any form of gut. Nutrients are absorbed across the general body surface which is densely covered in cilia and long branched microvilli[46] (see Fig. 2.28).

SYMBIOTE NUTRITION

Uni- and bidirectional nutrient flow between symbiotic partners has been analysed in many different categories of symbiotic association. In examples which concern two large animals, such as many cleaner symbioses, the experimental monitoring of food transfers is largely an ethological exercise and is considered elsewhere (Chapter 5, pp. 181 and 186). Most instructive in comparison with the trophic organization of host-parasite associations are the nutrient transfers which take place in symbioses which include microorganisms. These have proved amenable to the analytic methods of the biochemist, and in several classes of symbiosis we now have a reasonable, broad-based understanding of the dynamics of nutrient flow. As always, however, increased comprehension of the outlines of a biological system by the use of new technology, serves also to pose whole new sets of unanswered questions.

Lichen symbioses

Recent work on the nutritional interactions of the algal and fungal components of lichen partnerships will be used as an example of the way in which modern methodology has transformed our understanding of particular symbiotic partnerships. Since D. C. Smith and his associates began, in the early 1960's, to apply isotopic tracer techniques to the physiology of lichens, the strengths of these powerful experimental tools have become increasingly apparent.[131,321] Also important in this area of conceptual advance was the development of a standardized lichen system which was manipulatable in the laboratory. Circular, 7 mm diameter discs were punched from cleaned thallus lobes of the large foliose lichen *Peltigera polydactyla*. These constant-sized lichen pieces could be easily handled *in vitro* and enabled most of the basic techniques for the analysis of nutrient flow in lichens to be developed. Equally useful in providing evidence on the release of nutrients from the algal component has been the isolation of algae from the partnership and the study of their behaviour in the absence of the fungal hyphae which normally surround them.

Most of the information gained has concerned the way in which carbohydrates, photosynthesized by the algae, pass to the fungal parts of the lichen. The algal symbiote of *P. polydactyla* is the blue-green alga, *Nostoc*. Algal cells constitute only 5% of the dry weight of the thallus. As is common in lichens they are concentrated in a layer near the upper surface of the composite organism and come into intimate contact with the fungal hyphae.

The transfer of photosynthetically fixed ^{14}C from the autotrophic *Nostoc* cells to the heterotrophic fungus can be demonstrated by a technique that takes advantage of the restriction of algae to the upper half of the thallus. Discs of lichen are floated on a solution of $NaH^{14}CO_3$, which provides $^{14}CO_2$, and illuminated. At intervals discs are removed, split into fungal medullary and algal cortical zones and the halves assayed separately for ^{14}C (Fig. 3.18). Fixed ^{14}C appears in the algal zone almost immediately. It can be monitored in the fungal portion some ten minutes later. With continuing photosynthesis and

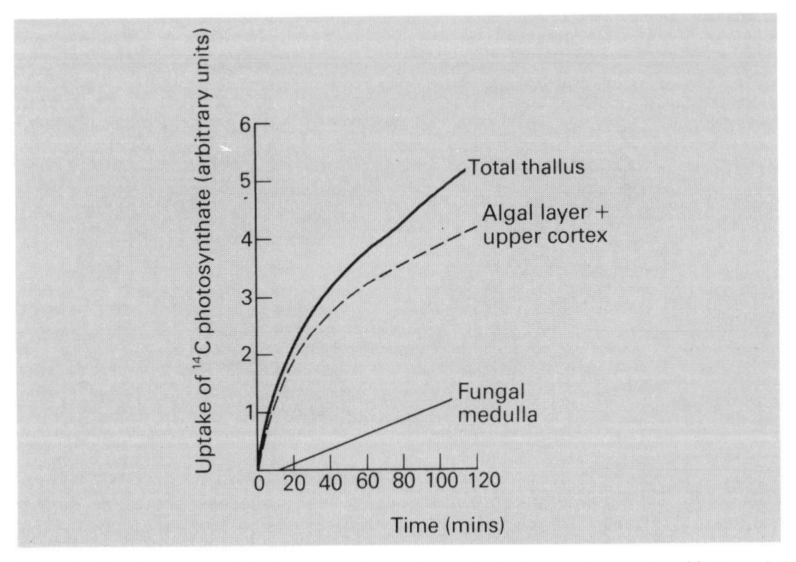

Fig. 3.18 Passage of ^{14}C labelled photosynthate from algal cells, *Nostoc*, in *Peltigera polydactyla* to the fungal medulla. (Redrawn from Smith.[321])

transfer of photosynthate, a dynamic equilibrium is soon set up with a considerable proportion of the total fixed ^{14}C being present in the medulla. Characterization of the substances containing ^{14}C after such incubations reveal interesting differences between the labelled molecules in algal and fungal symbiotes. Freshly isolated *Nostoc* cells, separated from the fungus, release ^{14}C-glucose when supplied with $NaH^{14}CO_3$, water and light. Analysis of the medulla, however, shows that the fixed ^{14}C has been incorporated into the polyhydric alcohol, mannitol. If separated *Peltigera* fungal hyphae are provided *in vitro* with ^{14}C-glucose they absorb it and convert it to ^{14}C-mannitol. These

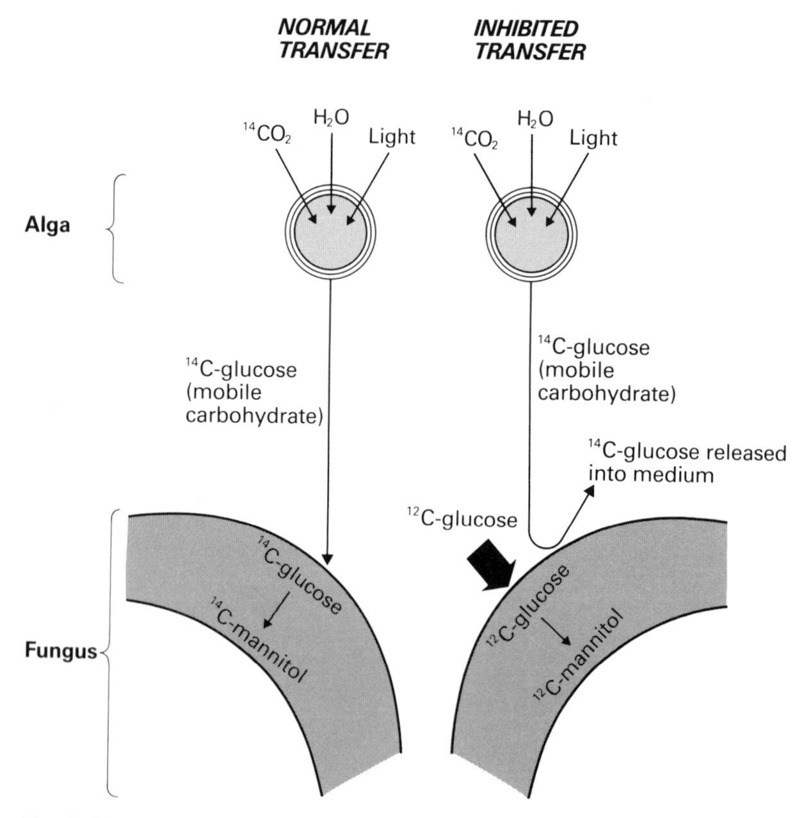

Fig. 3.19 Diagrammatic representation of the processes occurring during normal and inhibited transfer of photosynthate in a lichen where glucose is the mobile carbohydrate which after active uptake by the fungal hyphae is converted to mannitol. Inhibition is achieved by the presence of a large excess (large arrow) of unlabelled ^{12}C-glucose.

crucial findings suggest that in the entire *Peltigera* thallus the basic flow of fixed ^{14}C is as follows: $^{14}CO_2$ fixation by *Nostoc*, ^{14}C-glucose release by *Nostoc*, uptake of ^{14}C-glucose by the fungus and finally conversion of ^{14}C-glucose to ^{14}C-mannitol by the fungus. Interpretation of carbohydrate transfer results is made easier because *Nostoc* never synthesizes mannitol. This enables delicate probing of the speed of photosynthate transfer to be accomplished. The maximum time for transfer to take place can be established simply by assaying the rate of incorporation of ^{14}C into ^{14}C-mannitol in the entire thallus. For *Peltigera polydactyla* it is less than two minutes.

The nature of the photosynthate transfer mechanism has proved analysable by an ingenious experimental method termed the inhibition technique.[141,322] Some portion of the transfer process has adsorption characteristics, the mobile carbohydrate becoming bound to a receptor site. The site is probably an active transport locus in the plasma membrane of the fungus. Such sites imply the possibility of competitive inhibition and this potential is utilized in the inhibition technique. The following examples of its use illustrates the general principle. *P. polydactyla* is incubated with $NaH^{14}CO_3$ and illuminated in the presence of 0.06 M unlabelled ^{12}C-glucose. This glucose permeates all the extracellular spaces of the thallus and achieves concentrations adjacent to the transport loci very much higher than those of ^{14}C-glucose photosynthate. Assuming that the locus does not differentiate between ^{12}C- and ^{14}C-glucose, uptake of the two isotopic forms will be in proportion to their concentrations. Thus with a huge excess of externally applied ^{12}C-glucose, only a negligible quantity of ^{14}C-photosynthate enters the fungus. Instead, it mainly appears in the incubation medium, where it can be counted. The technique (Fig. 3.19) provides two central pieces of information, firstly the identity of the mobile carbohydrate (see Table 3.6) and secondly the rate of release of the mobile carbohydrate from the algal cells. In general, it has been found that lichens with blue-green algal symbiotes use glucose as the mobile carbohydrate, whereas those forms containing green algae utilize alcohols like ribitol, erythritol and sorbitol. Glucose is released relatively rapidly from blue-green algae like *Nostoc*, where the release of sugar alcohols from the green algae occurs more slowly.

In experimental attempts to unravel the pathway by which mobile carbohydrates are released from the alga, transferred to the fungus and then assimilated by the latter the effects of a wide variety of potential inhibitors have been examined to see if transport from alga to fungus could be prevented without blocking photosynthesis. Substances affecting protein synthesis (chloramphenicol and cycloheximide), fungiostatic antibiotics like griseofulvin and nystatin as well as

Table 3.6 Characteristics of mobile carbohydrate transfer in some lichens. (From data in Richardson, D. H. S. (1973). In *The Lichens*, Ahmadjian, V. and Hale, M. E. (eds). Academic Press, New York.)

	Lichen species	Algal symbiote	Mobile carbohydrate
Lichens with green algae	*Dermatocarpon hepaticum*	*Myrmecia*	Ribitol
	Umbilicaria pustulata	*Trebouxia*	Ribitol
	Rocella fuciformis	*Trentepohlia*	Erythritol
	Verrucaria hydrella	*Heterococcus*	Sorbitol
	Dermatocarpon fluviatile	*Hyalococcus*	Sorbitol
	Polyblastia hencheliana	*Trochiscia*	Sorbitol
Lichens with blue-green algae	*Peltigera polydactyla*	*Nostoc*	Glucose
	Coccocarpia spp.	*Scytonema*	Glucose
Lichens with both green and blue-green algae	*Lobaria amplissima*	*Myrmecia* and *Nostoc* in cephalodia	Ribitol and glucose
	Peltigera aphthosa	*Coccomyxa* and *Nostoc* in cephalodia	Ribitol and glucose

inhibitors of glucose uptake such as phloridzin are all without appreciable deleterious effect on transport in *Peltigera polydactyla*. So far, only four classes of compounds have been found to have such an inhibitory effect.[322] These are para-chloro-mercuri-benzo-sulphonate (PCMS) and related substances, sorbose, 2-deoxyglucose and digitonin. PCMS reduces ^{14}C incorporation into fungal mannitol. It appears to do this by a gross blocking action on active thiol groups in the fungal cell membrane. The other three types of effective inhibitors provide more clues in relation to the nature of transport mechanisms. 1% sorbose in the incubation medium reduces ^{14}C incorporation into mannitol by about 33%. It is conjectured that it has this effect because it binds to algal membrane carriers which bring about the outward movement of glucose photosynthate. Such sorbose-sensitive carriers have already been demonstrated in yeasts where they are implicated in a carrier-mediated facilitated diffusion system for carbohydrates.[73] It has been suggested[322] that a similar system is operating in the cell membrane of *Nostoc* in lichens.

Pretreatment of lichen experimental discs with 2% 2-deoxyglucose almost completely abolishes ^{14}C accumulation in fungal mannitol. Its operation as an inhibitor of transport appears to be multi-faceted, as it probably competes with glucose for fungal membrane uptake loci and could also interfere with the conversion of glucose to mannitol in the fungal cytoplasm. Its action certainly suggests strongly that the fungal uptake of algal glucose is an active process. Pretreatment of the lichen with 0.01–0.05% digitonin again reduces ^{14}C incorporation into mannitol to insiginificant levels without any concomitant reduction in the efflux of glucose from the algal cells. It is believed that the digitonin acts by binding to sterols in the fungal cell membrane. Its action demonstrates that the release of mobile carbohydrate from *Nostoc* in *Peltigera* is not dependent on the uptake of glucose by the fungal hyphae.

Some of the lichens containing the blue-green alga, *Nostoc*, can fix atmospheric nitrogen.[241] The nitrogenase activity within the algal cells converts molecular N_2 into ammonia which can be incorporated into small organic molecules. The inhibitor digitonin has proved useful in the analysis of ammonia transfer just as it has in relation to carbohydrate movements. In the case of *Peltigera canina*,[340] in the presence of 0.01% digitonin and nitrogen, ammonia is liberated into the incubation medium. In the absence of digitonin no ammonia can be identified and the supposition is that it is passing into the fungal hyphae rather than the medium.

Interesting triangular trophic relationships exist in lichens containing a fungus, a green algal species and a blue-green algae. *Peltigera aphthosa*,[173] to take a specific example, contains *Nostoc* in special regions called cephalodia, and the green alga, *Coccomyxa*, in the main part of the thallus. *Nostoc* passes glucose and fixed nitrogen compounds to the fungus. *Coccomyxa* supplies ribitol, which the fungal symbiote assimilates. None of the fixed nitrogen passes from the fungus to *Coccomyxa*.

Other symbioses

The algal-fungal nutritional relationships in lichens provide to some extent a good general model for such processes in the wide range of symbiotic associations in which algae figure as endosymbiotes. Such algal cells are principally found in flatworms, hydrozoan and anthozoan cnidarians, molluscs and protozoans but scattered records from other invertebrate groups do exist. It is likely that in most cases the animal macrosymbiote derives nutritional benefit from the algae in much the same way that the fungal hyphae in lichens do. In the animal-algal associations glucose, maltose, polyhydric alcohols and amino acids have all been identified as mobile organic compounds moving from plant to animal.[320]

Nitrogen-fixing blue-green algae, *Anabaena azollae*, have even been found in a tight symbiotic relationship with a floating aquatic fern, *Azolla*.[17] The algal cells are found in cavities in the dorsal lobes of the fern's leaves. Nitrogen-fixing carried out by the prokaryotic partner enables *Azolla* to colonize aquatic habitats of low nitrogen availability. This facility is perhaps part of the reason for the success of *Azolla* as an inconvenient aquatic 'weed' in some man-made reservoirs and as a very convenient source of fixed nitrogen in the paddy fields of Vietnam. It has recently been demonstrated[108] that the algal cells in the dorsal lobes are associated with specialized multicellular hairs which have a 'transfer cell' ultrastructure. The hairs have labyrinthine cell wall ingrowths and contain dense cytoplasm with many mitochondria and much endoplasmic reticulum (see also page 156). This organization has led to the hypothesis[108] that the hairs are implicated in the interchange of metabolites between *Azolla* and *Anabaena* (Fig. 3.20).

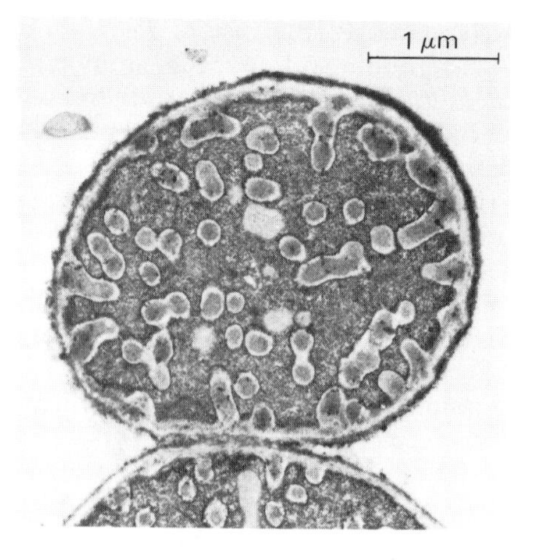

Fig. 3.20 Transverse section through a multicellular hair in a dorsal lobe cavity of the aquatic fern *Azolla*. The hairs are associated with the symbiotic blue-green alga filaments of *Anabaena azollae* within the cavities. The hair cells have a transfer cell-type ultrastructure with numerous inward protuberances of the cell wall into the cytoplasm. (From Duckett *et al.*[108])

Fig. 3.21 The *Elysia-Codium* system. (**a**) *Elysia viridis*, living animal. (**b**) Electron micrograph of part of a digestive tubule of *E. viridis* demonstrating the functional chloroplasts; c, within a gut cell. (Both from Trench.[349])

Trench[349] has recently, and surprisingly, demonstrated that some marine sea slugs (elysioid saccoglossans), notably *Elysia*, engage in what can only be termed chloroplast symbiosis. The molluscs eat a seaweed, *Codium*, and their gut cells phagocytose intact chloroplasts from the partly digested plant material in the gut. In some unknown way the chloroplasts avoid being digested themselves. Eventually the chloroplasts lie free in the host's cells, photosynthesizing at rates comparable to those attainable in the intact seaweed (Fig. 3.21). Photosynthate moves from the chloroplasts to the mollusc and is used in, for instance, mucus production. The plant organelles can photosynthesize for three months in this bizarre relationship but can neither divide nor synthesize many important plastid components like chlorophyll, glycolipids, membrane proteins and nucleic acids.

Perhaps the best-known example of obligate nutritional exchanges between symbiotic partners is the case of that veritable menagerie of symbiotes, the termite.[47] The paunch region of the gut of a wood-eating termite is packed with dense populations of obligately anaerobic protozoans and bacteria (Fig. 3.22). These symbiotes can represent 33–50% of the termite's total weight. Some of the large hyper-

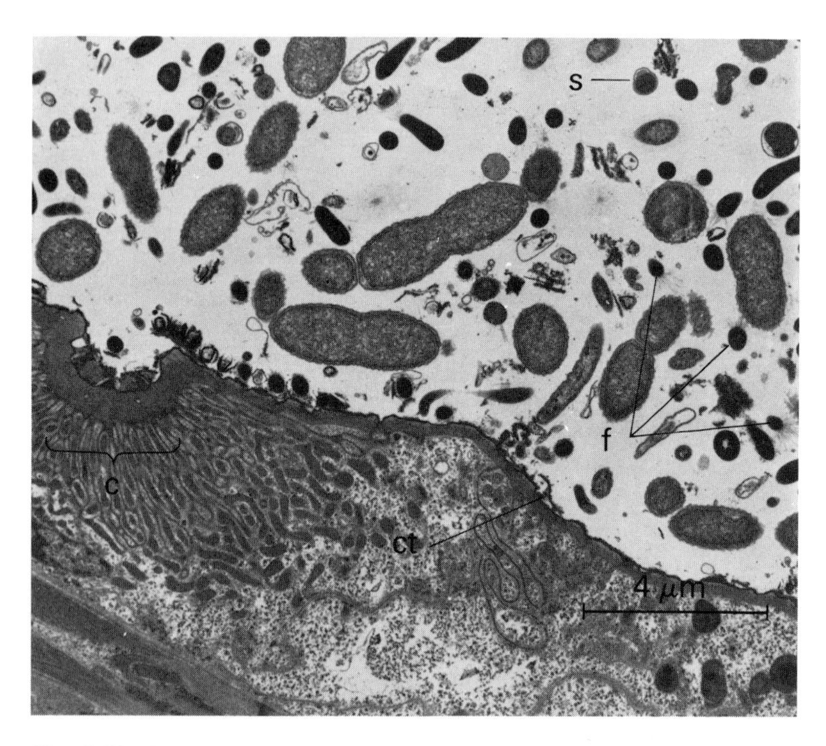

Fig. 3.22 Transmission electron micrograph of part of a cluster of bacteria in the paunch of the termite *Reticulotermes flavibes* adjacent to a depression in the paunch wall (c) which is lined with paunch epithelial cuticle (ct). The complex community of bacteria includes large rods (r), filament-bearing forms (f) and spirochaetes (s). (From Breznak.[47])

flagellates like *Trichonympha* have the ability to ingest and digest cellulose fragments, a faculty which enables the termite macro-symbiote to live on a diet of wood.[152] Without its flagellates a termite cannot hydrolyse this polysaccharide. The cellulose-cleaving protozoans ferment cellulose in anaerobic conditions to acetic acid, CO_2 and H_2. Acetic acid is taken up through the termite's gut wall and utilized aerobically by the termite tissues in respiratory metabolism.

The functions of the termite gut's bacterial flora are less well understood. Some of these microorganisms live attached to the termite's hind gut wall (Fig. 3.22), others are found free in the lumen. It is possible that some forms aid in maintaining anaerobic conditions in the hind gut, some certainly produce methane and yet others can fix atmospheric nitrogen. This latter discovery came only with the development of the acetylene reduction method for assaying nitrogenase activity. Nitrogenase can reduce acetylene to ethylene as well as performing its physiological function of reducing nitrogen to ammonia. The gas chromatographic assay for ethylene is a far more convenient method for monitoring nitrogenase activity than earlier techniques which relied upon the non-radioactive isotope of nitrogen, ^{15}N. All wood-eating termites possess N_2-fixing bacteria and if they are eliminated from a termite's gut by antibiotics the protozoan symbiotes eventually die. The output of fixed N_2 compounds from the bacteria seems to be reciprocally regulated by exogenous nitrogen sources. If ammonium salts or nitrates are fed to termites, their N_2-fixing activity declines. The significance of the N_2-fixing symbiotes is probably to be found in the low nitrogen content of the termite's cellulosic foods. Values of 0.03–0.05% by weight of nitrogen are not unusual. At times in the termite's life cycle when growth is maximal it is likely that the low intrinsic nitrogen content of the food could be a severe restraint. Supplementation by bacterially fixed N-compounds could be crucial in such instances. In this context, it is interesting that the highest specific N_2-fixing activities recorded for termites occur in young growing larvae.

It is obvious that without hyperflagellates and bacteria termites would not be able to exist on a cellulosic diet, and Ogden Nash could not have written:

> 'Some primal termite knocked on wood
> And tasted it, and found it good.
> And that is why your Cousin May
> Fell through the parlour floor today.'

4

Physiological and Regulatory Interactions between Associates

INTRODUCTION

The movements of nutrient molecules examined in the previous chapter are not the only molecular transfers that occur between the members of parasitic and symbiotic associations. Transfers take place which are quantitatively much smaller than the nutritional examples, but which have far-reaching qualitative effects out of all proportion to their absolute size. These interactions are characteristically mediated by molecules that can be said to have informational rather than trophic significance for the associating organisms. Molecules such as antigens, antibodies, lymphokines and hormones occupy the front of the stage in this area of scientific enquiry, but other, less well known substances synthesized by vertebrates, invertebrates, plants and microorganisms also play supporting yet crucial roles.

In a more general context, Whittaker and Feeny[377] have reviewed the significance of chemical messages between members of a species (pheromones) and between different species of organisms coexisting within a community. They initiated the phrase allelochemic interactions for those interactions that are mediated by chemical signals between members of different species. Their review also provides an extensive categorization of chemical signals on the basis of their functions.

The effects of the interplay of informational molecules are usually regulatory in character in parasitic and symbiotic associations. Substances produced by associate A may signal its presence and trigger a defensive response from associate B. Such defensive responses can themselves provide new informational stimuli for associate A which can be used to regulate counter-measures against the original defence. Even counter-measures to counter-measures may ensue! Interactions can involve the induction by one associate of specific changes in the behaviour, physiology or development of the other. Such changes usually have some obvious adaptive significance and must also be regarded as regulation within the bounds of the association.

DEFENSIVE RESPONSES

Almost all examples of well organized physiological defensive responses by one associate to another, occur in host-parasite associations. Due to its immense applied significance and also the enormous quantity of experimental evidence flowing from it, and despite the existence of many excellent books on the topic,[293–4] the field of vertebrate immunological responses must take up an important part of this chapter. At the present stage of knowledge, all other defensive responses must be assessed in the context of the immune responses of mouse and man. These are comparatively so much better understood that all other such regulatory interactions that they are, however arbitrarily, the only finely calibrated measure of these interactions that we have. Despite intense scientific activity since Jenner's first protective vaccinations in 1798, there remain enormous gaps in our comprehension of man's immunological make-up. This daunting realization must, at the very least, make all statements about similar processes in other associations extremely provisional.

Vertebrate defensive responses to parasites

Behind all the defensive responses against a parasite that an individual vertebrate is able to produce lies a genetic background of susceptibility. Some of us are genetically more available as hosts for specific parasites than others. The first suggestions of the complexity of this inherited state have come from medical studies on tissue-typing or histocompatibility antigens. In the mouse there is a genetic linkage between the histocompatibility complex H-2 and resistance to virus-induced leukaemia.[216] With a less firm experimental justification it appears that it can be said that people with particular patterns of tissue antigens are more likely to have a given disease than those with different patterns. Thus, at the very beginning of a defensive battle against a parasite some individuals have a strategic disadvantage. The important gene complex which has been identified for man in this context is the HL-A system of loci. HL-A genes code for glycoproteins in cell membranes and glycocalyces, and it is obvious that the precise conformations of such surface proteins will influence many types of interactions between host cells and parasites. In no case yet, however, is it possible to be specific about the mechanism of genetically determined susceptibility.

Human defensive responses can be generalized, non-specific ones which are operatives against a broad spectrum of genetically dissimilar parasites. Alternatively, the responses can be specific ones involving effector molecules such as antibodies which have high information contents.

Non-specific responses

Parasites attempting to colonize man's outer surfaces are subjected to a complex armoury of non-specific responses. The toxic effects of metabolites like lactic acid and fatty acids in sebum and sweat secretions can be directly detrimental to skin-dwelling bacterial parasites, as can the low pH produced by such substances and the compounds excreted from harmless skin-inhabiting yeasts and bacteria. These colonists must often be regarded as symbiotic partners.

At a biochemical level, the enzyme lysozyme represents a generalized weapon against most bacteria; it destroys their cell walls. Lysozyme is found in tears, saliva and gut secretions and must have a protective function on the surfaces of the eye, nasal passages and the alimentary canal.

Internal parasites are subjected to a range of more sophisticated, relatively non-specific weaponry including phagocytic cells. Perhaps the most complex non-specific effector is the complement system, a potentially interacting series of proteins, analogues of which can also be found in higher invertebrates. The separate complement components numbered C_1–C_9 exist in the serum at concentrations of up to 1200 $\mu g\,ml^{-1}$. They have molecular weights of 100000 to 400000 and constitute together a multifunctional defence system which is able to lyse the surfaces of microorganisms. Complement molecules are an extracellular defence of some complexity that evolved before the more specific, immunoglobulin (antibody)-mediated system. A complement defence can be triggered in a number of ways. Direct interaction with, for instance, bacterial surfaces can occur. Equally, linkage to antibodies that have attached to parasite surface antigens can be an initiation stimulus. This second type of activation is presumably a secondarily acquired responsiveness elaborated after the evolution of the immunoglobulins.

A complement response operates in a cascade fashion. C_1 binds to the target surface and activates a greater number of further complement enzymes which can activate other complementary enymes. The consequence of this expansion of activation is to localize on the target surface, in a tight array, thousands of activated complement molecules (C_8 and C_9) which have phospholipase activity. Each activated cluster can 'punch' a hole in the target surface. The damaging effects of thousands of these holes can kill a bacterial cell. Other parts of the complement sequence (C_{3a}) are chemoattractants for phagocytic cells that engulf the dead or damaged microorganisms.

A non-specific defensive response to viral infections is the production by cells infected with viruses of the substance interferon.[353] It is an inhibitor of virus replication and seems especially important in producing a local and immediate defence against viruses. A full-blown antibody response needs time to be mounted and constitutes a later phase of the total reaction to viral invasion.

All of the above non-specific defences play important and interrelated roles in resistance to parasitic establishment and population growth. None of them, however, are so vital that lack of them is incompatible with survival. Even the complement system can be impaired without enormous changes in susceptibility to parasitic infections. A quite different generalization has to be made about lymphocyte based, immunoglobulin-mediated immune mechanisms and cell mediated immunity, which respond with specificity to particular parasites. A lack of this defensive weaponry is frequently a fatal deficiency.

Parasite-specific responses

The immunoglobulins (antibodies) are proteins found in the γ-globulin electrophoretic fraction of serum. Their informational content concerns the structure of foreign, non-self macromolecules (usually other proteins). The information manifests itself when a particular immunoglobulin (Ig) reacts by specific attachment to its own appropriate spectrum of target macromolecules (antigens). This bonding reaction may itself provide a defensive response against a parasite bearing the antigen or a toxic antigen released by the parasite. More often, the antibody-antigen reaction is the trigger for further defensive measures that can be appropriately directed once the Ig has branded the antigens as relevant targets.

Secreted, free Ig molecules appear to have been a vertebrate invention. Their production only occurs in that taxon of animals and almost all present-day groups of vertebrates are able to produce such immunoglobulins.[226] Five basic classes of Ig have been identified in man, IgG, IgA, IgM, IgD and IgE (see Table 4.1). These Ig categories are not based on the proteins' specificities for particular classes of

Table 4.1 Properties of human immunoglobulin classes. (Based on data in Roitt.[293])

Name	IgG	IgA	IgM	IgD	IgE
Molecular Weight	150 000	160 000	900 000	185 000	200 000
No. of 4–polypeptide units	1	1 or 2	5	1	1
Concentration range in normal serum	8–16 mg ml^{-1}	1·4–4 mg ml^{-1}	0·5–2 mg ml^{-1}	0–0·4 mg ml^{-1}	17–450 ng ml^{-1}
% of total Ig	80	13	6	1	0·002
Principle characteristics	1 Commonest Ig in internal body fluids including extra-vascular sites	1 Principle Ig in sero-mucous secretions, providing protection for external surfaces of the body	1 High molecular weight pentamers of the 4-polypeptide unit	1 Present on the surfaces of some lymphocytes from neonate babies	1 Raised concentration in parasitic worm infections
	2 Crosses placental barrier		2 Multiple attachment sites to antigens make them efficient agglutinators of bacteria		2 Binds to mast cells
	3 Involved in complement fixation		3 Produced early on in infections as a first-line defence in the blood		
			4 Involved in complement fixation		

antigens, but on their own polypeptide structures. These structures determine the overall biological activity and functions of the different categories.

The basic Ig structure (IgG) is a paired Y-shaped molecule consisting of four polypeptide chains, two of which, the light chains, are shorter than the remaining pair of heavy chains (Fig. 4.1). The four polypeptides are linked by three disulphide (—S=S—) bonds.[59] Each arm of the Y includes a combining site, analogous to the active centre of an enzyme, which incorporates the specific binding activity for antigens. When these sites interact with antigen the arms of the Y swing open at a 'hinge' region. In any Ig class the combining sites are included in the first 100 or so amino acids at the N-terminal ends of the

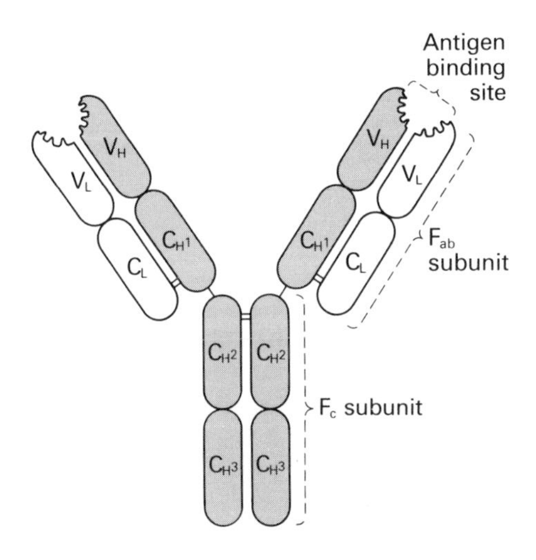

Fig. 4.1 Domain structure of IgG. A single IgG molecule consists of two heavy chain polypeptides (tinted) and two light chain polypeptides. Both heavy and light chains consists of globular polypeptide domains (represented by the ovoid outlines) defined by intrachain disulphide bonds. Each heavy chain possesses four domains (named V_H, C_{H^1}, C_{H^2} and C_{H^3}) each light chain two (V_L and C_L). In the domains designated C the amino-acid sequences are very invariant from IgG to IgG molecule in an individual, these are the constant domains. The V or variable domains have an amino-acid constitution which is in general less constant and in particular hypervariable or 'hot spot' subsequences, very variable from molecule to molecule. The hypervariable regions are indicated as spikes on the V_H and V_C domains. There are four hypervariable regions in each V_H and three in each V_L. Together the seven hypervariable zones help to determine the specificity of the antigen binding site. (Based on data in Capra and Edmundson.[59])

polypeptide chains.[59,279] Each combining site is constructed from parts of the two variable regions (domains) at the end of each arm of the Y. Within each variable region amino acids at particular sequence positions show variability from one Ig to another. Counting from the N-terminal end, at positions 31–37, 51–68, 84–91 and 101–110 in the V_H domain the amino-acid residues are hypervariable, showing profound differences between different specific Ig's. Three such regions exist in the V_L domain.[60] These hypervariable amino-acid transpositions (as well as deletions and additions) in V_L and V_H determine the geometry and binding activity of the combining site. In each variable subunit the polypeptide chain is bent into two irregular, roughly parallel sheets consisting of antiparallel polypeptides. The chains surround a subunit interior composed of hydrophobic amino-acid side chains. This basic and characteristic framework has been found at the combining ends of all Ig molecules yet examined. Each half of the composite ($V_H + V_L$) combining site is a slot incorporated into the sheet backbone surrounded by the hypervariable amino acids.

Heterogeneity of hypervariable amino-acid sequences is the basis for the different combining specificities of the 10^5–10^6 different antibodies that a man can produce. In any Ig class the most constant runs of amino acids, at the C-terminal ends of the heavy chains, constitute the shaft of the Y, part of which is termed the F_c region of the heavy chain backbones. The F_c region mediates most of the other biological activities of the entire Ig molecule which do not directly involve antigen binding. Its structure determines the types of bodily barriers that the molecule can cross. Complex allosteric changes can occur in the F_c portion as a consequence of antigen bonding at the other end of the molecule. These changes bring about significant and appropriate changes in the binding activities of the F_c region (see, for example, page 126).

Ig molecules are made in cells which develop from lymphocytes (Fig. 4.2). Lymphocytes arise from bone marrow stem cells, enter the blood system and exist there as a recirculating population. They pass from the blood into lymph nodes, spleen and other organs, reaching the blood system again via the lymphatic system and ultimately the thoracic duct. Two types of lymphocyte have been characterized which are dependent for their immunological programming on processing in different lymphoid organs. T-lymphocytes (T-cells) are processed in the thymus gland. B-lymphocytes (B-cells) are handled in birds by the Bursa of Fabricius, a lymphoid organ adjacent to the intestine. In each case the lymphoid organ is responsible for the development of immunocompetence in the lymphocytes. The identity of the mammalian equivalent of the bursa is not clear but tonsils,

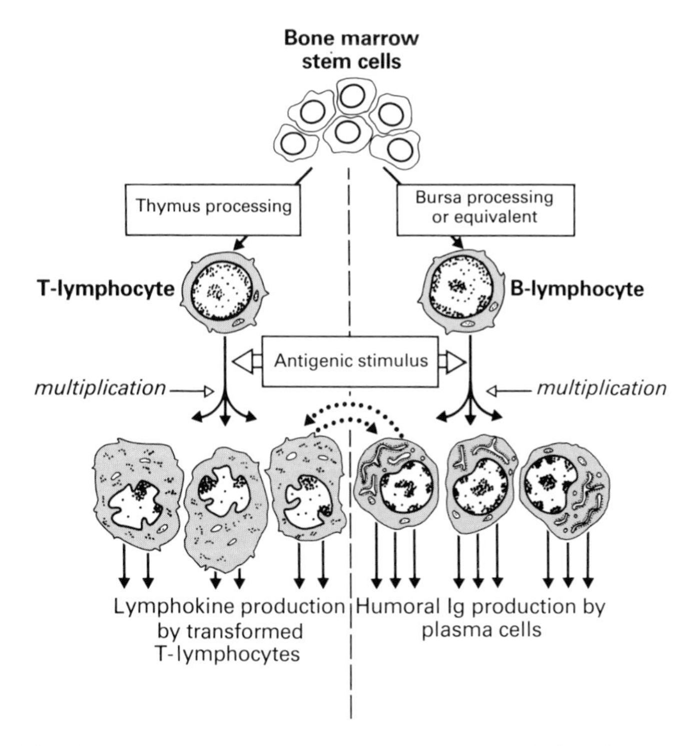

Fig. 4.2 The origin, processing and activities of T- and B-lymphocytes. The dotted arrow represents cooperative interactions between the progeny of the two cell types.

appendix and Peyer's patches in the small intestine all have the correct sorts of credentials.

Both T- and B-cells divide and differentiate by a process called blastogenesis when they counter an appropriate antigenic macromolecule. In fact recognition of antigen by lymphocytes of the immune system is the basis of the system's exquisite specificity of response. B-cells produce plasma cells. The latter have a cytoplasm packed with granular endoplasmic reticulum, typical of cells that are synthesizing protein for export from the cell. Indeed, each plasma cell synthesizes a particular class and specificity of Ig and secretes it into the extracellular environment. These secreted antibodies are termed humoral antibodies. Under antigenic stimulus T-cells undergo a cytodifferentiation and division process which transforms them into lymphoblasts (transformed lymphocytes). Lymphoblasts have a greater cytoplasmic volume than the untransformed T-cell and this cytoplasm stains with an intense basophilia due to large numbers of free, polysomal ribosomes. Ribosomes unassociated with endoplasmic reticulum membranes are normally found in cells producing protein that will remain within the synthesizing cell. Lymphoblasts in fact do not secrete free antibodies. Such cells, however, can respond directly to antigens. They are important in the operation of cell-mediated immunity responses (CMI).

The receptor on the surface of B-cells which responds to an antigen is known to be an immunoglobulin. The nature of the analogous receptor on T-cells is an incompletely resolved problem; it appears not to be a normal Ig molecule. Present opinion on the matter has been influenced by the discovery that in many forms of immunity T- and B-cells must co-operate (T–B cell collaboration). For many antigens both T- and B-cells must recognize the antigen, the B-cell response (humoral Ig production) depending on help from the T-cells. The assistance appears to be due to a factor produced by T-cells that crucially has been shown to have antigenic properties determined by genes in the histocompatibility complex. It is possible that this T-cell factor, an expression of such genes, is the elusive receptor.[20] The conceptual difficulty relating to this finding is that a non-immunoglobulin appears to have binding specificity of an Ig type. Interestingly, though, it has recently been shown[272] that proteins bearing the antigenic specificities of mouse and human histocompatibility gene products possess the same structural folding as the immunoglobulins. So the conceptual difficulty is probably more apparent than genuine as it is conceivable that histocompatibility antigens and Ig have a common genetic origin. Even more exciting is the discovery[289] that superoxide dismutase, a primitive metabolic protective enzyme found even in prokaryotes, also has this same basic structure. It seems that the immunoglobulin type of molecule might have been in existence for over 3000 million years!

How are the correct antibodies for corresponding antigens produced appropriately when a parasite establishes itself in a vertebrate? It seems likely that the process is genetically a selective one in the sense that the genetic propensities for the production of all possible antibodies are always present in the gene pool. In a selective process, interaction of lymphocytes with a single antigen selects, or induces, the production of the relevant type or types of Ig. It must be emphasized, in contrast to earlier ideas of a one antibody-one antigen conceptual framework, that a single purified antigen will induce from a vertebrate a degenerate antibody response. That is, several different antibodies will be produced, each of them including the antigen in their range of combining abilities. This degeneracy has profound genetic advantages compared with a one-for-one response.[279] Degeneracy provides a comprehensive insurance against the possibility that genetic drift or spontaneous mutations could remove the ability to respond to a particular virulent antigen.

A widely accepted hypothesis to account for the induction process is the clonal selection theory of Burnet.[55] It suggests (considering only B-cells, for simplicity) that each single lymphocyte is committed genetically to the potential production of a single specific Ig. Thus, within the millions of lymphocytes in the body's total pool are a vast range of different antibody specificities. Each lymphocyte is assumed to have its own marker antibody built into its plasma membrane. If a single antigen type is introduced into the body it will eventually combine with those cell surfaces containing Ig molecules whose binding specificities encompass that antigen. Such lymphocytes, by this means, receive a stimulus which signals the binding of antigen. The stimulus switches on cell division and Ig synthesis in such cells. Each one produces a clone of many cells, all synthesizing the same appropriate Ig. Clones of different lineages, however, will be producing different Ig's which nonetheless can all combine with the antigen.

It is not yet known how the original broad spectrum of genetic differentiation is obtained in terms of genes specifying different antibodies. Whatever mechanism is operating to generate this diversity it must explain the ability to produce 10^5–10^6 different antibodies. One could simply hypothesize that there are very many different specific genes in the human genome coding for the various Ig's. An alternative theory suggests that precursors of lymphocytes have a foundational Ig gene which undergoes randomized mutations especially in the base-pair regions corresponding to the variable sections of the Ig polypeptide chains. It is not possible at present to differentiate between the likelihoods of these two theories corresponding with reality. It should be noted, though, that the multiple gene theory might not need as many genetic determinants as might be supposed. As each combining site is a composite one formed from two different variable regions, random assortment of n different chains will give n^2 different combining sites.[279] Thus theoretically, $\sqrt{10^5}$ or 316 genes for chains could code for 10^5 different combining abilities.

Defensive responses to bacteria involving Ig

The initial step in the phagocytosis of bacteria by neutrophil leucocytes and macrophages is a tight adherence between the bacterial surface and the plasma membrane of the phagocytic cell. Many bacteria have coats and capsules around them which impede this adhesion. Such microorganisms are poorly phagocytized in the absence of Ig. If, however, antibodies in the form of specific IgG or IgM are present, they can bind to surface antigens on the bacterial surface and macrophages ingest the previously non-adhering bacteria avidly. Much of this enhancement is caused by one of the allosteric effects that occur in the F_c portion of IgG when it reacts with antigen and the arms of the Y molecule hinge open. Amino-acid configurations are produced which can bind to specific receptor sites on macrophage surfaces. Thus, the previously refractory bacterial surface is stuck to the macrophage via an immunological adhesive. This method of increasing the effectiveness of phagocytosis is termed opsonization.

The complement system, interacting with Ig, can also enhance phagocytic activity. In a process termed immune adherence the C_3 portion of the complement cascade when bonded to a bacterial surface can also bind to specific receptors on the macrophage. The helper effects between immunoglobulins and complement components are reciprocal because the complement sequence is often initiated by the binding of C_1 to yet another reactive region on the multifunctional F_c region of an antigen–IgG complex.

IgA is important in immunological resistance to bacterial parasites in sero-mucous secretions like saliva, tears, nasal fluid, lung secretions and the mucous lining of the gut. The IgA is synthesized locally by plasma cells. IgM (macroglobulin) is a phylogenetically primitive, first-line attack against bacteria. It is produced early on during a bacterial infection and is restricted to the blood stream. Humoral antibodies of more than one class have defensive regulatory functions with regard to toxins released from bacteria as well as to bacterial cells themselves. Many of the more pathological of these exotoxins are macromolecular and enzymic. They can thus act as antigens against which antibodies are produced by plasma cells.

As described in chapter 3 (page 69), some bacterial parasites like the mycobacteria can survive and reproduce inside macrophages which have ingested them. The eventual killing of these impudent parasites in immune hosts, and the production of immunity against further infections is a function of CMI. Sensitized T-cells which encounter mycobacterial antigens trigger the death of bacteria within macrophages. The stimulus that effects this alteration in macrophage potency is one or several of the soluble effector molecules released by sensitized T-cells. These substances are called lymphokines.

Immune responses to eukaryotic parasites involving Ig and CMI

Compared with the bacterial situation, far less is known about Ig-mediated defensive responses against protozoan, helminth and arthropod parasites of man. In protozoan infections antibodies can be isolated which show specificities for surface and internal antigens and sometimes exoantigens released from the parasitic cells. Even in infections where the parasites inhabit the circulatory system as is the case with malaria, trypanosomiasis and piroplasmosis, CMI as well as humoral antibody responses can be produced. Complex patterns of heterologous immunity exist in small mammals,[80] and doubtless in man as well, to protozoan infections. Heterologous immunity is immunity that extends to species other than the one that induced the primary immunological response. In the small mammal models which have been examined, previous experience with malaria parasites of the genus *Plasmodium*, for instance, confers increased protection against later infections with piroplasms of the genus *Babesia*. Such complex interactions of consecutive or simultaneous infections need not always be due to the obvious mechanism of shared antigen classes. Non-specific production of lymphokines could eliminate a number of coexisting species of parasite in a particular host. In another example, after the recovery of a mouse from an erythrocytic rickettsial infection such as *Haemobartonella*, some later infections of malarial and piroplasmic parasites are ameliorated. This effect is not Ig mediated but happens because the rickettsias cause anaemia and a consequent increase in the proportion of erythrocyte precursors (reticulocytes) in the blood. This change in blood cell population structure stops some of the protozoan parasites from developing in their preferred mature red blood cells. Mouse oncogenic (tumour-inducing) viruses such as the Rowson–Parr virus makes the host's response to a malarial infection less effective. This change is a direct immunodepressive one by the viruses on the host's antibody production system.

Helminth parasites in vertebrates seem to be one of the most potent stimuli known for the production of IgE antibodies. Normal concentrations of IgE lie in the range 17–450 $ng\,ml^{-1}$. In helminth infections these values can rise to 10000 $ng\,ml^{-1}$, and it has seemed likely that these antibody molecules must have some role to play in eliminating worms, reducing their reproductive output or conferring immunity to subsequent infection attempts. It has proved remarkably difficult, however, to demonstrate these functions rigorously. IgE certainly binds to mast cells and causes them to release histamine and 5-hydroxytryptamine when the antibody molecule also reacts with antigenic substances on the surfaces of parasitic worms.

There appear to be better experimental grounds for ascribing a real protective function to CMI in worm infections. Larsh and Weatherly,[187] in a recent comprehensive review paper, have collected together experimental findings that appear to provide evidence for the active participation of CMI in the defensive responses of vertebrates to nematodes, tapeworms and digeneans.

There is contention about the mode of action of T-cells in ejecting worms. It is possible that lymphoblasts with the appropriate binding specificities combine directly with antigens on the worms, disabling them in the process. Larsh and Weatherly,[187] however, relegate this immunologically specific interaction to a mere triggering role. They believe it initiates a delayed hypersensitivity reaction (DH). This then produces host-tissue injury that makes the local environment untenable by the worms. In their models of the rejection phenomenon, specifically sensitized T-cells arrive randomly at the parasite location and their surface receptors combine with parasite antigens. This activation causes them to release the lymphokines which are the direct effectors of the DH response. Chemotactic lymphokines attract macrophages, lymphocytes and polymorphs to the parasite's locale, macrophage inhibition factor (MIF) serves to induce phagocytic macrophages to congregate near the parasite, and cytotoxic substances can be released. The final result is tissue injury and non-specific allergic inflammation which changes the environment to the parasite's disadvantage. Figure 4.3 shows macrophages congregating around a helminth parasite; in this instance, *Fasciola hepatica*.

Little work has been done on the immunological component of human defensive responses to arthropod parasites. A certain amount seems to be known about scabies.[238,269] The mite *Sarcoptes scabei* lays its eggs in burrows in the keratinized layer of human skin. Starting with a pair of adult mites, unimpeded reproduction would produce a population of 3×10^6 acarines in seven weeks, this total representing close to the theoretical carrying capacity of the human skin surface. Luckily, typical scabies infections rarely comprise more than twenty burrowing adult females at one time. What regulates the parasite population to this bearable, extremely non-maximal level? The onset of intense itching (*Sarcoptes* is called the itch mite) about four weeks after a primary infection is thought to be due to a DH reaction. IgE has also been implicated. Such a response could impair the mites' environment directly but the indirect effect of scratching induced by the itch stimulus cannot be ignored. Scratching excavates mites from their burrows, produces inflammation and can introduce secondary bacterial infections. All of these factors can reduce acarine reproductive efficiency. Interestingly, there exists a series of natural and iatrogenically induced experiments where the defensive responses to scabies have broken down. This series is comprised by the reported cases of so-called Norwegian scabies.[269] In this form of the disease an

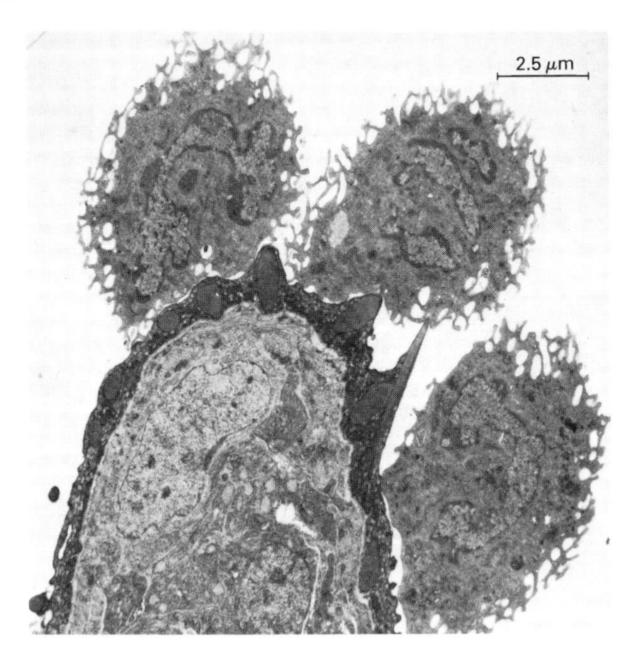

4.3 Transmission electron micrograph of macrophages attached to the surface of a young specimen of *Fasciola hepatica* during its migration through the peritoneal cavity of the final host. (From Davies.[96])

overwhelming infection of 0.2–2 × 10^6 mites takes over the whole skin surface. It can occur if cutaneous sensation is lost as happens in leprosy and in the tabes dorsalis complication of syphilis. In these instances, the stimulus to scratch is not perceived. It has also occurred in the case of a patient given immunosupressive drugs after a renal transplant operation. Mercaptopurine, the drug used, when given at the same time as an antigen, inhibits the development of a primary immunological response. Obviously in the absence of that, no hypersensitivity or consequent itching occurred and the mites could multiply unhindered in the skin.

COUNTER-MEASURES TO VERTEBRATE DEFENCES

Behavioural patterns, genetic resistance, and cellular and macromolecular defensive responses can combine to hinder the establishment of a parasite in a vertebrate. In the face of this welter of adverse selection pressures, parasites have responded in an evolutionary context by developing a number of strategies which negate some or all of these responses. If one was looking for effective examples of such counter-measures one might begin one's search where the total defensive response might be expected to be most multifunctional and specific, namely within the blood system itself of the host. This expectation is easily fulfilled. The two most distinctive examples of evasion of the vertebrate immunological response concern parasites that are blood-dwelling in their habits. They are the schistosome digeneans, typified by *Schistosoma mansoni* and the salivarian trypanosomes, exemplified by *Trypanosoma brucei*.

Schistosoma mansoni[75,345]

The paired male and female worms of *S. mansoni* in the blood vessels of man, monkeys or mice (Fig. 4.4) certainly induce a powerful immune response. Humoral antibodies are produced in a primary infection against surface antigens of the worms. The importance of humoral factors in the total response is shown by the fact that serum taken from mice 12–15 weeks after a primary infection transfers to naive recipients a partial resistance to schistosome challenge.[314] It is likely that worms are killed by a combined attack by IgG and eosinophils. In addition, an over-enthusiastic cell-mediated immune response produces granulomata around schistosome eggs in the tissues. In the monkey[324] antibodies to worms prevent further infections by cercariae after the primary one. Paradoxically, then, most of the adults derived from a primary infection can persist in the blood system for many years in the face of a powerful defensive response which can thwart subsequent cercarial infection attempts. This situation, in which a parasite is unaffected by an effective immune response which it has itself generated, is called concomitant immunity.[345] There is now partial evidence that this type of immunity exists in human populations infected with schistosomes [42,215] just as it does in experimentally infected rhesus monkeys. How is the trick achieved? The answer is that the worms possess host antigens associated with their outer surfaces and these appear to deflect the host's immunological attack in an, as yet, unexplained way. This antigenic disguise has been elegantly demonstrated in a series of experiments[325] in which schistosomes grown in the hepatic portal system of mice were transferred surgically to the same blood vessels in rhesus monkeys. If control monkeys were used, about 80% of the worms survived the hazards of transfer and re-establishment. Egg production halted for about 20 days but then resumed at the original level. So worms experiencing and overcoming an immune response in one species of host could, after some original difficulty, overcome a

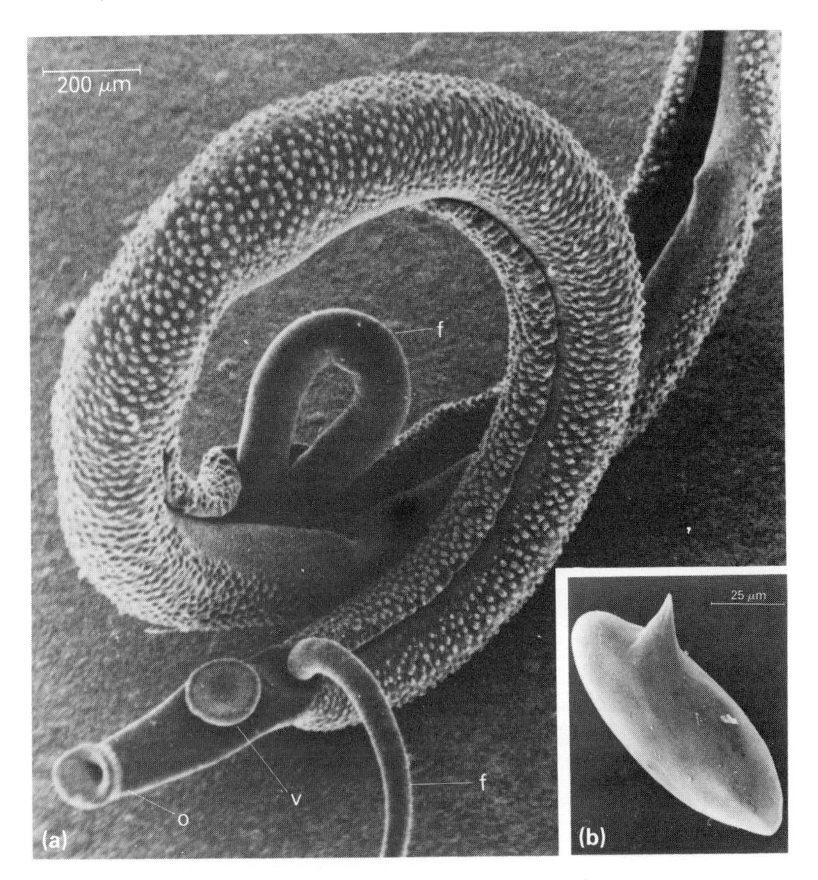

Fig. 4.4 *Schistosoma mansoni* (**a**) A male and female pair of worms, 4 months old, in permanent copulatory union. The female lies in the ventral gynaecophoric canal of the male. f, female worm; o, oral sucker of male worm; v, ventral sucker of male worm. (**b**) A spined egg from the female worm. The spine aids the penetration of host tissue barriers. This penetration is necessary to allow the eggs to pass from the veins of the hepatic portal system into the host's gut lumen for ultimate passage into the outside world. (Both scanning electron micrographs supplied by Dr. H. D. Blankespoor, Museum of Zoology, University of Michigan, U.S.A.)

new response in a different host species. A quite different result was obtained when worms were switched from mice to monkeys previously immunized against mouse red blood cells, a choice of antigen that was fortunately an extremely apposite one for the understanding of concomitant immunity in schistosomes. In such transfers, all the worms were normally destroyed by an intense immune response which produced catastrophic damage to the worm's surface teguments. This result suggests forcibly that the worms from mice had incorporated host antigens (probably from host red cells) into their surface to evade attack in the mouse. When they are transferred to monkeys that can attack mouse erythrocytic antigens their disguise becomes a fatal characteristic. These indications have been confirmed by the location of mouse antigens in the plasma membrane of the tegument using an electron-microscopical cytochemical technique involving a complex Ig-bridge linking the mouse antigens to horseradish peroxidase[222] (Fig. 4.5).

Other experiments[75] have shown that host blood-grouping antigens, which are glycolipids in red blood cell membranes, are very important among the host antigens involved in concomitant immunity. Monkeys were immunized against purified human eryth-

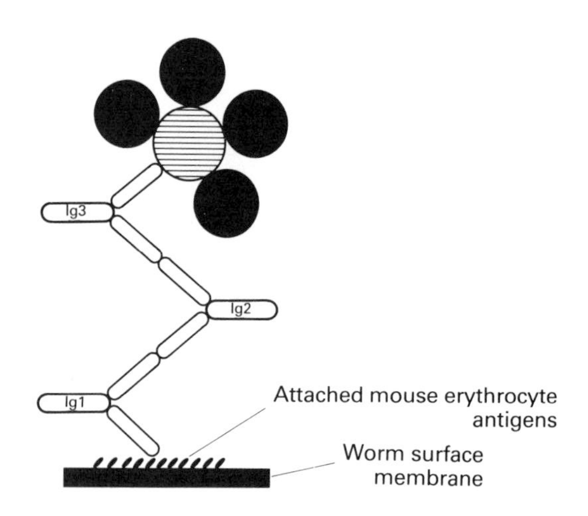

Fig. 4.5 The Ig bridge technique used by McLaren *et al.*[222] to demonstrate mouse erythrocyte antigens attached to the surface membrane of schistosomes. Ig1 (monkey anti-mouse erythrocyte antibody) binds to the mouse antigens on the worm surface. Ig2 (rabbit anti-monkey Ig antibody) binds to Ig1 which is a monkey Ig. Ig2 will also bind to Ig3 (monkey anti-horseradish peroxidase antibody) because that too is a monkey Ig. Ig3 will bind to horseradish peroxidase itself (hatched circle) which can be visualized by an electron microscopical cytochemical reaction which produces osmium deposits (solid spheres). Mouse antigens can be demonstrated on the worms by this technique four days after entry into the mouse but not at 3 hours. Correspondingly monkey anti-worm Ig can be demonstrated to bind to 3-hour worms but not to worms after four days in the mouse.

rocyte glycoproteins covalently linked to a protein carrier molecule. Very young adult schistosomes, termed schistosomula, were grown *in vitro* for 15 days in an incubation medium containing human red cells and serum. They were then transferred surgically to the blood system of the anti-human glycolipid monkeys. Almost all of them were destroyed by an immune response. Control schistosomula grown in the same culture medium survived perfectly well both in normal non-immunized monkeys and, importantly, in monkeys immunized against the carrier protein molecule.

The surfaces of schistosomula have been tested using mixed agglutination techniques to ascertain which glycolipids (glycosphingolipids) were being absorbed from various culture media.[98] In the tests, indicator cells (human red blood cells of known blood grouping types) were bound to the surfaces of schistosomula by antisera specific for particular glycolipid antigens. For example, type A, but not type B red cells were bound by anti-A serum to worms which had type A glycolipid on their surface. Results were quantified by measuring the percentage area of each worm covered with indicator cells. In experiments where schistosomula were grown with human red blood cells possessing glycolipid antigen types A, B, H, C, c, d, E, e, M, N, Duffy^a and P, only types A and B were ever shown to be transferred from red cells to schistosome surfaces. Schistosomula in culture could also take up these two glycolipids from alcohol extracts of red cell membranes and the saliva of a person known to be a secretor of soluble blood group A substance. The passive nature of the absorption, and the route of uptake, were both neatly demonstrated in a simple experiment in which dead schistosomes fixed in cold 0.2% formalin were shown to be able to acquire glycolipids A and B from culture in a medium containing red blood cells of these two types. Dead schistosomes do not ingest red blood cells so the absorptive route must be the direct one from red cell to surface via the medium. The absorptive process must be a passive one as it is most unlikely that any energy-generating metabolic pathways are still operating in the tegument of a fixed worm.

On the basis of this experimental evidence, it seems likely that in a human host glycolipids A and B can transfer from serum or from red cells via the serum directly to the surface membrane of the parasite. In an elegant confirmation of such processes Goldring *et al*[126] immunized monkeys with purified A group substance and showed that such monkeys rejected schistosomes grown in A group blood but not a similar group of worms grown in blood of group B. As shown in Fig. 4.6, glycosphingolipid molecules have a highly hydrophobic end consisting of sphingosine linked to a fatty acid. The other end of the

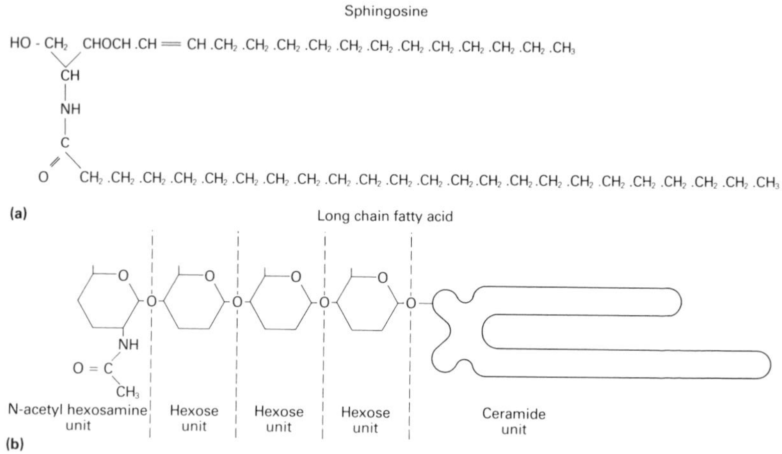

Fig. 4.6 Red blood cell glycolipid structure. These glycolipids are constructed from a hydrophobic ceramide molecule (**a**) (itself formed by a linkage between sphingosine and a fatty acid) linked to a hydrophilic oligosaccharide chain. In the total glycolipid molecule (**b**) the links between hexose units and N-acetyl hexosamine units are schematic in position.

molecule consists of a variable chain of sugar units such as glucose, galactose and N-acetyl galactose, which are responsible for the hydrophilic and antigenic properties of the glycolipid.[67,75] Such molecules link freely to existing cell membranes by interpolating their long paraffinic chains into the double layer of lipid molecules of the membrane. This mechanism does not pose the need to postulate specific receptors of glycolipid antigens on the schistosome membrane. On its own a glycolipid is too small to act as an effective antigen, but bonded to a carrier protein or incorporated into a membrane it can act as a hapten and show immunogenic properties.

When a schistosome cercaria penetrates the host and transforms into a schistosomulum it takes some time to acquire a detectable complement of host glycolipid antigens.[76,222] By day 7 of an infection in mice, not all schistosomula have acquired enough antigen to be killed by transfer to an anti-mouse red cell monkey. Presumably this delay is the basis of the concomitant immunity phenomenon. In a host containing a primary infection of adult schistosomes, subsequent invading schistosomula are damaged by the immune response before they have been able to acquire sufficient protective host antigens. How the host-derived glycolipids actually forestall immunological attack has not been determined with any certainty, but forestall it they certainly do. If schistosomula are cultured in normal monkey serum plus red cells for some days before being challenged with hyper-immune monkey serum they are protected. Schistosomes without this pretreatment have their growth inhibited, or are killed, by the same serum. As a useful working hypothesis, it seems that host glycolipids which have slotted into the bilayer of membrane phospholipids in the worm plasma membrane, or are associated with the hydrophobic regions of exposed surface proteins, act to deflect an immune attack sterically. They can do this because their antigenic, hydrophilic ends are regarded as self rather than non-self by the host's immune system. To the host's lymphocytes an adult schistosome appears to be nothing more threatening than a red blood cell which happens to be 2 cm long!

The above discussion has concentrated on the now well-substantiated incorporation of host-produced antigens into the schistosome surface. Schistosomes also seem to synthesize antigens themselves which show immunological similarities with host antigens; that is, they elaborate a 'home-made' disguise. An example is the substance produced by adults of *S. mansoni* which is antigenically very similar to mouse α_2-macroglobulin.[93] Such schistosome products are probably best regarded as a strategy for ensuring low overall levels of antigenicity throughout a range of different host species. The host glycolipid incorporation system is perhaps a 'fine-tuning' mechanism enabling a schistosome to become specifically protected within a particular host species. Probably the specialized heptalaminate surface of adult schistosomes can be thought of as a membranous sponge for adsorbing host glycolipids (see page 57).

Trypanosomes

A typical infection of a pathogenic salivarian trypanosome in a mammalian host consists of a repeated series of immunologically induced recoveries, each of which is characterized by a rapidly declining trypanosome population. Between these population crashes occur population expansions (relapses from the patient's point of view) which are difficult to explain. In these relapses trypanosome parasites are surviving and flourishing in a blood environment which a few days before was destroying them. Clues to the nature of this incongruous situation came when it became obvious that trypanosomes isolated from successive relapses in a single infection differ in their surface antigenic properties.[128] They are said to display antigenic variation and as one variant is being destroyed by the host another is increasing in numbers to take its place. What is more, the sequential pattern of transiently successful antigenic variants is, to some extent, predict-able. After passage through tsetse flies, new infections in the vertebrate

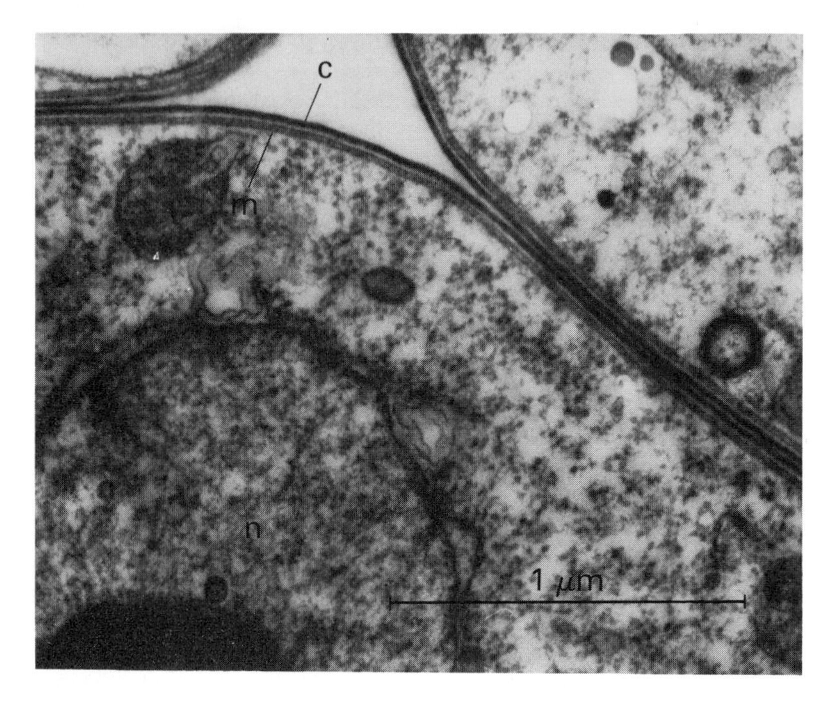

Fig. 4.7 The surface coat of *Trypanosoma brucei.* Portions of three individual bloodstream inhabiting forms are shown each with a dense compact cell coat. c, cell coat; m, microtubule; n, nucleus. (Transmission electron micrograph supplied by G. J. Morgan, King's College, London.)

usually start with a basic antigen (A) which is then eliminated to be replaced by a temporal sequence of new, but usually predictable, antigens (B, C, D, etc.). Until recently, the largely stereotyped nature of this antigenic gamut suggested to most workers that the antigen changes could not be the result of a mutation/selection mechanism acting on the parasite population's intrinsic genetic variability during that particular infection. Antigenic variation appeared to be more a case of the separate induction, perhaps by the effects of the host's specific immune responses, of a number of genetically predetermined antigenic capacities.

What is the actual nature of the variable surface antigens? Many different types of evidence from ultrastructural, biochemical and immunological investigations point to their being glycoproteins (protein-oligosaccharide complexes) attached to the outer surface of the trypanosome's plasma membrane. They constitute 5–10% of the total protein of a trypanosome cell and are probably the sole structural component of the compact and uniform surface coat (glycocalyx) that characterizes *T. brucei* and related trypanosomes (Fig. 4.7). Each sequential variant in a trypanosome infection is entirely covered with a unique glycoprotein which differs in amino-acid sequence and probably carbohydrate components from all other variants in the

Amino-acid sequence numbering from N-terminal

Variant	1	2	3	4	5	6	7	8	9	10
I	Thr -	Asn -	Asn -	His -	Gly -	Leu -	Lys -	Leu -	Gln -	Lys -
II	Ala -	Lys -	Glu -	Ala -	Leu -	Glu -	Tyr -	Lys -	Thr -	Trp -
III	Thr -	Asp -	Lys -	Gly -	Ala -	Ile -	Lys -	Phe -	Glu -	Thr -
IV	Ala -	Glu -	Ala -	Lys -	Ser -	Asp -	Thr -	Ala -	Ser -	Gly -

↑ N-terminal amino acid

Fig. 4.8 First ten amino acids from the N-terminal end of variant-specific surface glycoproteins isolated from four successive variants of a cloned strain (427) of *Trypanosoma brucei.* (From Bridgen *et al.*[48])

infection[48,90] (see Fig. 4.8). Cross[90] has estimated that there are between seven and ten million specific glycoprotein molecules in the coat of a single trypanosome and has produced a structural model for the nature of the association between glycoproteins and cell membrane (Fig. 4.9).

The production of a compact glycocalyx formed from the variant-specific glycoproteins is characteristic only of the bloodstream trypanosomes and the so-called metacyclic forms in the salivary glands of the insect intermediate host. The metacyclic stages are those that are destined to reinfect the vertebrate host. Insect host midgut forms or trypanosomes grown in culture at 25°C do not possess a compact cell coat. Vickerman[362] has suggested that the antigen-containing glycocalyx associated with newly constructed membrane is elaborated in the trypanosome's Golgi apparatus and is perhaps incorporated into the existing surface at the flagellar pocket. No detailed information,

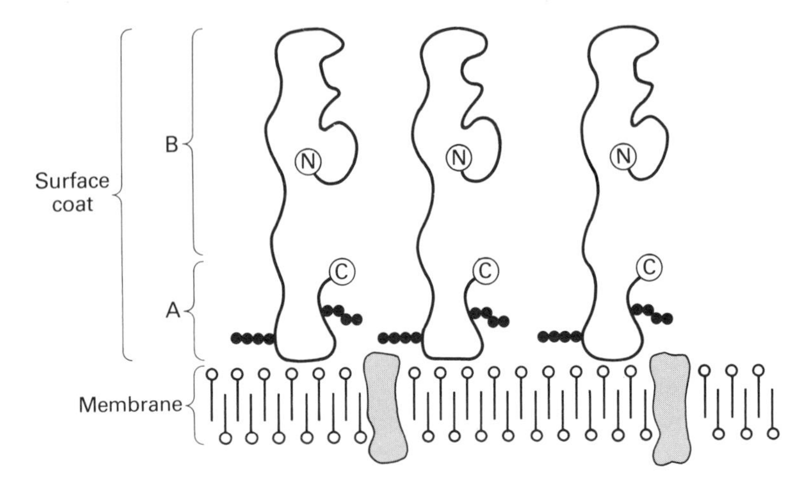

Fig. 4.9 Structural model of the surface coat of *T. brucei* (based on Cross[90]). The cell membrane is shown conventionally as a phospholipid bimolecular leaflet. The glycoprotein molecules are shown attached to the membrane's outer surface. Each glycoprotein has a basal (C-terminal) domain (A) which contains most of the carbohydrate (●●●●), a high concentration of lysine residues and presumably a somewhat conservative structure. This end of the glycoprotein binds it to the cell surface. The outer (N-terminal) domain (B) is highly variable in amino-acid sequence between different variant-specific glycoproteins. The whole glycoprotein molecule has a molecular weight of about 65000.

however, exists on the rate of turnover of the surface glycoproteins. It would be interesting to know, for instance, if the formation of an antibody-glycoprotein complex resulted in the detachment of that complex from the parasite surface membrane.

The mechanism and expression of antigenic variation is being actively investigated by a number of workers. From this work there seems to be a consensus of opinion that the genetic basis of the different phenotypic antigens consists of a number of different genes each coding for a separate variant antigen. Any one gene can be expressed at a particular stage of an infection by an individual parasite. The hypothesis that there might be a single gene for the surface glycoproteins and that it could mutate frequently has been dismissed. The rejection is based on the following pieces of evidence: (i) the variant antigens appear in a preferred sequence, (ii) the parasites revert to a basic antigenic type at the metacyclic insect stage, and (iii) the N-terminal regions of successive variant glycoproteins have very many amino-acid substitutions from point to point along the amino-acid sequence.

There is less of a consensus concerning the processes that generate switching from one antigen to the next. Work in this area depends to a large extent on recent developments in immunofluorescence methodology. These enable the variant types of individual trypanosomes to be ascertained in dried blood smears or suspensions of whole washed parasites. Such determinations enable the detailed variant structure of trypanosome populations to be examined in different conditions.

The results from these and other techniques relating to the switching mechanism point simultaneously in at least two ways. Some results indicate a potent role for the host's immune responses in inducing antigenic variation. It can be shown, for instance, that cloned populations of trypanosomes maintained in irradiated mice, that cannot mount an immune response, can maintain their antigenic homogeneity during repeated experimental passage from mouse to mouse. In other words, in the apparent absence of an immune response, variation itself is absent. Equally, trypanosomes of one variant (D.11), when incubated *in vitro* with homologous (anti D.11) antiserum at particular dilutions, responded by producing a significant proportion of D.12 variants, that is, the next variant in the preferred sequence. With greater dilutions of homologous antiserum, in antiserum against a different variant or in homologous antiserum inactivated with 2-mercaptoethanol no D.12 variants were induced[90] (Table 4.2).

Table 4.2 Results of an experiment in which trypanosomes (variant D.11) were incubated *in vitro* with various dilutions of homologous (anti-D.11) or heterologous (anti-variant 103) antisera then injected into lethally irradiated, hence immunosuppressed, mice. The population structure of the parasites in terms of variants D.11 and D.12 was assessed 72 hours after multiplication in the mice. The initial population contained no detectable individuals of variant D.12. (From Cross.[90])

Antiserum dilution (reciprocal)	Percentage of each variant 72 hours after infection with D11 variant	
	D11	D12
160 (anti-D11)	77 ± 9	30 ± 5
640 (anti-D11)	85 ± 9	21 ± 5
1280 (anti-D11)	100	0
160 (anti-D11 plus 2-mercaptoethanol)	100	0
160 (anti-103)	100	0

Other findings point to variant heterogeneity in the apparent absence of an immune response. Some of the results of Van Meirvenne and his collaborators[359–60] have been interpreted as showing that the initial population of trypanosomes in a vertebrate does not simply contain parasites with the basic antigen variant but also shows small proportions of other variants. The implication would seem to be that heterogeneity can exist before the immune response starts. Indeed, there is also some preliminary evidence that the parasite population in the salivary glands of the tsetse fly is also antigenically heterogeneous.

Probably the actual mechanism for the expression of antigenic variation in trypanosomes is multiple in character, and contains processes compatible with both of the apparently contradictory sets of findings described above. It would be possible for the parasites to have a background polymorphism of antigenic variants at the beginning of an infection and also to demonstrate variant induction by the host's immune response.

Antigenic variation appears to be only one aspect of the trypanosome's ability to side-step the immune responses mounted against them by vertebrate hosts. They can also hide from the response or depress it.

Hiding seems to involve survival in locations within the host where the response cannot operate at full efficiency. Such locations are usually termed privileged sites. Viens *et al*[366] have shown that the stercorarian trypanosome, *Trypanosoma musculi*, can survive in a mouse host which is immune to that parasite. In such cases there are no parasites detectable in the blood and new challenges do not produce a parasitaemia. Nonetheless parasites persist, not in the blood, but in the vasa recta of the kidneys. Presumably neither humoral antibodies nor CMI responses can destroy all the trypanosomes in this location.

Depression of the host's immune responsiveness in trypanosome infections is a well documented phenomenon. Mice harbouring a trypanosome infection show a reduced ability to respond immunologically to both artificial antigens like sheep red blood cells[248] and other parasites like the gut nematode *Nippostrongylus brasiliensis*.[356] The immunodepression is paradoxically characterized by high IgM levels and the presence of antibodies to antigens in no way related to trypanosomes. One hypothesis which seeks to explain this type of trypanosome induced immunodepression suggests that some aspect of a trypanosome infection activates B-lymphocytes polyclonally, that is unspecifically.[154] If this were so, a chronic infection would progressively deplete antigen-reactive B-lymphocytes because these would have been previously converted into antibody secreting cells.

INVERTEBRATE DEFENSIVE RESPONSES TO PARASITES

When collating information on vertebrate defensive responses one is in the happy position of being able to select illustrative examples of experimental work which demonstrate particular defence mechanisms. Findings are often independently verified by many different techniques performed on a range of different organisms. Unfortunately, but inevitably, information from the field of invertebrate defensive responses is more anecdotal in character, less experimentally supported and often derived from work on a single species using a restricted choice of methodologies. Some relationships between invertebrates and parasites, however, have economic importance because the parasites concerned are either potential biological control agents or pathogens of man or his domestic animals. These types of practical significance have encouraged experimental work on such systems.

Encapsulation

Many invertebrates respond to invading parasites or, with special relevance to insects, parasitoids, by patterns of cellular retaliation termed encapsulation. The encapsulation response implies some initial recognition of the foreign nature of the invader and consists of the production of a host-produced capsule around the intruder. Almost always the effector cells are types which are mobile or can become so. The cells most often implicated are free-floating haemocytes in insect haemolymph, or phagocytic amoebocytes and fibroblastic cells that can secrete extracellular collagen or similar structural proteins.

In many instances the encapsulation response is not a specific one directed against parasites, but rather a generalized reaction against introduced foreign material. Encapsulation by haemocytes in insects has been extensively reviewed by Salt.[304] In molluscs,[382] amoebocytes and fibroblasts play more important roles in encapsulation. The parasite itself, or local damage caused to host tissues, can act as initiators of the response. In infections of the snail *Biomphalaria* (*Australorbis*) *glabrata* by *Schistosoma mansoni*, there is a rapid increase in the numbers of dividing fibroblasts and large amoebocytes as cercariae begin to be produced. Where cercariae become immobilized in the snail tissues they become invested with a layer of amoebocytes.[262] The innermost cells remain amoeboid and phagocytic, the outer ones differentiate into fibroblasts and form an outer capsule. Within this, dissolution of cercariae occurs, the remains being ingested by the inner phagocytic cells. It is intriguing that this violent and effective cellular response against some cercariae does not appear earlier in response to the antecedent sporocysts. The latter are, in fact, attacked during the cercarial production phase as though the thresholds for reaction have been lowered.

Other invertebrate defences

It is possible that cells capable of phagocytozing parasitic bacteria and other microorganisms are present in all metazoans. In contrast, macromolecule-mediated defences analogous to the vertebrate immune system appear to be less common. It must be remembered, though, that this apparent deficiency is just as likely to be caused by a lack of relevant investigations as by a genuine gap in defensive abilities. This under-investigation of the field coupled with the novelty shown by some of the examples discussed below, hint strongly that many unusual defensive reactions are waiting to be discovered among the more obscure invertebrate groups.

Substances termed bactericidins can be produced in the haemolymph of large decapod crustaceans like lobsters as a response to infections by, or injections of, bacteria. The bactericidins have a broad spectrum of effectiveness, in that they can kill bacterial types unrelated to the original invading species. Recent work on the lobster, *Homarus americanus*,[339] seems to suggest that induced bactericidin is a result of an interaction between soluble components of the haemolymph and material within haemocytes. There appear to be haemolymph fractions which are present in an inactive form until activated by haemocyte-derived substances.

The sipunculid, *Sipunculus nudus*, has in its coelomic fluid remarkable 'urn' cells which secrete mucus that is sticky for foreign cells but not for autologous cells.[25] Each urn is a bicellular structure which detaches itself from the peritoneal epithelium. One cell of the urn is vesicular; the other is a ciliated mucus secreting cell. Thousands of urns swim in the coelomic fluid removing foreign and cellular debris by gathering it onto their sticky mucus tails. If pathogenic bacteria are introduced the urns produce long tails of mucus which trap the invading parasites. Such urns are said to be hypersecreting. Their mucus will trap the foreign cells but not other urns nor the additional types of coelomic cells in the fluid. Urns with a haul of attached parasites are eventually incorporated into encapsulated 'brown bodies' in the coelom or discharged through the nephridia.

COUNTER-MEASURES TO INVERTEBRATE DEFENCES

Most experimentally investigated examples of effective parasitic counter-measures to invertebrate defensive reactions involve methods for the evasion of encapsulation. In all instances the counter-measures seem only to work in the correct, commonly inhabited, hosts. The experimental introduction of the parasite into an abnormal host usually results in encapsulation and parasitic death, so the evasion mechanisms must be specifically designed to operate in very constrained host environments.

Moniliformis in *Periplaneta*[183]

When the ovoid egg of the acanthocephalan *Moniliformis dubius* hatches in the midgut of a cockroach (*Periplaneta americana*) an active hooked larva, the acanthor, is released. This larva, by repetitive inversion and eversion of its anterior rostellar hooks,[373] penetrates the

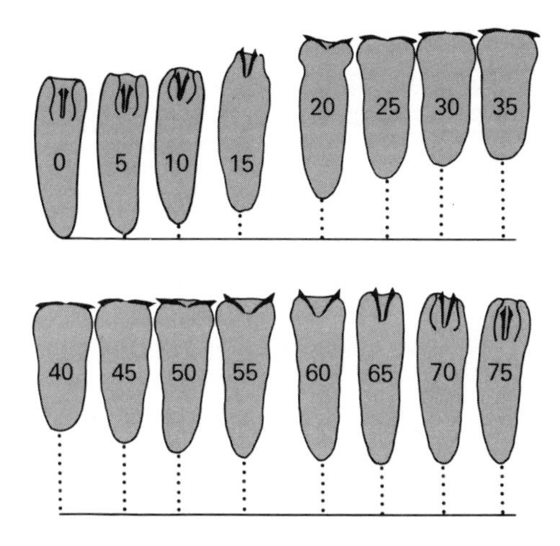

Fig. 4.10 The penetrative locomotory method of a *Moniliformis dubius* acanthor larva. Using this means of progression the acanthor can bore through the gut wall of its cockroach intermediate host. The drawings are taken from the frames of a microcinematographic film, the numbers in each tracing referring to frame numbers (film speed 18 frames per second). The dotted line illustrates the effective anterior progression of the posterior tip of the larva from its original position in frame 0. (From Whitfield.[373])

gut wall (Fig. 4.10) and enters the cockroach's haemocoele. Almost immediately it is densely covered by a classical haemocytic encapsulation (Fig. 4.11a). Layer upon layer of haemocytes flatten themselves against the acanthor's tegument, bonding themselves together with septate desmosomes. At some stage between day 2 and

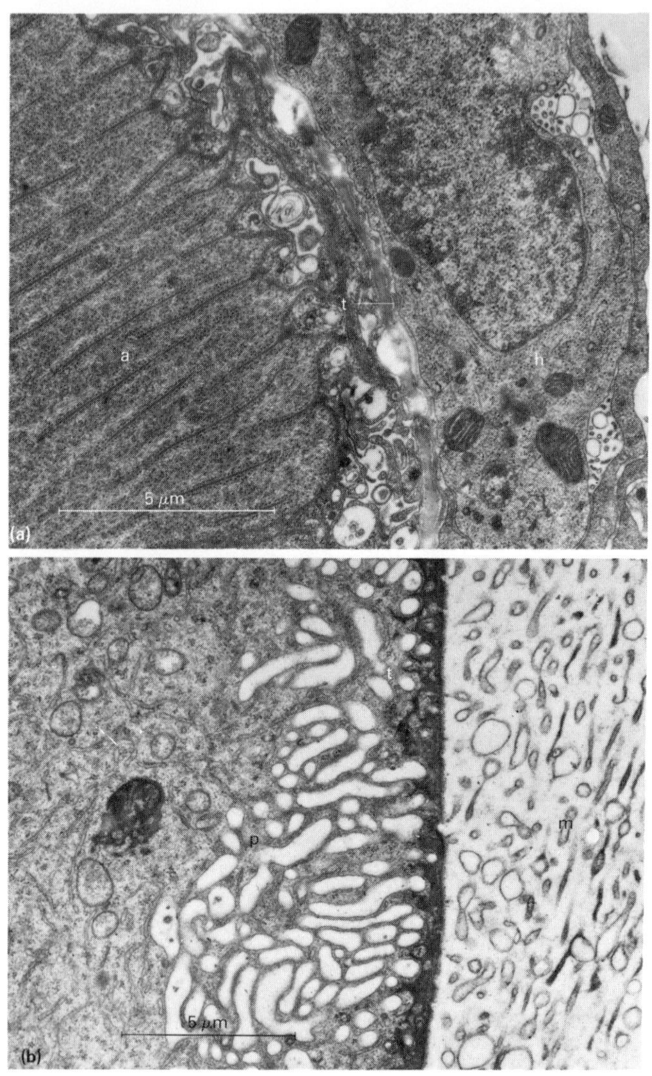

Fig. 4.11 The haemocytic response by the cockroach host to a *Moniliformis* acanthor and the parasite's microvillous counter-measure. (**a**) 3 day old acanthor in the insect haemocoele. The larva was closely attached to the gut and is tightly invested with cockroach haemocytes. a, acanthor muscle; h, cockroach haemocyte; t, outermost layer of acanthor tegument. (**b**) 8 day old acanthor in the insect haemocoele. The larva was loosely attached to a trachea on the serosal surface of the intestine. Tegumental microvilli have been produced and no investing haemocytes are visible. m, tegumental microvilli; t, outermost layer of the sub-microvillous tegument; p, developing tegumental pore canal. (Transmission electron micrographs supplied by Dr. S. Rotherham.)

day 8 of a cockroach infection this potentially disastrous state of affairs for the parasite is dramatically reversed. Long thin microvilli grow out from the parasite's tegumentary plasma membrane and form an interwoven membranous layer (Fig. 4.11b). The microvilli appear physically to push the investing haemocytes away and cause them to disaggregate.

The few haemocytes that do remain in contact with the outer surface of the membranous layer do not adhere to it strongly.[296] The microvillous envelope then elevates from the parasite's surface, losing cytoplasmic connections with it. The mechanism of this process is unknown. As the parasite grows through acanthellae stages to the much larger cystacanth phase the envelope continues to surround the parasite completely, but in the absence of cytoplasmic connections it does not increase in bulk. Instead, it stretches so that the envelope round the cystacanth is much thinner (1.15 μm) than that around the stage 1 acanthella (8.39 μm). Even the tenuous envelope around the cystacanth successfully inhibits haemocyte flattening and no complete, host-produced capsule is ever present after day 2 of an infection.[184] Mercer and Nicholas[239] in an early description of the microvillous membranous layer in the *Moniliformis*-cockroach interaction interpreted it as a product of host haemocytes. Lackie and Rotherham,[184] however, in their more recent study, have shown that similar microvilli can be produced by acanthors maintained in *in vitro* culture where no insect haemocytes are present. This finding of course does not mean that host produced substances need be entirely absent from the envelope. The inhibition of haemocyte flattening and haemocyte-haemocyte fusions is still to be explained and host macromolecules associated with the envelope could provide the answer. It does seem clear, though, that the elevation of the microvillous envelope is not an invariant phase of a successful larval acanthocephalan response to encapsulation. In the case of another acanthocephalan, *Pomphorhynchus laevis*, this separation does not occur in the haemocoele of the amphipod intermediate host, *Gammarus pulex*. The microvillous layer forms but stays closely adjacent to the parasite right through larval development as far as, and including, the infective cystacanth stage.[231]

PLANT DEFENSIVE RESPONSES[377]

Plant-parasites like fungi and nematodes are only one part of the range of organisms that cause extensive damage to, or kill, plants. Plants also have to contend with pathogenic bacteria and viruses, herbivorous vertebrates and insects as well as competing for space, light and soil nutrients with other plants. In the face of this multi-faceted attack, the immobile plants have evolved a wonderfully diverse selection of chemical defences.[29, 119, 377] These molecules usually come in the category of secondary substances, that is, organic molecules that are not of widespread distribution throughout all living organisms but are restricted to particular taxa. The range of molecules is vast but six major groups can be recognized namely phenylpropanes, acetogenins, terpenoids, steroids, alkaloids and uncommon amino acids (see Fig. 4.12).

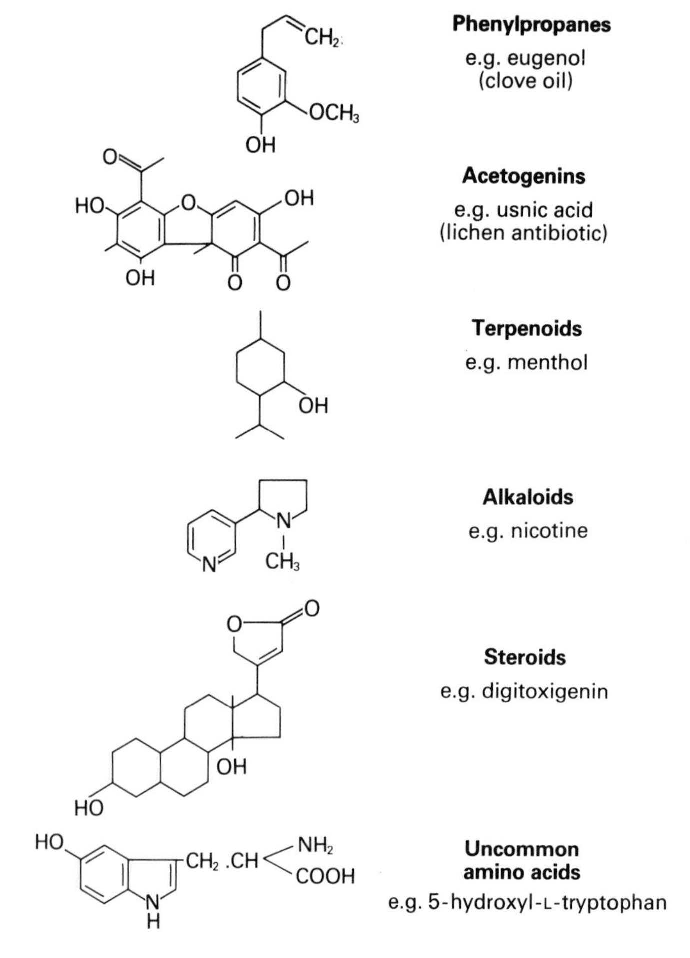

Fig. 4.12 Representatives of the main classes of secondary substances.

Before considering some of the ways in which defensive chemicals and cellular responses protect plants against parasites it will be instructive to consider some examples from the wider range of allelochemic interactions involving higher plants. These examples can serve as a chemical and ecological context for the more specifically anti-parasitic responses. They will also demonstrate the broad relevance of allelochemic interactions.

(1) Several gymnosperms[377] contain substantial amounts of the insect moulting hormone, ecdysone. They also contain analogues of these hormones which are even more active than the originals. Normal growth and development in insects is a function of precisely changing internal levels of ecdysone and juvenile hormone. When insect larvae feed on the foliage of the gymnosperms their metamorphosis can be fatally accelerated with an obvious advantage to the plant.

(2) Larkspurs (*Delphinium* spp.) contain potent neurotoxic alkaloids such as delphinine.[377] These alkaloids are quite capable of killing large herbivores like cows when ingested in relatively small quantities.

(3) The grass, *Aristida oligantha*, releases phenolic acids into the soil which inhibit the establishment and development of nitrogen-fixing bacteria and blue-green algae.[377] *Aristida* is resistant to the resulting low levels of available soil nitrogen but this lack impedes the invasion and replacement of the grass community by many other plant species.

(4) A staggering 14% of the dry weight of the seed of the West African legume, *Griffonia simplicifolia*, consists of the uncommon amino acid 5-hydroxyl-L-tryptophan (5-HTP). This substance is extremely toxic to caterpillars and weevil larvae when they attempt to feed on the seeds,[30] presumably due to some derangement of chemical transmitter function.

Responses to parasites

When defences to parasites are considered a similar diversity of defensive strategies is apparent. Glycoproteins called agglutinins (lectins), like WGA (wheat germ agglutinin), have been found to be present in plants as well as in invertebrates and fish.[312] Recent work[242] has shown that WGA will bind specifically to the growing tips of the hyphae of parasitic fungi. Associated with this binding is an inhibition of ^{3}H-acetate uptake by the fungus and a correlated decrease in hyphal growth. The binding abilities of the different agglutinins relate to various types of substrate polysaccharides and these specificities could be regarded as responsiveness to different bacterial and fungal phytopathogens with chitin, glucans, galactans, mannans and heteropolysaccharides in their cell and hyphal walls.

It is intriguing to note that the specificity of the saccharide binding abilities of lectins has also been implicated in symbiotic associations. Bohlool and Schmidt[37] have demonstrated that a lectin which can be isolated from soybeans will combine specifically with 22 out of 25 different strains of the soybean-nodulating bacterium *Rhizobium japonicum*. On the other hand it did not combine with any of 23 other *Rhizobium* strains which would not induce nitrogen-fixing nodules in soybean roots. This evidence suggests that the original specific binding of appropriate *Rhizobium* strains to root hairs is mediated by plant surface lectins.

Potatoes which are resistant to the larvae of the plant parasitic nematode, *Heterodera rostochiensis*, owe part of their resistance to the production of phenolic glucosides.[371] In *Nicotiana glutinosa*,[23] diterpenes are the effective protection against fungal rust parasites. In this example, 10% of a surface exudate of the surfaces of the plant's leaves consist of an epimeric mixture of the diterpenes sclareol and 13-epi-sclareol. These substances appear to be highly specific inhibitors of the development of rust fungus uredospores. The substances can be transferred experimentally to the leaves of plants of other species where they can effectively protect against rusts. A $100\,\mu g\,ml^{-1}$ solution of sclareol sprayed onto broad bean leaves produces an almost complete inhibition of the germination of the broad bean rust, *Uromyces viciae-fabae*. Examples like the latter one offer hopeful evidence that some of the defensive secondary chemicals used by plants might be utilized by man to his advantage in the protection of crop plants.

Perhaps associated with the production of substances harmful to pathogens and parasites of plants is the so-called hypersensitivity reaction (HR) which has been demonstrated in many plants.[371] It is found in resistant varieties of plant after invasion by parasitic nematode larvae. The latter usually die as a consequence of the response. The HR produced against nematodes is similar to the responses elicited by pathogenic bacteria in, for instance, the bean, *Phaseolus vulgaris*.[205] When a form of tomato plant (variety Nematex) which is resistant to the nematode, *Meloidogyne incognita*, is infected with larvae of this parasite, a characteristic series of HR-associated ultrastructural changes occurs.[270] At first, migrating larvae cause cell wall destruction and produce cellular debris in the cortex of the root. Within 8–12 hours of infection, indicators of increased protein production are apparent in the cells of the root tip provascular tissue. Also, so-called hypersensitive cells around the nematodes lose the contents of dense osmiophilic inclusions that characterize their cytoplasm. Such changes appear to be correlated with rapid cell breakdown around the nematodes and the release of substances damaging to them. It is also possible that these processes have analogies with lysosomal autolysis, as nematode secretions appear to induce the release of lytic substances from plant lysosomes. The tactics of the plant in all these HR responses appear to be similar to a 'scorched earth' military strategy. The point of such a response is to deny sustenance to the attacker by controlled self-destruction. The nematode larvae locked in a shell of HR-destroyed cells find no intact cells to penetrate with their feeding stylets and, like Napoleon's troops in Russia, die of starvation after an apparently successful invasion. The HR is so rapidly mounted in a resistant host that the induction of nurse or giant cells (see page 155 below) necessary for the feeding of several plant endoparasitic nematodes does not have time to occur.

Little is known about the ways in which plant defences can be evaded by parasites; in most cases all that is observed is an infection running an apparently uninhibited course in a susceptible host. If defences are operating in such instances they are only partially effective. Very often the differences between susceptibility (compatibility) and resistance (incompatibility) between a plant and a fungal or nematode parasite have been shown to have a distinct genetic basis. The patterns of susceptibility and resistance are labile in space and time.[371] Geographical varieties of the same host species show great differences in susceptibility to a particular parasite and in time new 'resistance-breaking pathotypes' of the parasite can emerge. One is here observing the genetic basis of evolutionary selection and the efforts of plant breeders to produce resistant crop strains have probably increased the intensity of the selection pressures on parasites for the production of resistance-breaking abilities. Man has thus escalated an already existing genetic conflict.

Cell pathology and the consequences of the genetic differences between compatible and incompatible host/parasite associations have been investigated in the fungus-plant interaction at an ultrastructural level[79] (see Fig. 4.13 and 4.14). The vegetative spore stage of the obligately parasitic rusts is the uredospore. After germination of this spore on the surface of a host leaf, the leaf's waxy cuticle acts as a trigger for the production of a specialized infection hypha that penetrates a leaf stoma. Thereafter hyphal lobes called haustoria (Fig. 4.13) invaginate mesophyll cell plasma membranes and nutrient transfers occur from plant to fungus. Haustoria are produced within 48 hours of infection even in highly incompatible pairings of host and parasite strains. In the latter associations, however, subsequent rapid

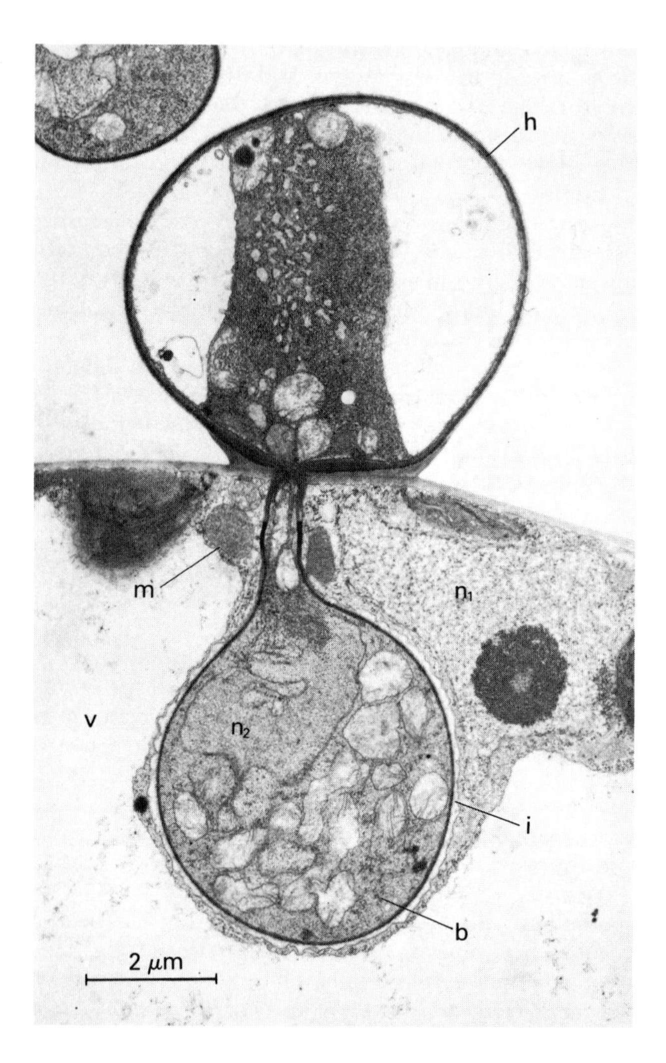

Fig. 4.13 Median section through the haustorial apparatus of the flax rust; *Melampsora lini*, in a 5 day old infection in a susceptible host plant. The haustorial body with a nucleus has 'penetrated' the host cell cytoplasm, but there are no obvious pathological changes in the host cell. b, haustorial body; h, haustorial mother cell; i, invaginated host plasma membrane; m, host microbody adjacent to neck ring of haustorium; n_1, host cell nucleus; n_2, haustorium nucleus; v, vacuole of host cell. (A more extensive version of a transmission electron micrograph first published in Coffey.[79])

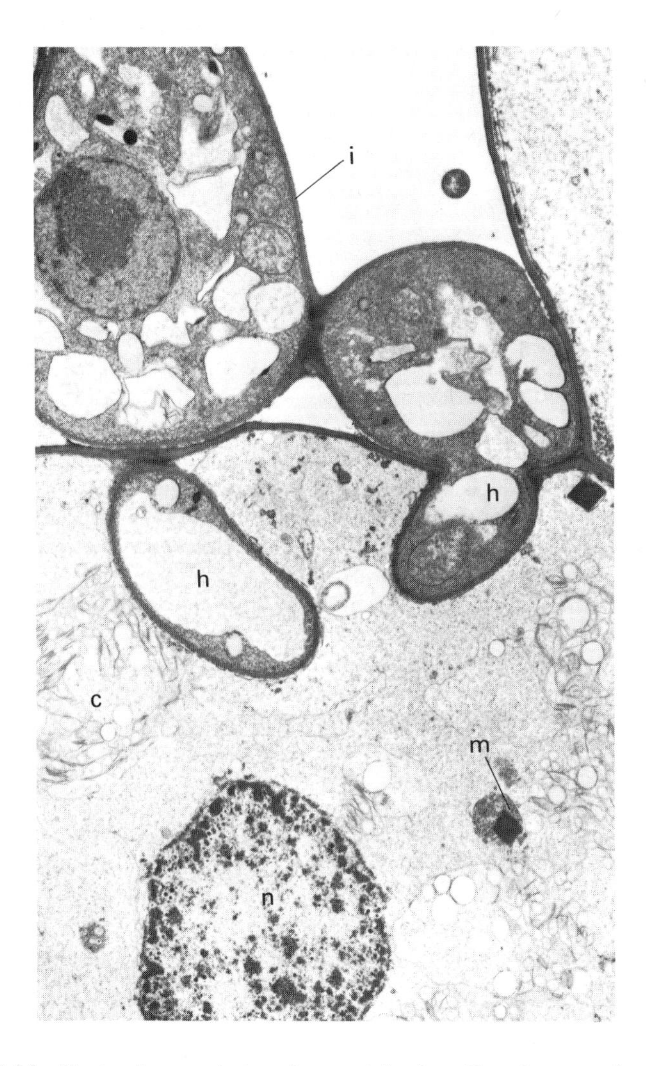

Fig. 4.14 Host cell necrosis in a fungus infection. The electron micrograph shows two haustoria (h) of the pathogen *Phytophthora infestans* (causative agent of potato blight) penetrating a potato leaf mesophyll cell. The haustoria arise from an intercellular hypha (i) and appear to have caused the complete necrosis of the plant cell. Only remnants of chloroplast (c), nucleus (n) and crystal-containing microbodies (m) are visible. (Micrograph supplied by Dr. M. D. Coffey, Trinity College, Dublin.)

disorganization of host cell cytoplasm destroys both cell and haustorium. Genetical evidence suggests that both host and parasite gene products interact to produce incompatibility. Susceptibility may represent the absence or feeble level of such specific interactions. It is interesting that in the rust-plant interaction, recognition and destruction occur only after intimate cell contacts have been established.

NON-DEFENSIVE PHYSIOLOGICAL RESPONSES IN ASSOCIATIONS

In both host-parasite associations and symbioses the physical and chemical presence of one partner can induce changes in the other which are not concerned with defensive regulation. Such induced alterations, sometimes termed physiological responses, often have considerable significance in the integration of the relationship between the organisms. It has usually been assumed that the mechanism of these reactive changes is chemical in nature. As a working hypothesis this has proved a useful assumption. In several cases it can now be shown that changes in metabolism, with consequent perturbations of development or behaviour, occur in associations after the reception by one partner of chemical triggers from the other. In the remaining sections of this chapter a variety of such examples will be discussed. In some, the inductive molecules, with information content for the relationship, have been specifically characterized; in others, attempts at characterization continue.

Induction of reproduction

Fascinating examples of close integrative interactions of associating organisms exist in situations where synchronization of parasitic breeding effort is achieved with some appropriate segment of the host's life cycle. This integration is often a consequence of the parasite's responsiveness to hormones produced by the host. Evidence for this type of intermeshing of development of host and parasite has been found for fleas parasitizing rabbits, the monogenean *Polystoma* which inhabits the bladders of frogs and the protozoan *Opalina* in frogs and toads. A description of the work relating to the flea example will give some impression of the amazing degree of integration to be found in such circumstances.

*The rabbit flea (*Spillopsyllus cuniculi*)*

Fleas are endopterygote insects and have larval forms which are quite different in organization from the blood-sucking adults. Flea eggs are not attached to the host, although they are normally laid while the female flea is on a host animal. The eggs drop to the floor of the host's nest or burrow. There after a period they hatch to free legless, worm-like larvae. The coordination of the reproductive development and behaviour of the rabbit flea, *S. cuniculi*, has been shown to depend crucially on the changing patterns of rabbit hormones that the fleas ingest along with their blood meals. This coordination involves a synchrony between rabbit and flea reproduction such that young fleas are available to infect each new generation of rabbits.

It was shown over 15 years ago that the ovaries of *S. cuniculi* females will not develop until the fleas have fed on pregnant doe rabbits or their newborn young.[235] Later Rothschild and Ford,[298–9] in a series of experiments in which they injected rabbits with different hormone preparations, attempted to unravel the component effects of different host hormones. During the last 10 days of a rabbit pregnancy, blood levels of corticosteroids and oestrogens increase rapidly. These substances appear to stimulate ovarian maturation in the flea. Immediately after parturition male and female fleas leave the mother rabbit and move onto the surface of her newborn offspring. They feed avidly on the blood of the young rabbits. Flea copulation and egg laying on the young rapidly follows. On average 11 days later the fleas once more change hosts. They leave the young rabbit and return to the lactating doe. If any gravid females remain on the female rabbit after parturition, or return, still gravid, from the nestling to the doe, ovarian regression occurs.

The stimuli which induce flea copulation and egg laying on the newborn rabbit are complex. It does seem though that growth hormone in their blood is one contributory influence. Certainly when nestlings 11 days old or more are injected with human growth hormone, fleas upon them begin to copulate. It has also been suggested[299] that a pheromone-like factor produced by nestling rabbits plays a part in stimulating the same aspects of flea reproductive behaviour.

Nutritional release from symbiotes and hosts

In many symbiotic and parasitic associations changes in the physiology or even gross structure of one partner are induced by the other with the effect of facilitating nutrient exchanges in the relationship. Examples from the range of known parasitic and symbiotic instances will illustrate the profound ways in which associate organization can be remoulded in such inductions.

One of the central problems of lichen physiology is why the algal component should behave so 'altruistically' in releasing photosynthate as it does.[322] Typically the green algal cells pass about half the carbon that they fix to the fungal component and blue-green algae similarly donate almost all the nitrogen that they fix reductively.

The fungus-related influences that induce carbohydrate release in the lichen association are not known. We are closer, however, to an understanding of the regulation of the nitrogen-containing substances released from the nitrogen-fixing blue-green algae in some lichens. Work on nitrogen-fixing bacteria in leguminous root nodules seems to provide indications of the sort of metabolic control that might be involved in the analogous lichen associations. *Rhizobium*, the bacterial partner in the legume symbiosis, can fix atmospheric nitrogen. Investigations, reviewed recently by Postgate,[280] have shown that the legume provides the bacteria with both a pentose sugar and a dicarboxylic acid such as succinate. Without the exogenous provision of these two substances the bacterial cells alone will not fix nitrogen in culture; with them, they will. It is known that the blue-green alga, *Nostoc*, ceases its nitrogen fixation almost immediately after it is dissociated from a lichen thallus. Perhaps it is a lack of one or both of these necessary substrates, normally supplied by the fungus, that causes the cessation.

Plant-parasitic nematodes and their hosts can be used to provide an informative series of examples which demonstrate an increasing tendency for the induction of novel morphology and physiology in the host[34] (Table 4.3). The nematode-induced alterations are basically related to the feeding habits of the parasites. At one end of the range of interaction types described in Table 4.3 are ectoparasitic nematodes that cause no obvious long-term changes in the cells upon which they feed. Forms like *Paratylenchus projectus*, for instance, insert their feeding stylet into a root hair cell. Despite the removal of cytoplasm, the plant cell survives the feeding attack. Indeed, cyclosis, that is cytoplasmic streaming, persists through the feeding encounter.

Fig. 4.15 The induction of transfer cell-like giant cells in plants by nematodes in the genus *Meloidogyne*. (**a**) Portions of two giant cells in a root of *Helianthemum* caused by *M. javanica*. The cells abut both xylem and sieve elements. They are syncytial in nature containing several nuclei and the inner surface of the giant cell cell wall is thrown into tortuous protuberances. n, giant cell nucleus; p, cell wall protuberances; s, sieve element; x, xylem element. (**b**) As in (**a**), detail of cell wall protuberances in the giant cell. p, cell wall protuberance. (**c**) Scanning electron micrograph of the inner cell wall surface of a giant cell induced in the root of *Impatiens balsamina* by *M. incognita*. Note the branched tortuous protuberances. (Transmission electron micrographs in (**a**) and (**b**) from Jones, M. G. K. and Gunning, B. E. S. (1976). *Protoplasma*, **87**, 273–9; scanning electron micrograph in (**c**) from Jones, M. G. K. and Dropkin, V. H. (1976). *Cytobios*, **15**, 149–61.)

At the other end of the spectrum are endoparasitic nematodes, members of the family Heteroderidae, which induce the formation of giant cells or multinucleate syncytia in plant tissues. The nematodes (usually from the genera *Heterodera* and *Meloidogyne*), then feed on

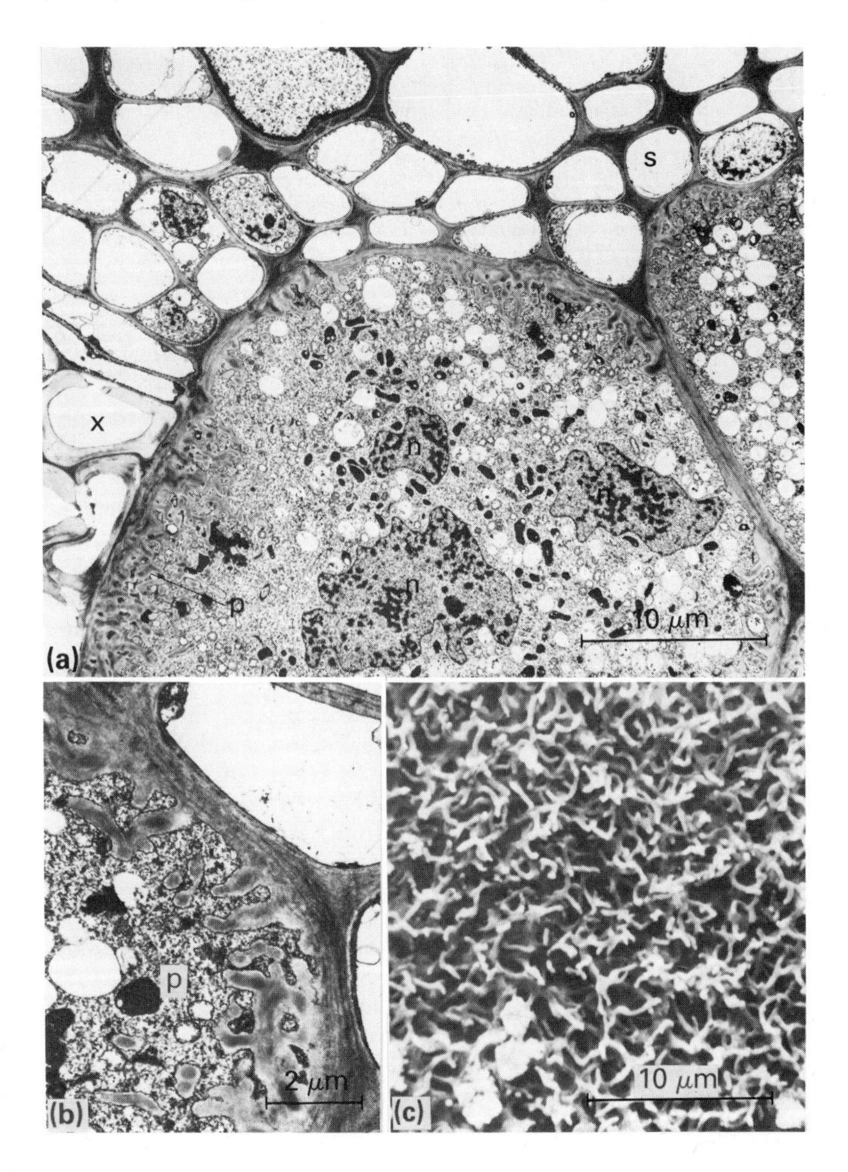

Table 4.3 Examples of host cell responses, in order of increasing complexity, to feeding by plant parasitic nematodes. (From Bird.[34])

Response	*Example*
1 Cyclosis normal, cell survives	*Paratylenchus projectus* feeding on root hair cell
2 Cyclosis decreases, cytoplasm coagulates, cells die gradually	*Tylenchorhynchus dubius* feeding on epidermal or root hair cells
3 Cytoplasmic coagulation and death immediately after feeding	*Trichodorus similis* feeding on epidermal on root hair cells
4 Hypertrophy of cells, loss of chloroplasts, lysis, cavity formation	*Ditylenchus dipsaci* feeding on cortical parenchyma cells
5 Nuclear and nucleolar enlargement, increased cytoplasmic density, thickened cell walls, nurse cells, 6–10 of which make up a feeding site	*Tylenchulus semipenetrans* feeding on cortical parenchyma cells
6 Nuclear and nucleolar enlargement, increased cytoplasmic density, synchronous mitosis of host cells without cytokinesis, cell fusion	*Meloidogyne javanica* feeding on cells in the stele

the contents of these induced structures. One of the most detailed ultrastructural studies on such inductions has been carried out for the potato cyst-nematode, *Heterodera rostochiensis*, which produces syncytia in the cortex of potato roots.[163] Larval nematodes of this species cut through cell walls with their stylet after invading a root. They migrate intracellularly until they take up a permanent feeding position in the cortex.[107] Here they induce plant cell syncytium formation. If the syncytia are large the larvae develop into females, if they are smaller they form male worms.[351] Jones and Northcote[163] found that where the larva is feeding plant cell wall dissolution begins and neighbouring cells are incorporated into a growing syncytium. The syncytium enlarges towards the central vascular tissue of the root and this centripetal growth is only halted when lignified xylem cells are reached. At that point syncytial spread occurs laterally with the involvement of xylem parenchyma and pericycle cells. Cell wall protuberances develop on the syncytium walls which are in contact with conducting elements in the vascular bundle. The presence of these protuberances led Jones and Northcote to suggest that the syncytium is a multinucleate transfer cell. It could function in the translocation of nutrients from a vascular vessel source to an intrasyncytial nematode nutrient sink (Fig. 4.15). The actual mechanism of syncytial induction is unknown. What is apparent, however, is a widespread ability among plant parasitic nematodes for interfering with the concentrations and activities of plant growth promoting substances like cytokinins,[54] gibberellins[54] and indol-3yl-acetic acid (IAA)[367] This ability might well be a partial explanation of the syncytium-forming activity.

5

Behavioural Aspects of Organism Associations

INTRODUCTION

Behaviour is an important facet of the dynamics of organism associations such as parasitism. Parasites and their hosts, commensals and symbiotes all interact behaviourally with their respective partners. Both spontaneous behaviour and elicited behavioural responses are crucial to the integrity of many associations. For the obligate partners in relationships they are often the means by which those relationships are maintained. It is usually possible, by reference to the behaviour of related animals that are free-living, to distinguish non-specific activities from behaviour that is association-specific. The latter behaviour patterns are only meaningful and comprehensible in the context of an association. The non-specific ones have a biological relevance that would exist outside of a parasitic, symbiotic or commensal relationship. For example, the unique ability of the cuckoo to lay one of its eggs in as little as nine seconds[311] is understandable only in the light of the brood-parasitic habits of this bird. Related but non-parasitic birds take much longer to lay. In contrast, cuckoos fly like any other bird, such behaviour being non-specific in relation to brood-parasitic associations.

Purely by convention, the behavioural dimension becomes less well defined as the organisms involved become simpler. The responses of bacteriophages to host bacterial cells, or of algae to fungi in a lichen, are behavioural in a sense, but are more usually described in the language of biochemical, cellular or physiological events. The same can be said of plant activities. Most studies of the behavioural content of associations have involved relationships between metazoan animals with nervous systems. It is from this accumulation of findings that a review of the behavioural aspects of parasitism and related associations must proceed.

THE SENSORY BASIS OF ASSOCIATION-SPECIFIC BEHAVIOUR

When vertebrates, or higher invertebrates like insects, are involved in intimate associations, the fact that behavioural responses are a vital component in the relationships does not cause surprise. A plethora of information exists on the sensory and nervous physiology of animals like mammals, fish, birds and arthropods. This knowledge points to the conclusion that such animals, in associations, will be capable of reacting to subtle sensory information with equally subtle responses. By contrast, facts about the physical foundations of behaviour in parasitic groups like the monogeneans, digeneans and cestodes were, until very recently, fragmentary and few. Consequently, the behaviour of these helminths was assumed to be simple, and their receptor abilities generalized and insensitive. This assumption helped to maintain the belief that 'parasites are degenerate'. In fact, the lack of information on the sensory and neural apparatus of parasitic worms was more rooted in the technical problems posed by small, aquatic invertebrates as experimental animals, than in any degeneracy on the parasite's part.

Many of the best studied examples of association-specific behaviour relate to parasitic worms. A survey of their recently revealed sensory capabilities must therefore precede an analysis of the behaviour itself.

Investigations in recent years, especially with the electron microscope, have uncovered an unsuspected diversity of presumed sensory structures in parasitic worms. There is now little doubt that almost all these animals can receive and utilize a considerable range of sensory inputs.

Many supposed sense organs have been described morphologically, but good neurophysiological or behavioural evidence for the sensory modalities that they mediate is almost entirely absent. The functions tentatively ascribed to the receptors are deduced from the ultra-structural characters and types of locations that they share with sense organs of proven function in other, non-parasitic, animals.

Monogenean sensory systems

The transformation in our understanding of helminth sensory equipment is well illustrated by the monogeneans. Until 1960 we knew little more of monogenean sensory biology than that some of these

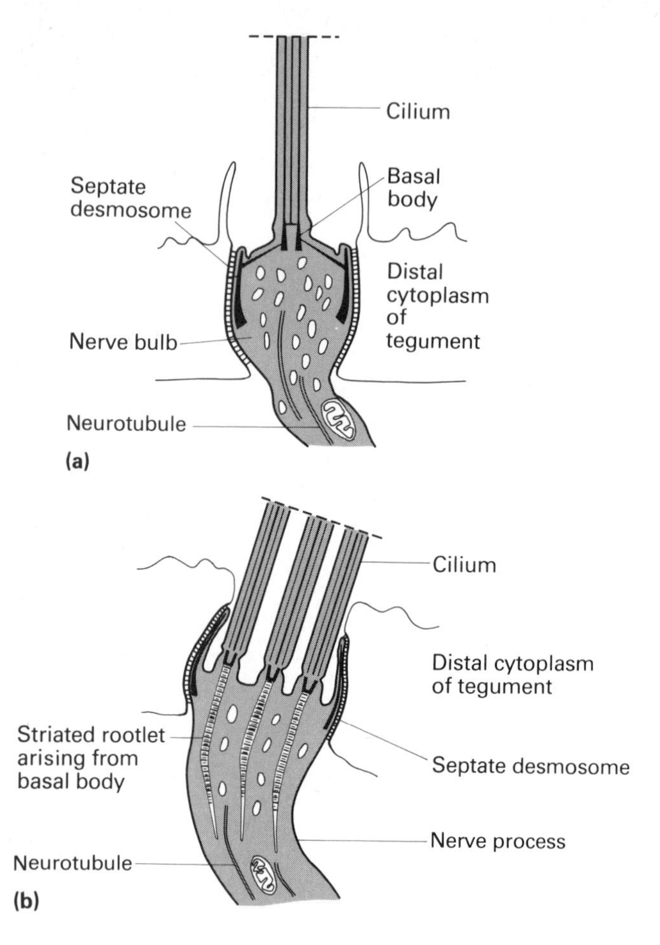

Fig. 5.1 Monogenean ciliated receptor endings. (**a**) Single receptor with a 9+2 cilium. Probably a tangoreceptor (mechanoreceptor). (**b**) Compound (multiciliate) receptor of the 'pit' type, with several 9+2 cilia arising from a single nerve process. Probably a chemoreceptor. (Redrawn from Lyons.[208])

flukes had eyes and that 'sensilla' were present on the surface of the worms. Now, mainly due to the work of K. M. Lyons,[206–211] we have an ever-lengthening inventory of the sense organs present in the larval and adult phases of monogeneans such as *Gyrodactylus*, *Amphibdella* and *Entobdella soleae* (see Fig. 5.1).

The ciliated, free-swimming oncomiracidium larva of the latter

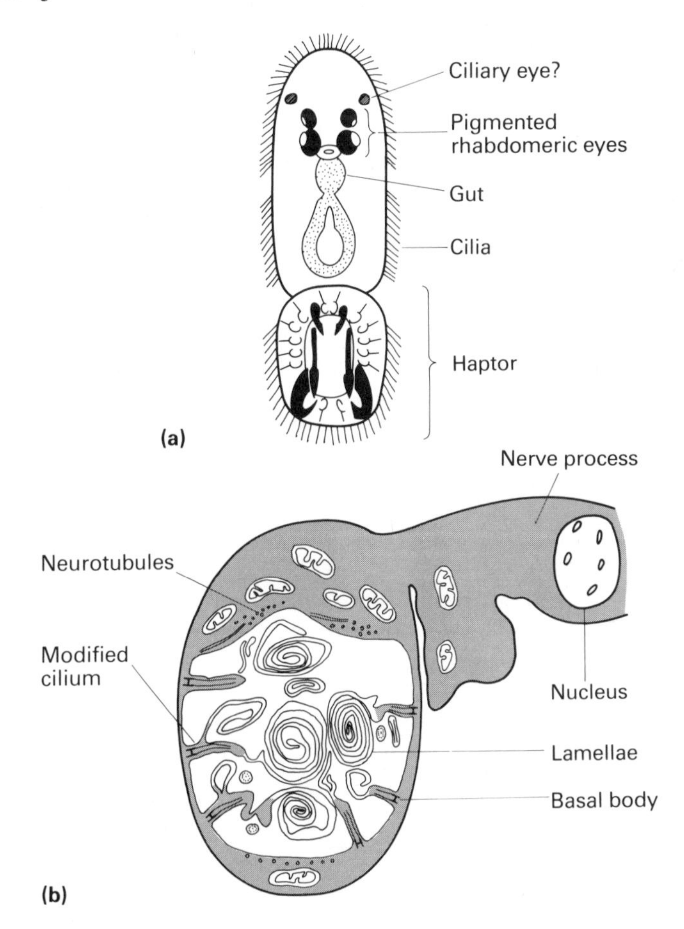

Fig. 5.2 The eyes of the oncomiracidium of the monogenean *Entobdella soleae.* (**a**) The locations and orientations of the pigment-cup, rhabdomeric eyes and the presumed multiciliate photoreceptors. (A composite drawing from Kearn[169] and Lyons.[208]) (**b**) An interpretation of the ultrastructure of the presumed multiciliate photoreceptor. (Redrawn from Lyons.[208])

species has a formidable battery of photoreceptors. It appears to possess both rhabdomeric and ciliary eyes. Rhabdomeric eyes, like those of most insects, have their photoreceptor pigment associated with the membranes of multiple microvilli extending from the receptor cell. Ciliary eyes, like the rod and cone eyes of vertebrates, have the photoreceptor pigment in membrane expansions of highly modified cilia. In the *Entobdella* oncomiracidium four rhabdomeric eyes with pigment cups, and at least two presumed ciliary photoreceptors without masking pigment are present. The rhabdomeric eyes are organized in two pairs, the posterior pair with a light acceptance angle directed anterolaterally, the anterior eyes pointing posterolaterally (Fig. 5.2a). This patterning of pigment cups suggests strongly that these eyes are directional photoreceptors. The ciliary photoreceptors (Fig. 5.2b), with no pigment cups, can accept light from all directions. In the molluscs *Pecten* and *Cardium*, ciliary eyes produce bursts of nerve impulses when previous illumination is shut off. They are important in shadow responses. It is possible that the ciliary receptors of the oncomiracidium also signal sudden decreases in light intensity. They could also be long-term light receptors, connected with the phase setting of rhythmical behaviour.

In the oncomiracidial epidermis are embedded a variety of sense cells bearing non-motile cilia. In each case a nerve bulb of sensory cell cytoplasm is sealed into the epidermis by a circular septate desmosome. Solitary receptors have a single, long cilium with a 9+2 microtubule organization (Fig. 5.1a). They are thought to be mechanoreceptors signalling information either about contacts (tangoreception) or water currents (rheoreception). Such sense cells occur over much of the larval surface. Groups of similar uniciliate receptors are found on the head of the oncomiracidium. In these, each nerve bulb has its own nerve process and the periphery of each bulb extends outwards as a tall collar. The function of these sense cell clusters is unknown.

Positioned on each side of the head, near to the openings of adhesive glands, is a group of three ciliated pits, each of which bears 15–18 highly modified cilia (see Fig. 5.1b). The ultrastructure and location of these pits is suggestive of a chemosensory role.

Adult *Entobdella* also possess solitary, uniciliate receptors and clusters of uniciliate cells. The latter groups occur in front of the adhesive head lappets. The ventral surface of the posterior sucker of the adult worm is covered with more than 800 sensory papillae (Fig. 5.3). Each papilla represents a hypertrophied zone of sub-epidermal fibrous material penetrated by nerves.[211] These form a stack of dendritic lamellae running parallel to the surface of the sucker. The

Fig. 5.3 The opisthaptor of *Entobdella soleae*. (**a**) Scanning electron micrograph of the attachment surface of the opisthaptor. a, accessory sclerite; m, marginal valve; p, posterior hamulus; s, sensory papillae. (From Lyons.[211]) (**b**) Drawing of a transverse section through a sensory papilla on the opisthaptor. (Based on an electron micrograph in Lyons.[211])

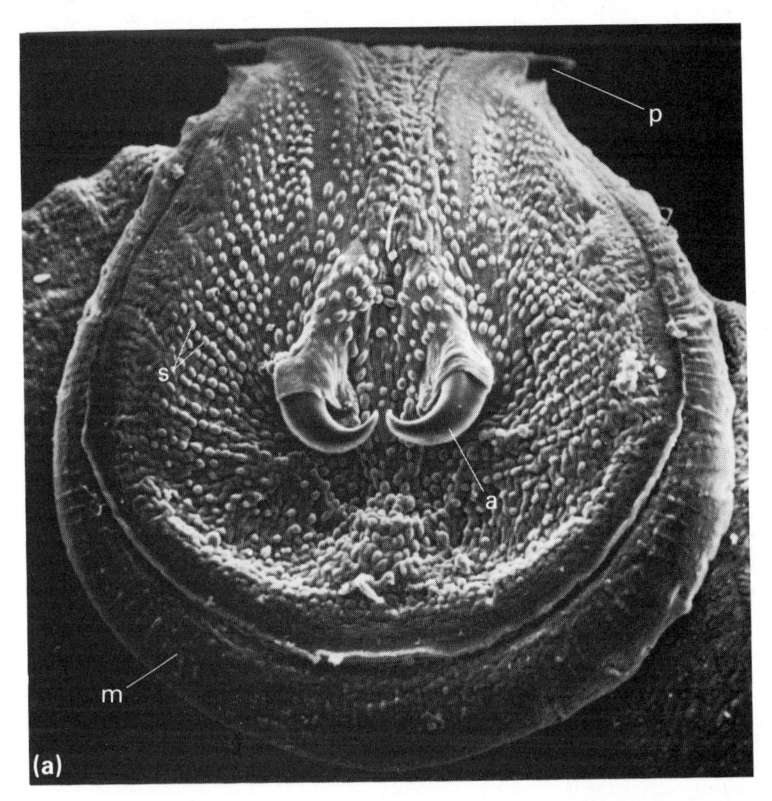
p
s
a
m
(a)

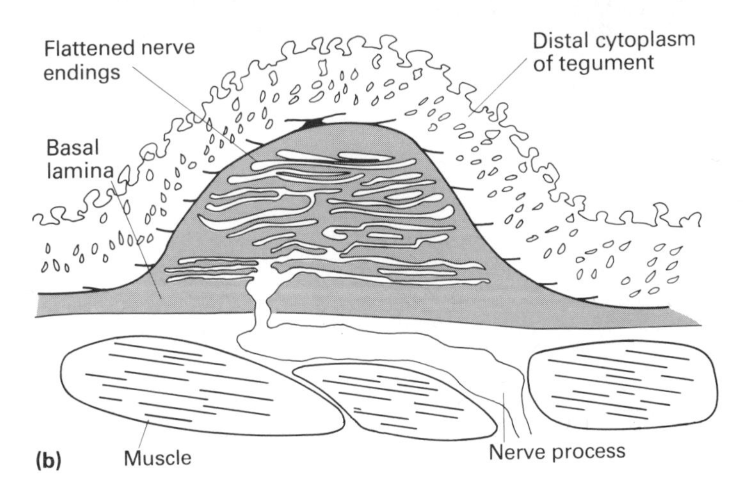
Flattened nerve endings
Distal cytoplasm of tegument
Basal lamina
(b)
Muscle
Nerve process

field of papillae might constitute a receptor array for monitoring either the details of sucker contact with the host's scales or a pattern of strains in the attached sucker. The papillae must develop after contact with the fish host (the sole, *Solea solea*), because the functional posterior sucker of the oncomiracidium does not possess such receptors.

Digenean sensory systems

Eyes, and ciliated and non-ciliated sensory endings, have been investigated ultrastructurally in digeneans at most developmental stages. Uniciliate 'tangoreceptive' sense cells and 'chemoreceptive' multiciliate pits are the most common. Uniciliate sensory endings have been described from the miracidia of *Schistosoma mansoni*,[52] sporocysts of the fellodistomatid, *Bacciger bacciger*,[232] cercariae from at least six species including *S. mansoni* as well as the sexually reproducing adults of several digenean species including *Cyathocotyle bushiensis*,[114] *S. mansoni*[243] and *Transversotrema patialense* (P. J. Whitfield, unpublished observations). Ciliated pits have been demonstrated in the miracidia of *Fasciola hepatica*[381] and *S. mansoni*[52] and also in the cercariae of *Echinostoma paraensei*[232] and *S. mansoni*.[258]

A variety of techniques have been used in attempts to locate and visualize sensory endings in these digenean parasites. Scanning and

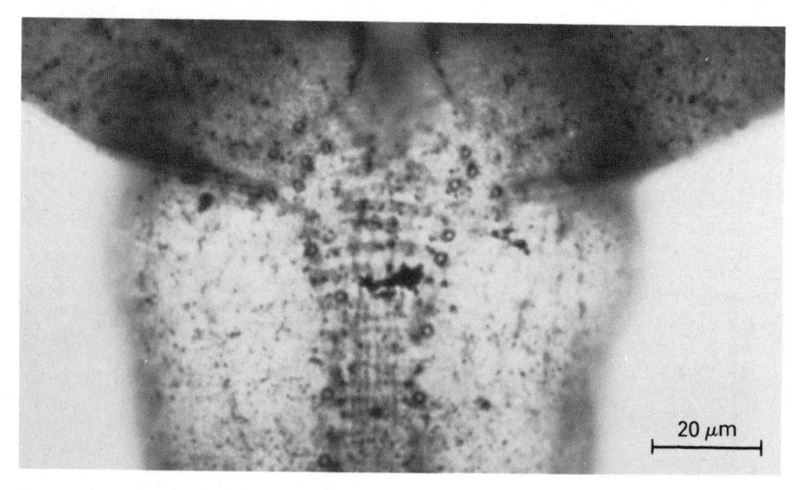

Fig. 5.4 Photomicrograph of a silver-stained whole mount prepration of the base of the tail stem of a *Transversotrema patialense* cercaria. The dorsal surface is in focus and shows two longitudinal rows of ciliated receptors, apparent as dense annuli.

transmission electron microscopy have given complementary information on the surface and internal architecture of the receptors. Histochemical methods for cholinesterases have been used with some success to map the total nervous systems of parasitic flatworms (see for instance Halton and Morris[132]). The histochemical techniques, however, do not usually possess sufficient spatial resolution to allow the individual staining of single sensory endings. Silver staining techniques, though, do appear to have the necessary resolving power. Using them, sensory endings that penetrate the tegument, can usually be stained with metallic silver. In many cases, nerve bulbs stained in this way appear as dark annuli (Fig. 5.4) as though silver had been deposited in the septate desmosome area around each bulb.

Ciliated receptors often occur in distinct patterns which are thought to have functional significance. Dense arrays are found on the surface of the gynaecophoric canal of male schistosomes (Fig. 5.5a), where they can perhaps signal the presence and position of a female worm. In the ectoparasitic digeneans of fish, the transversotrematids, interesting and apparently conservatively maintained patterns of ciliated receptors are found in a number of locations. Two rings of endings are situated on the attachment surface of the ventral sucker, where they presumably provide information concerning sucker contact with the fish host (Fig. 5.5d). Similarly, clusters of tall mammiform receptors protrude from the tips of the arm processes of transversotrematid cercariae[375] (Figs 5.5b and c). The arms appear to be the organs responsible for recognition of the fish host surface and subsequent attachment to it, so it seems that the clustered receptors are involved in contact chemoreception.

Miracidia[52,381] seem particularly well endowed with sense cells. The apical papilla and the gaps between ciliated epidermal cells bear uniciliate solitary receptors as well as ciliated pits. As in the oncomiracidium, rhabdomeric eyes with pigment cups and presumed ciliary photoreceptors are both present. The commonest arrangement of the pigment cup eyes consists of two adjacent cups each enveloping the neatly stacked rhabdomeric microvilli of two retinular cells. A fifth retinular cell with more irregularly disposed microvilli makes contact with the posterior surface of the left pigment cup. The mitochondria of each retinular cell are concentrated in a dense mass in the opening of the cup. This mass is probably the refractile region described as a crystalline lens in light microscopical publications concerning eye spots (Fig. 5.6). In *Diplostomulum spathaceum*,[52] the dendrites of the retinular cells of each pigmented eye make contact with the surface of a ciliary photoreceptor. Thus, in this species there is indirect evidence of neurally mediated integration between the functioning of rhabdomeric and ciliary eyes. Perhaps one adjusts the response thresholds of the other.

The miracidia of most digeneans possess a pair of lateral papillae located between the first and second tiers of external ciliated cells.

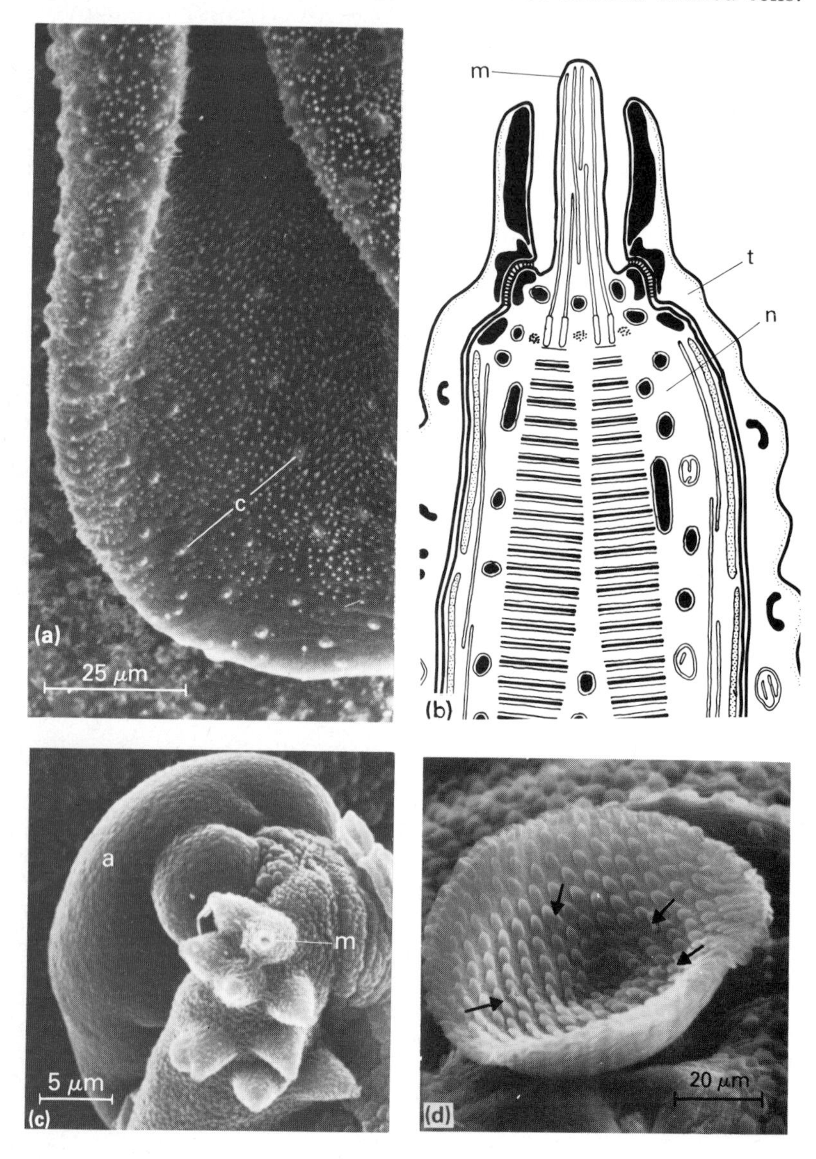

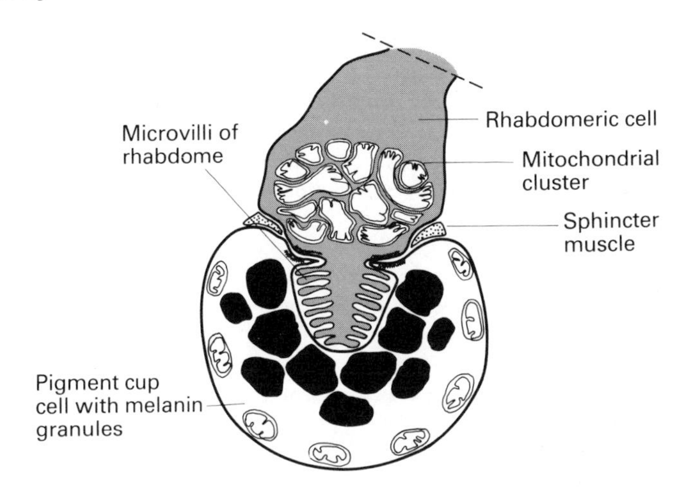

Fig. 5.6 A miracidial pigment cup (rhabdomeric) eye. The drawing is based on an electron micrograph of an eye in the miracidium of *Diplostomum spathaceum*. (From Brooker.[52]) The section only passes through one of the rhabdomeric cells. Encircling the junctional area between pigment cup and rhabdomeric cell is a circular spincter muscle. This might operate to close the aperture of the pigment cup or to alter the geometry of the mitochondrial zone of the rhabdomeric cell.

These papillae are linked to the central nerve mass of the miracidium by a neural connective. In *S. mansoni* and *D. spathaceum*, each papilla contains a single bulbous nerve ending filled with small vesicles, but without cilia. The nerve ending has no direct contact with the external environment, being covered with non-ciliated epidermis. Each lateral

Fig. 5.5 Patterned groups of ciliated receptors in digeneans. (**a**) Ventral view of the posterior end of the gynaecophoric canal of a 4 month old male *Schistosoma mansoni* showing an array of ciliated receptors (c) in the canal tegument. (Scanning electron micrograph supplied by Dr. H. D. Blankespoor, Museum of Zoology, University of Michigan, U.S.A.) (**b**) Interpretation of the internal ultrastructure of a mammiform receptor from the cercaria of *Transversotrema patialense* (see (**c**)). m, macrocilium formed from the microtubules from several basal bodies; n, nerve bulb; t, tegument of the arm process. (Based on a drawing in Whitfield *et al.*[375]) (**c**) Array of nine mammiform receptors at the tip of an arm process of the cercaria of *Transversotrema patialense*. The receptors are thought to be chemoreceptors signalling a contact with a fish surface. Immediately after contact has been achieved adhesive droplets are extruded from the adhesive pad, sticking the cercaria to the fish. a, adhesive pad; m, mammiform receptor. (Based on a scanning electron micrograph in Whitfield *et al.*[375]) (**d**) Ciliated receptor array on the spined surface of the ventral sucker of an adult digenean (*Transversotrema patialense*). (Scanning electron micrograph supplied by D. A. P. Bundy, King's College, London.)

papilla lies immediately posterior to a solitary, uniciliate receptor (Fig. 5.7). Brooker[52] has suggested that physical interaction between the adjacent sensory structures might provide the miracidium with information about its orientation with respect to gravity. If the two components had different flexing rigidity, they could equally well function as a directional accelerometer sensor for antero-posterior accelerations.

Some miracidia also possess dorsal sensory papillae that lie between the ciliated cells of the most anterior tier. In *D. spathaceum*, each papilla consists of a nerve bulb that penetrates the epidermis as a hermispherical bulge bearing a ring of short cilia that lie parallel to, and slightly above, the surrounding epidermal plasma membrane. The function of these intriguingly shaped receptors is unknown. Chemosensitivity has been postulated but not proven.

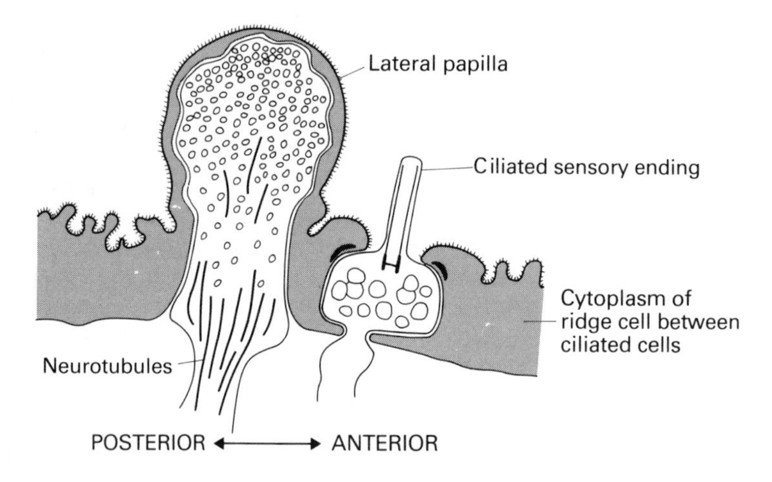

Fig. 5.7 A miracidial lateral papilla complex, consisting of an aciliate papilla immediately behind a uniciliated sensory bulb. (Drawing based on an electron micrograph of the complex in *Diplostomum sphathaceum* produced by Brooker.[52])

Cestodes sensory systems

Few investigators have searched for sense organs in tapeworms. Those results which have been gathered, however, begin to fit into a consistent pattern (Fig. 5.8). The scolex attachment region usually possess a high density of uniciliate sense cells and the proglottids have non-ciliate as well as ciliate types. In *Raillietina cesticilllus*,[35] many sensory endings are found on the scolex, especially on the sides of the terminal rostellar region. These receptors are all of one type. Each nerve bulb carries a single, stout cilium which is approximately the same length as the surrounding tegumentary microtriches. The cilium has an unusual microtubular organization, namely 9 + 6 + 1 (see Fig. 5.8). Broadly similar ciliated sense cells have been demonstrated on the rostellum of *Echinoccoccus granulosus*[161] and the posterior proglottids of *Taenia hydatigena*.[117] With only ultrastructural evidence available, the receptors could equally well be mechano- or chemosensory.

The non-ciliated sense cells of cestode proglottids do not penetrate

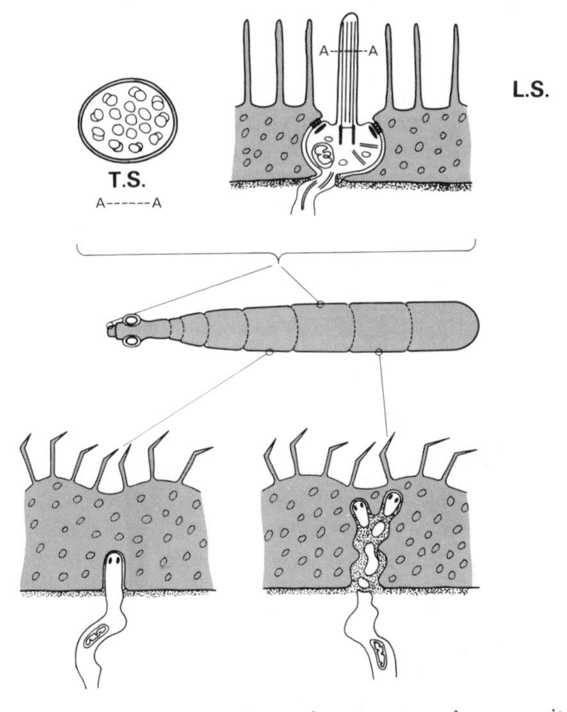

Fig. 5.8 Tegumental sensory endings of tapeworms. A composite diagram showing the structure and locations of sensory nervous terminations in a hypothetical cestode. The upper two diagrams show the ciliary microtubule arrangement (9+6+1) and nerve bulb ultrastructure of a rostellar receptor (similar receptors are found on the proglottids). The lower two diagrams show non-ciliated nerve endings in the tegumental cytoplasm of proglottids. The left hand example only penetrates the cytoplasm shallowly. The right hand example illustrates the situation where the nerve process penetrates deeply into the cytoplasm surrounded by basal lamina material.

the outer surface of the tegumentary epidermis. They lie enclosed in its distal cytoplasm. There has been some confusion in the literature concerning the relationship between these receptors and so-called 'pore-canals'. Most results, however, are understandable if it is assumed that there are two types of non-ciliate sense cells in the proglottids (Fig. 5.8). Both types have a bulbous, distal termination which contains a dense annulus; they differ in the ways in which the nerve processes from the terminations associate with the tegumentary cytoplasm. The first type of receptor does not enter deeply into the cytoplasm, but is closely enveloped by the latter's basal plasma membrane.[117,244] The second type, however, is located at the distal ends of broad evaginations of the fibrous, subtegumentary, basement lamella into the tegumentary cytoplasm. Sensory nerve processes weave an irregular course through such channels to the bulbous terminations.[35,117,200,244] Particular sectioning planes or poor fixation can sometimes obscure the nerve processes in such channels, giving rise to some of the reports of non-nervous 'pore-canals'. Both types of non-ciliate ending are shielded from direct contact with the external environment of the tapeworm, so that tango- rheo- or chemosensitivity seems unlikely. Their location and structure suggest far more that they monitor deformations of the cestode body wall. Such deformations could arise from external pressures, caused by gut peristalsis, for instance, or from the activities of the worm itself.

The sensory systems of parasitic nematodes

Nematodes are eutelic animals. Their somatic organs tend to demonstrate species-specific numbers of cells or nuclei. There is, thus, every likelihood that, in a particular species, the numbers, specializations and positions of sensory neurones will be invariant. This fact has prompted ambitious attempts to correlate the genetics and behaviour of nematodes with their precise neural 'wiring diagram' as determined by serial-section electron microscopy.[44–5,369] Almost all of this research activity has been based on a small, free-living, self-fertilizing hermaphroditic nematode called *Caenorhabditis elegans*. Luckily, however, all nematodes have a broadly similar sensory organization. This means that the extensive work on *C. elegans* can be considered together with that performed specifically on parasitic nematodes. A synthetic viewpoint of this type provides an unusually thorough picture of the extent of the sensory equipment available to these helminths. In *C. elegans*, for instance, it is possible to state categorically that there are exactly 58 sensory neurons in the anterior head region, 52 of which are arranged in sensilla, that is, are surrounded by specialized supporting cells[370] (Fig. 5.9).

Despite this sort of information on the numbers and distributions of sensory endings we are far from understanding the modalities and functions of all of the receptors. Most is known about the amphids (Fig. 5.9). These sense organs are located laterally on the head and usually contain a greater number of sensory neurones than other sensilla. In *C. elegans*, for instance, each one contains twelve. In this species, work with behavioural mutants[369] has been directed towards correlating the precise conformation of the sensory endings of amphidial neurons with observable defects in chemotaxis behaviour.

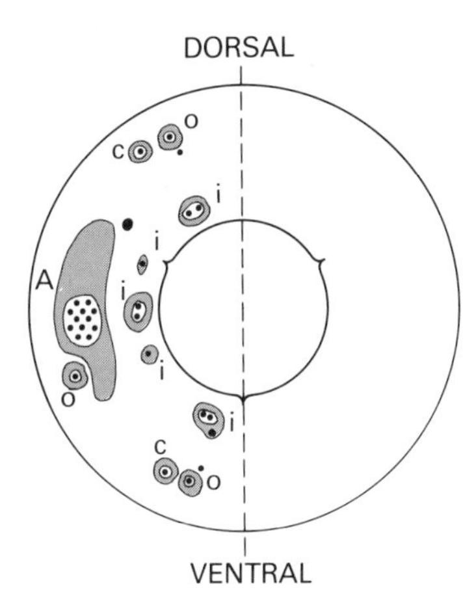

Fig. 5.9 Plan of the sensory neurones in the head of *Caenorhabditis elegans*. The plan is highly diagrammatic and shows the arrangement of neurones and sensilla around the mouth. Only the right hand, lateral half of the sensory array is shown, the other half being its mirror image. In the plan the spots represent individual sensory neurone endings and the shaded areas represent the supporting cell portions of sensilla. i, inner labial sensilla; o, outer labial sensilla; c, cephalic sensilla; A, amphid. All of the sensory endings except the three not in sensilla and the ending within the supporting structure of the ventral, inner labial sensillum are ciliated endings. (A redrawn and simplified version of a detailed plan produced by Ward *et al.*[370])

In several cases it has been possible to demonstrate that a nematode individual with a genetic constitution that produced specific changes in sensory ending ultrastructure in the amphid also had specific sorts of chemotaxis impairment. This is quite strong circumstantial evidence that at least one of the amphid's roles is chemoreceptive.

Amphidial ultrastructure has been extensively investigated in the larval and adult forms of a number of parasitic species by McLaren.[220]

One of the interesting generalizations that can be made about her findings is that the sensory endings in amphids (and indeed in certain other nematode sense organs as well) have an intimate structural relationship with a secretory cell (Fig. 5.10). In all cases, most of the ciliated sensory endings are in contact with the external environment via an amphidial channel and pore which are both cuticle-lined. The channel itself is surrounded by a supporting cell filled with fibres. As this cell appears to be implicated in the production of the channel's cuticular lining after each moult it is probably a specialized hypodermal cell.

Each sensory neurone, before it terminates in a ciliary ending, is enclosed by another cell, termed the secretory cell, and forms extensive membrane contacts with it. Earlier light microscopical examinations of the sense organ resulted in this cell being termed the amphidial gland. It usually shows some ultrastructural evidence of secretory activity such as granular endoplasmic reticulum, Golgi

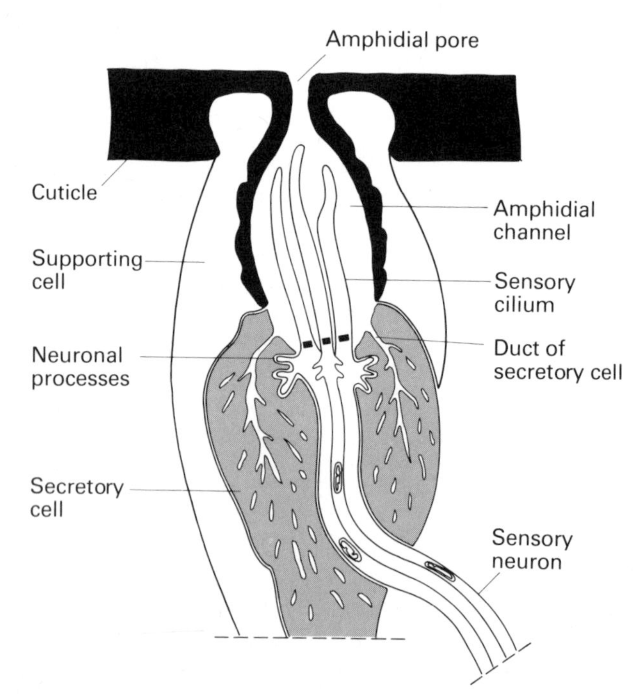

Fig. 5.10 Generalized diagram of amphid ultrastructure. (Based on the interpretations of examples from several parasitic nematode species produced by McLaren.[220])

apparatus and secretion droplets. The phase and degree of secretory activity, however, vary with nematode species and developmental stage. The most elaborate amphidial glands occur in the hookworms. Those of *Necator americanus* extend posteriorly for 25% of the length of the worm in close association with the lateral hypodermal cords.[219] In large secretory cells of this type secretion droplets pass out into intracellular secretory ducts that open into the amphidial channel.

At present the functional interrelationships of sensory endings and secretory cell are difficult to unravel. Some sensory cilia may monitor the external conditions that are the appropriate stimuli for secretion synthesis or release. Others may monitor and control secretion flow itself. Alternatively, the sensory cilia may require a local environment containing the secretory products in order to carry out their sensory roles. Our speculations on the nature of these interactions are vitiated by the fact that it is only in the case of hookworms and filarial nematodes that we have any information about the chemical nature of amphidial gland secretions. In larval and adult examples from both of these groups, the secretions have been shown to have acetylcholinesterase activity.[220–1]

TYPES OF ASSOCIATION-SPECIFIC BEHAVIOUR

The behaviour that takes place within any animal association is a continuous and coordinated sequence of activities. Within this continuum of behaviour, however, it is possible and convenient to demarcate areas of activity that apparently have different types of biological significance. One useful, if arbitrary, way of subdividing association-specific behaviour patterns is on the basis of the phase of association establishment in which the behaviour appears. Four such phases, and types of behaviour, are recognized here (Table 5.1). They will form the framework for subsequent descriptions of behaviour in parasitic, symbiotic and commensal relationships.

Partner-finding behaviour

For some animals involved in intimate associations, the problem of locating a reciprocal partner arises only infrequently because the relationship involved are semi-permanent ones. Many generations of human body lice, *Pediculus humanus*, or itch mites, *Sarcoptes scabiei*, for instance, can flourish, one after the other, on and in the skin of a single man. For these populations of larval and adult parasites the human skin is the sole habitat requirement for continuous life-cycle propagation. When host to host transfer does occur in such parasites it is unlikely to be due to specific, host-finding behaviour. *Sarcoptes*

Table 5.1 Behavioural phases in the initiation and perpetuation of intimate associations.

Phase	Main features of phase: described in terms of two associating organisms, A and B
Associate finding	Associate A matches microhabitat with that of B usually by means of responses to environmental variables
Associate recognition	Associate A perceives and responds to specific stimuli emanating from B
Establishment behaviour	Associate A usually makes physical contact with B
Maintenance behaviour	Associate A (and sometimes B) acts so as to continue physical contact despite disruptive perturbations

cross-infection only takes place when human bodies are in close contact for long periods, as happens in bed.[237] The mites crawl inadvertently from one warm, humid skin surface to the next.

In contrast to the situation illustrated above, in most associations there are regular intervals when one partner species must re-establish contact with the other. In the face of this general problem, two significantly different strategies have evolved. They differ in the relative importance of the behavioural dimension during their operation.

1) Non-behavioural strategies involve the establishment of associations by random, undirected processes. Typical is the case of fungal rust spores. These are disseminated into the atmosphere with vanishingly low possibilities of settlement on the correct host plant at some later date. The path of a successful spore is the result of essentially random air currents. Many helminth parasites use an essentially comparable strategy in that the infective stages, such as resistant eggs, make the transfer from host to host passively.

2) Associate-finding behaviour is the first phase in a sequence of behavioural stages that can be recognized in the establishment and maintenance of most intimate associations between animals. Behaviour of this type usually involves migration into, or remaining within, a micro-habitat where contacts with the appropriate associate are most probable. It is best understood in terms of a response to habitat characteristics which have nothing directly to do with the target animals. The latter will have specific habitat preferences which may alter diurnally and seasonally; the efficient searching animal will closely mimic these changing habitat preferences both in space and

time. Such a mirroring of responses will normally ensure a sympatric and synchronous distribution of searching and sought animals.

The simpler the nervous system of the searching animal, the more generalized and stereotyped will be the parameters by which the correct micro-habitats are recognized. Conversely, in animals with advanced nervous systems, habitat recognition by complex pattern recognition is the rule and learned behaviour becomes important. Most aspects of associate-finding behaviour can be illustrated by the hatching and initial locomotion of both monogenean oncomiracidia and digenean miracidia. These examples probably provide good general models for this phase of association establishment as it is displayed in simpler, invertebrate animals. Larval nematodes, as well as fish and shrimps that act as cleaner symbiotes, will be used as illustrations of the more complex behaviour patterns.

Oncomiracidia

The polygonal eggs of *Entobdella soleae* are found among sand grains in marine bottom deposits where soles, their host fish, lie just buried during daylight hours. The eggs are anchored to the sand by a sticky filament that extends out from each egg shell (Fig. 5.11). Once the enclosed oncomiracidium larva is mature, hatching is possible. The hatching process is an active one and appears to be due to the motile larva pressing against the operculum of the egg shell. Hatching

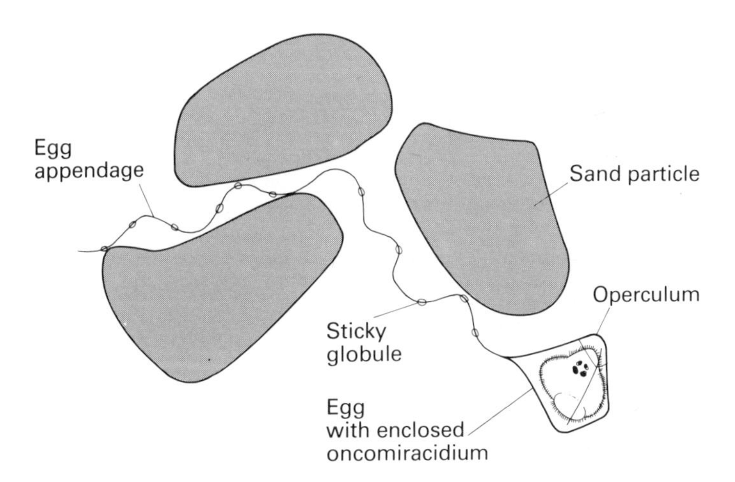

Fig. 5.11 Embryonated egg of the monogenean *Entobdella soleae* attached by sticky globules on the egg appendage to sand grains. (Redrawn from Kearn, G. C. (1963). *Parasitology*, **53**, 435–47.)

behaviour, however, is not a random process in time: it is modulated so as to increase the chances of free-swimming larvae achieving contacts with host fish.[170] Most hatching occurs at dawn when the night-feeding soles are returning to their daytime resting places. The circadian hatching rhythm is endogenous in that it persists even when the eggs are kept in constant darkness or continuous light (Fig. 5.12). The eggs, however, need to have had some previous experience of

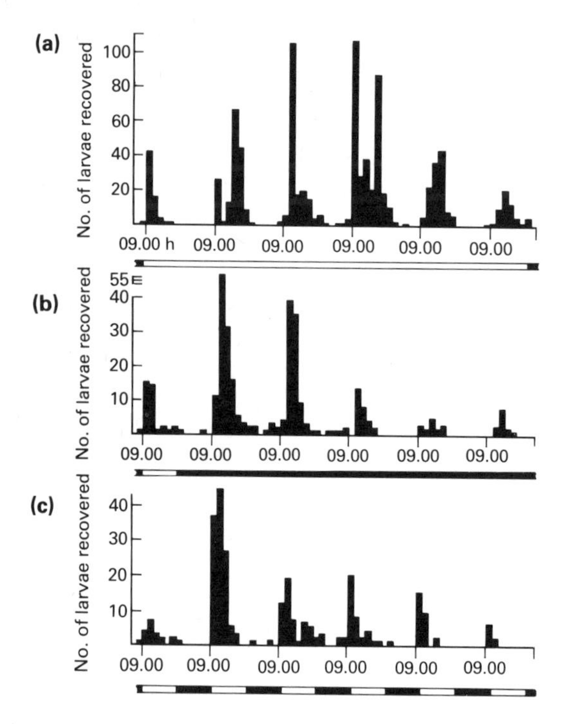

Fig. 5.12 The hatching of *Entobdella soleae* eggs in constant light or constant dark. (**a**) Continuous light following LD 12:12. (**b**) Continuous darkness following LD 12:12. (**c**) Control, LD 12:12. Larvae were collected at 2 hour intervals. solid bars: darkness; open bars: light. (From Kearn.[170])

diurnally fluctuating light intensity (LD 12:12) for the behaviour to manifest itself. It is possible that the ciliary photoreceptors (Fig. 5.2b) of the *E. soleae* larvae are the sense organs responsible for processing information about long-term light intensity changes.

The hatched oncomiracidia of several monogenean species show directional responsiveness to light, presumably mediated by the pigment cup eyes (Fig. 5.2a). Often, as in *Diplozoon paradoxum*[38] and *Discocotyle sagittata*,[261] there is an initial positive phototaxis followed by a negative or unresponsive phase. The first period appears to have a dispersive significance and spreads the bottom hatching larvae into the same location as middle- and upper-water fish. The second represents a phase when host-specific clues, such as chemosensory ones, can predominate.

Miracidia

The eggs of the bloodstream-inhabiting, schistosome digeneans are deposited by female worms at a stage when they already contain fully-developed miracidia. In *S. mansoni*, the eggs leave the blood vessels in the gut mesenteries, penetrate into the gut lumen and are voided with the faeces. Hatching behaviour is keyed to environmental parameters that ensure that the miracidia mainly emerge in situations which favour contacts with intermediate host snails like *Biomphalaria* (*Australorbis*) *glabrata*.[334] This sensory control of hatching depends on inhibitory as well as excitatory stimuli. High salt concentrations, darkness and high temperature all inhibit hatching, whereas low salt molarities, light and temperatures in the 25–30°C range stimulate hatching. These stimulatory conditions will only occur naturally when the infected faeces have been diluted in freshwater, that is, in the probable location of the molluscan hosts of the miracidia. The nature of the inhibitory stimuli makes hatching of the eggs in man, the final host, extremely unlikely. Such hatching would not promote continuation of the life cycle. The stimulatory effect of light on hatching behaviour is presumed to be a function of the responsiveness of either the ciliary or pigment cup eyes in the schistosome miracidium (Fig. 5.6). It is not known which of the other sense organs are responsible for sensitivity to osmotic pressure changes and temperature.

The general behavioural responses of free-swimming miracidia have been the subject of many investigations. Most findings confirm the hypothesis that initial behaviour patterns mimic those of the target snails and thus encourage contacts between miracidia and their hosts. The actual interactions of responses in any particular species of miracidium can be complicated, with one type of response being reversed or becoming subsidiary as another environmental factor passes some threshold value. The miracidia of *Schistosoma japonicum*, for instance, will move towards light of any intensity at 15°C, whereas at higher temperatures, high light intensities induce negative phototaxis and low intensities positive phototaxis. The same larvae move upwards against gravity in the dark. At high ambient temperatures, above 20°C, this response becomes subordinate to the

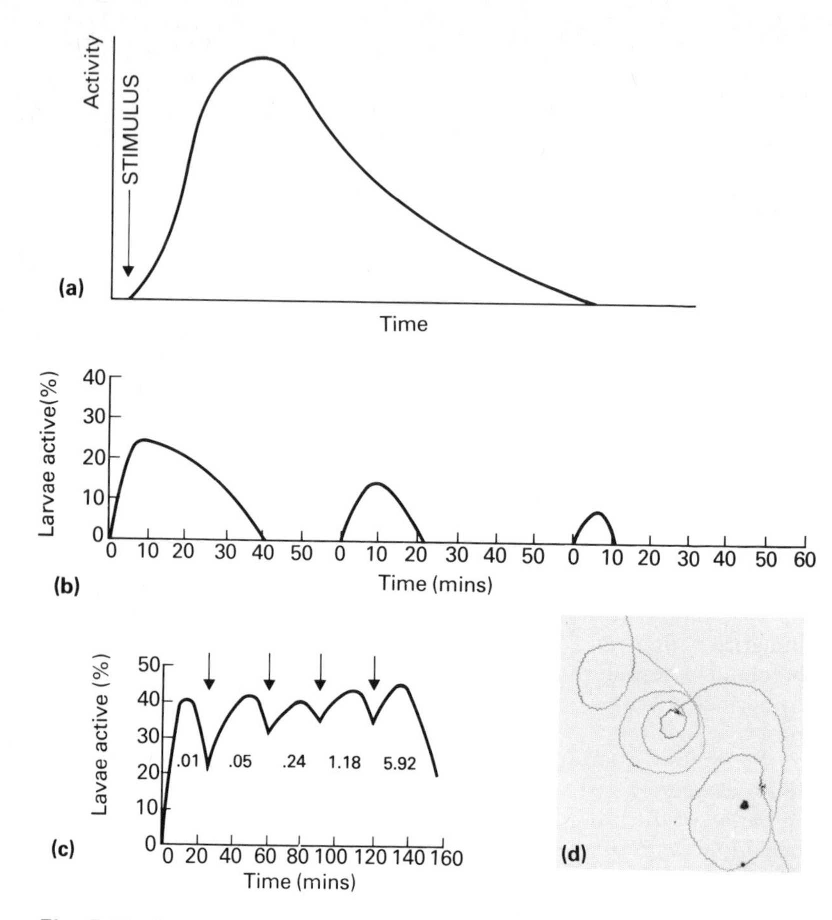

Fig. 5.13 The behaviour of hookworm larvae. (**a**) The 'quantized' stimulus response curve of infective nematode larvae. The stimulus is a change in environmental conditions. Activity may be measured as the proportion of a population exhibiting activity of some sort or the actual rate of movement. (**b**) Larvae of *Ancylostoma tubaeforme* were kept in constant conditions in darkness for 20 hours and then exposed to a sequence of alternating one hour periods of light and darkness. Each 60 minute period in the graph shows the successive activity responses in the first three light stimulus periods. With each repetition of the stimulus the response diminishes in intensity and duration. (**c**) The graph illustrates the activity level of dark-adapted *A. tubaeforme* larvae which has been maintained at an approximately constant level by successive five-fold increases in light intensity. The intensities are shown as the numbers (lamberts) under the curve, and the times of illumination alteration are indicated by arrows. (All based on Croll and Al-Hadithi.[84]) (**d**) Photograph of the physical track of an L_3 larva of *Ancylostoma tubaeforme* on an agar surface in a Petri dish. Note the basic sinusoidal wave form and the consistent sense of the large scale spiralling. (Photograph supplied by Professor N. A. Croll, McGill University, Canada.)

negative phototaxis shown at this temperature if light intensities are high. These involved interactions produce an integrated behaviour on the part of the miracidia that is very similar to the responses of the molluscan host, *Ocomelania nosophora*, to the same environmental conditions.[342]

Infective nematode larvae and cercariae

One aspect of host-finding behaviour appears to have special relevance to the infective larvae of parasitic nematodes. These are usually free-living larval stages that have stopped feeding. They show host-finding behaviour in the form of bouts of locomotion (Fig. 5.13d) that can arise in response to a number of general stimuli like mechanical disturbance, light intensity changes and changes in temperature. Croll and his collaborators[83–4] have shown, for instance, that the spontaneous activity of *Trichonema* and *Ancylostoma* infective larvae is minimal in constant environmental conditions. Greater activity is only elicited by novel external stimulation, and when elicited, it occurs in what Croll calls a 'quantum' of exogenously produced movement to which the larvae are committed, once activity has started (Fig. 5.13a). After this packet of movement is finished, a repetition of the original stimulus produces a much feebler response by the population of larvae under test (Fig. 5.13b). With repeated photic stimuli, a five-fold increase in intensity at each stimulus change is required to produce an undiminished response, measured as a percentage of larvae that become active (Fig. 5.13c).

This rapid habituation of the responsiveness of infective larval nematodes is thought to be a tactic for favouring economy in the use of energy. The non-feeding larvae have a life span dictated by the size of their finite nutrient reserves, which exist mainly in the form of lipids. Activity that might bring about contact with hosts and subsequent penetration is reserved for occasions when novel environmental changes occur. Unchanging patterns of environmental stimuli usually mean that no active hosts are nearby, so rapid habituation carries an adaptive advantage. In this way nutrients are conserved and a longer infective period is achieved. The habituation itself is due to sensory and possibly neurosecretory adaptation. This conclusion has been reached because potent cholinesterase inhibitors like neostigmine bromide uncouple larval habituation and allow constant movement in unchanging environmental conditions.

Similar energy-conserving strategies in associate-finding behaviour have been demonstrable in other host-parasite systems where infective transmission stages are non-feeding. Many digenean cercariae fit into this category. In the case of *Transversotrema*

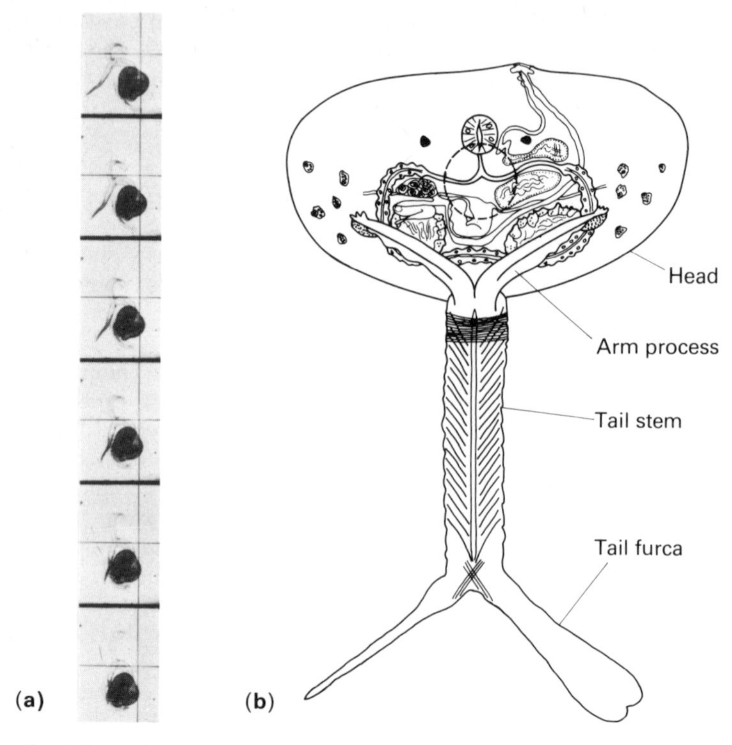

Fig. 5.14 *Transversotrema patialense* cercaria. (**a**) The cercaria performing tail-first swimming. The cinemicrographic sequence should be read from the top. Each frame shows the changing cercarial configuration at millisecond intervals. The cercaria is moving from the bottom to the top of the frame, the tail stem is beating from right to left with respect to the head. (A part of a cinemicrographic sequence produced by D. A. P. Bundy, King's College, London.) (**b**) Drawing of the cercaria to show internal organ systems and location of the arm processes (see also Fig. 5.5). (From Whitfield *et al.*[375])

patialense, event-recording techniques used throughout the entire active life-spans of individual cercariae have enabled a detailed picture of changing activity patterns to be described.[374]

These furcocercous (fork-tailed) (Fig. 5.14b) cercariae spend their time after emergence from the snail host either (i) actively swimming in a tail-first configuration (Fig. 5.14a). (ii) dropping passively using the spread tail furcae to retard the dropping rate or (iii) resting motionless on the bottom of the aquatic environment. Of these three types of activity, only the first utilizes significant amounts of energy and significantly reduces remaining survival time. The pattern of overall activity is dictated by the temporal pattern of active swimming bursts. These, it appears, do not change in duration as the larvae age but their frequency of occurrence does (Fig. 5.15). Over a range of different water temperatures, in constant environmental conditions, the first few hours of free-living life are spent with a high and constant frequency of swimming bursts. Thereafter, the frequency declines markedly and progressively. Probably the first, more active, phase represents a period when the cercariae are kept off the bottom by energetically expensive swimming which can enhance the likelihood of infective encounters with fish. If such contacts do not rapidly occur, progressively lower levels of spontaneous activity are exhibited and only novel stimuli such as shadowing can induce higher swimming burst frequencies. The overall strategy must prolong the period over which the cercariae are potentially active.

The sensory adaptation exhibited by infective nematode larvae is a simple type of learned behaviour. In higher taxa than the nematodes, learned responses become increasingly important in associate-finding and other behaviour.

Cleaner symbiotes

Many marine fish, especially free-swimming, littoral forms, actively seek the attention of cleaner symbiotes.[118,121] These are small fish and shrimps that obtain much of their food on the surface of larger fish in the form of ectoparasites and damaged or diseased tissues. Associations between the cleaners and the cleaned may be brief or prolonged, but elaborate behavioural interplay is almost always characteristic of the relationship. Partner-finding behaviour seems to be shown principally by the larger fish. Most cleaners show territoriality in the form of a 'cleaning station' where they await 'customers'. This station, for most cleaner fish, is a prominent bottom feature in the littoral habitat. Outcropping coral growths, or even, off the coast of the United States, discarded car tyres, have been noted as cleaning sites. Pederson's cleaning shrimp, *Periclimenes pedersoni*, is always found on, or next to, an individual of the large sea anemone, *Bartholomea annulata*. The fish in a particular littoral area learn the positions of cleaning stations and give the appearance of congregating there for cleaning. Underwater naturalists have described fish 'jostling for position' in a shoal awaiting the attentions of a cleaner shrimp at its station (see also page 186).

Associate-recognition

Once associate-finding behaviour has provided the chances for contact with associate animals, recognition of the partner must occur

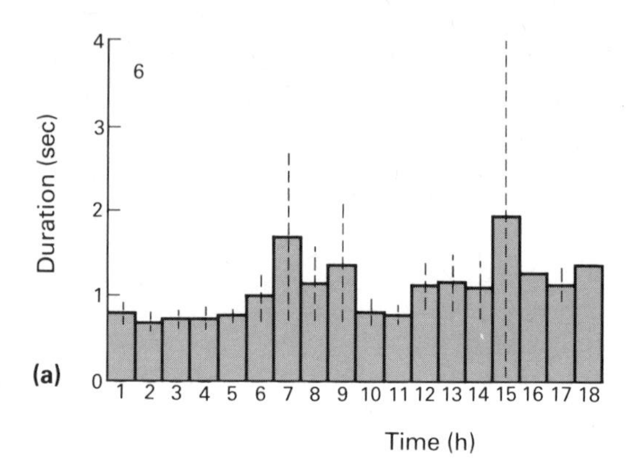

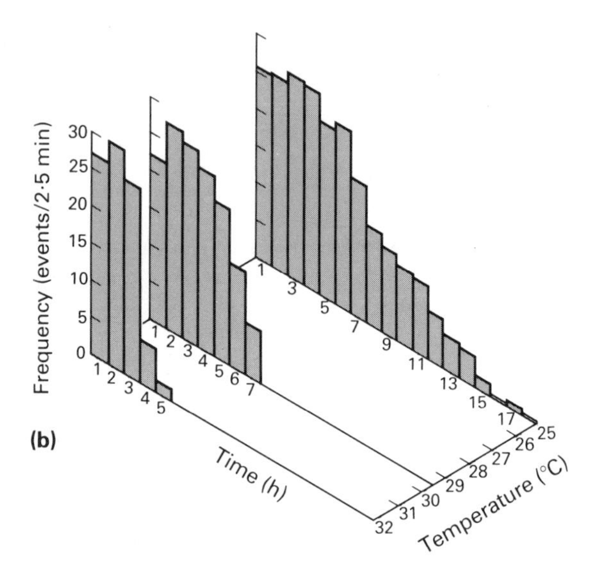

Fig. 5.15 Activity patterns in the host-finding behaviour of a transverotrematid cercaria, *Transversotrema patialense*. (**a**) Mean durations of swimming bursts throughout the spontaneously active life span of cercariae at 25 °C. The dashed vertical bars show 95% confidence limits. Analysis of variance shows that there is no significant change in swimming burst duration during the 18 hours. (**b**) Mean frequencies of swimming burst production during aging at three different water temperatures. In each case swimming bursts are produced initially at a high constant rate. Thereafter the frequency of bursts declines. (Both from Whitfield *et al.*[374])

before host-parasite, symbiotic or commensal relationships can be established. This crucial phase in the initiation of an association often cannot be clearly differentiated from associate-finding activities on the one hand or actual establishment on the other. What characterizes it above all else, though, is a positive response to stimuli which are specific to the partner.

Only in rare circumstances do unambiguous, partner-specific stimuli reveal the location of a partner at long range. These circumstances arise when sophisticated nervous systems are in action. It is very likely, for instance, that some mosquitoes which have distinct host preferences are able to distinguish between potential hosts at long range by sensing odours downwind.[124–5] Any terrestrial vertebrate produces a plume of contaminated air that stretches downwind perhaps hundreds of metres before it is dispersed by turbulence. The plume will contain both an excess of CO_2 and host-specific odours. CO_2 is often a general attractant for blood-sucking invertebrates. For the mosquitoes, the odours provide detailed information at long range about the identity of the smell producer.

Only in a few instances do we have chemical information about the nature of compounds that enable relatively long-distance identification of associates to take place. One interesting and complex example where such information is available concerns the braconid hymenopteran parasitoid, *Diaeretiella rapae*.[287] This insect uses crucifer-feeding aphids as its hosts. It locates them not by reaction to substances produced by the homopterans themselves but by its attraction to secondary compounds (see page 146) produced by the aphids' host plants in the family Cruciferae. Crucifers demonstrate a great adaptive radiation in the ability to synthesize mustard oils and their glycosides as potent defensive chemicals against most herbivorous insects. *Diaeretiella* has been shown to locate and identify plants likely to bear appropriate aphids by responding to one of these mustard oils, allyl isothiocyanate (Fig. 5.16).

In most situations, however, partner recognition occurs after an earlier phase of partner-finding activity. This more usual and hierarchical organization of the behaviour involves specific 'homing' responses after the correct micro-habitat for contacts has been reached.

A hierarchical structure is imposed on the behaviour of most associating animals when long-range reception of partner-specific stimuli is not feasible. Homing activity at close quarters requires sensory and locomotor structures which can be simpler, and therefore more economical to produce in energetic terms, than those necessary for long-distance responsiveness.

In all types of association, examples have been analysed in which the determining stimuli for recognition can be deduced. Most of the instances that have been subjected to experimental investigation have been found to involve either visual, chemosensory or auditory clues.

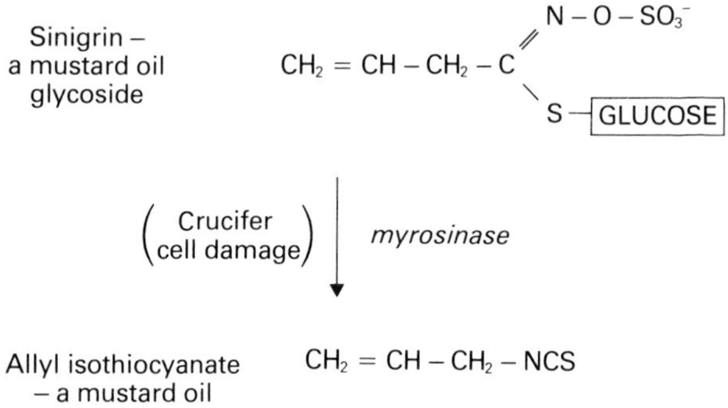

Sinigrin – a mustard oil glycoside

$$CH_2 = CH - CH_2 - C(=N - O - SO_3^-)(-S - \text{GLUCOSE})$$

(Crucifer cell damage) ↓ *myrosinase*

Allyl isothiocyanate – a mustard oil

$$CH_2 = CH - CH_2 - NCS$$

Fig. 5.16 The production of allyl isothicyanate after cell damage in crucifer plants.

Visual recognition

The relationship between certain orchids and the hymenopteran insects that pollinate them passes beyond the normal interactions that occur between insect-pollinated plants and their pollinators. Several of these relationships appear to arise from the adaptation of orchid pollination mechanisms based on the deception of the hymenopteran pollinator. Dobson[103] has estimated that more than half of existing orchid species do not provide food in the customary sense for their pollinators. Instead, bees and wasps are deceived into responding to orchid flowers as though they were prey insects, territorial competitors, food flowers, or even potential copulatory partners! In most of these cases some of the sensory clues used in the identificatory phase of the behavioural interaction are visual ones.

In a number of cases, particular orchids are fertilized quite specifically by males of a single species of bee or wasp in a process termed pseudocopulation. The looking-glass orchid, *Ophrys speculum*, of southern Europe and Algeria, for instance, is always pollinated by the wasp *Scolia ciliata*. The wasps cross-pollinate the orchid flowers by attempting to copulate with them. In the case of *Scolia* and its orchid, the orchid floret is a close visual mimic of the resting female wasp. The blue centre of the lip looks like the half-crossed wings of the female and the red hairs round the lip imitate similar hairs round the insect's abdomen. Dark, thread-like upper petals closely resemble the antennae of the female wasp.[236] (Fig. 5.17). The array of visual cues presented by the flower released copulatory behaviour by the male wasp. The mimicry produces a distinct reproductive advantage for the plant; for the wasp, quite the reverse. It is difficult to fit such a relationship into any neat category, but on present evidence it would seem that the orchid parasitizes the hymenopteran. Some work has suggested[181] that orchids in such associations also mimic hymenopteran sex pheromones. If this were true, it would mean that partner recognition in the hymenopteran-orchid system depends on both visual and chemosensory stimuli.

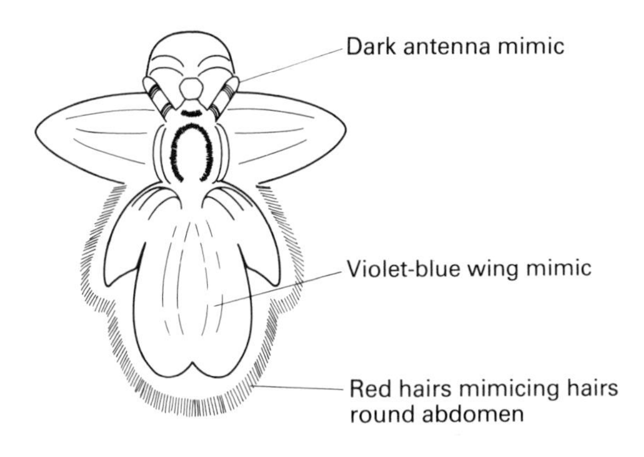

Fig. 5.17 A single flower of the looking glass orchid (*Ophrys speculum*) which is a visual mimic of the females of the wasp *Scolia ciliata.* (Based on a drawing in Meeuse.[236])

Perhaps the most extreme and refined example of the importance of orchid-hymenopteran interactions in orchid reproductive biology has been provided by Nierenberg.[256] On Grand Bahama Island, in the Caribbean, sympatric populations of two orchid species, *Oncidium bahamense* and *O. lucayanum*, are reproductively isolated by differential patterns of behavioural interaction with the male and female bees of a single species, *Centris versicolor*. *O. bahamense* flowers attract male bees who respond to them as though they were territorial competitors to be driven away. Conversely, *O. lucayanum* attracts female bees because they mimic the flowers of an important food plant of the females namely *Malpighia glabra*. The two orchid species are reproductively compatible by artificial fertilization in the laboratory but because of the extraordinary utilization of the two sexes of *Centris* natural hybrids between the orchids have never been discovered.

Marine cleaner symbiosis is usually initiated by a fish that requires cleaning. Once it arrives at a cleaning station, a behavioural balancing act of some delicacy must be performed. Cleaner fish and shrimps are of a size that would normally be regarded as food by the fish that are cleaned. Somehow the cleaners must suppress the predatory responses of their clients in order that they may feed on the latter's skin surfaces. This inhibition, and consequently recognition of the cleaner's status, is achieved by visual stimuli that the cleaners present to the larger fish. Colour patterns and showy behaviour make the cleaners conspicuous and easily recognized. There is some evidence that different cleaner species help to reinforce this recognition by sharing similar conspicuous patterns termed 'guild marks'.[113] Several small cleaner fish, for instance, share longitudinal black stripes, others are brown with a black spot on the tail. The common currency of cleaner fish

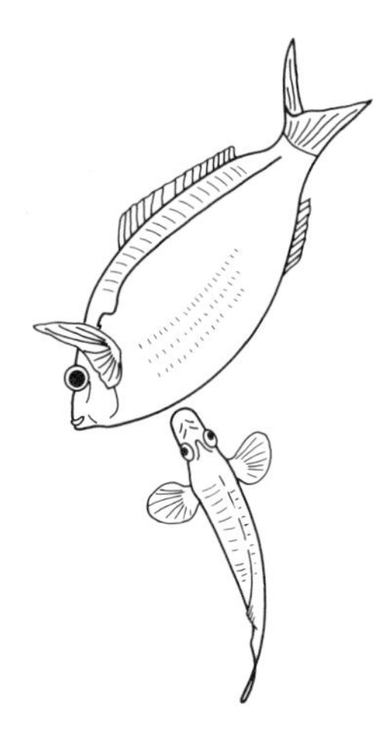

Fig. 5.18 Cleaner symbiosis: a pink bream (the larger fish) adopts a head-down invitation posture to a corkwing wrasse which removes ectoparasites from the ventral surface of the bream. (Based on a photograph published by Potts.[281])

guild marking has been exploited by a sabre-toothed blenny, *Aspidontus taeniatus*. This accomplished mimic copies the colour patterning of small cleaner wrasse like *Labroides dimidiatus*. It employs this disguise to approach larger fish in the privileged and unmolested way that cleaner fish can. Having got within range,

Aspidontus uses its specialized and formidable dentition to chop a mouthful of fin rays and surrounding tissue from the duped fish.

Many of the cleaner shrimps are marked with red and white stripes. These are especially prominent on the long antennae, whose waving appears to attract fish. Cleaner fish also have elaborate, ritualized displays that serve the same function. Once such visual stimuli have allowed initial recognition to occur, the host fish might have to learn to accept the close proximity of the cleaner and its attentions to its skin.[281] This phase of the relationship can entail special 'invitation postures' on the part of the larger fish (Fig. 5.18) which enable the cleaner to approach areas which are normally concealed. If such behaviour is not entirely stereotyped, it may be postulated that the removal of epidermal irritation in the form of skin parasites acts as a conditioning stimulus that reinforces cooperation by the host fish in the association. The stomach contents of Californian moray eels, however, bear witness to the fragility of the behavioural equilibrium of cleaner symbiosis. A shrimp, *Hippolysmata californica*, cleans the surface of these formidable predatory fish, yet sometimes is regarded by its host not so much as a symbiote, but more a crustacean food.

Chemosensory recognition

The recognition of partners by chemical clues seems especially common in aquatic organisms. Davenport,[95] in a series of pioneering experimental studies, has examined the chemosensory stimuli, which he terms chemical signs, by which commensal marine crabs and polychaetes recognize their larger partners.

Among parasites the oncomiracidia of *Entobdella soleae* show considerable discrimination in their ability to recognize the surface of their host, the sole.[168] When swimming larvae are presented with mixtures of sole scales and scales from other flatfish, they show marked preferences, as shown by non-random attachment behaviour, for the sole scales (Table 5.2). This discriminatory behaviour can occur in complete darkness, so it is evident that chemosensory clues are very likely to be those responsible for recognition. This conclusion is confirmed by the demonstration that the larvae can distinguish between control agar circles and circles that have been placed in contact with sole skin for 30 minutes. Preferential attachment to the latter almost always occurs. Only skin areas containing mucus glands possess this ability to cause oncomiracidial attachment. Corneal epidermis, for instance, which has no gland cells is unattractive. It seems likely, therefore, that some component of the secretions of these glands is the host-specific chemical stimulus for attachment. It is not clear whether distant- or contact-chemoreception is involved, but in

Table 5.2 Results of experiments in which oncomiracidia of *Entobdella soleae* were allowed to settle on mixtures of scales removed from *Solea solea* and also *Limanda limanda, Solea variegata, Pleuronectes platessa* and *Buglossidium luteum.* In each experiment the larvae had equal opportunities for settlement on each of the two scale types. (From data in Kearn.[168])

Scale origins	Numbers of oncomiracidia attached to scales
S. solea	49
L. limanda	2
S. solea	30
S. variegata	8
S. solea	20
P. platessa	1
S. solea	15
B. luteum	4

either case it is probable that the ciliary pit receptors of the larvae (see Fig. 5.1b) are implicated.

Auditory recognition

Evidence concerning associate recognition by sound stimuli is sparse. The most likely candidates for consideration in this respect are brood-parasitic birds. Vocal mimicry is an important adaptive feature in the behaviour of the young of several brood parasites such as the great spotted cuckoo, *Clamator glandarius*, in sub-Saharan Africa and the striped crested cuckoo, *C. levaillantii*, in southern Africa.[247] Fledglings of the former species, for instance, give a begging call for food remarkably similar to that produced by the natural young of the commonest fostering bird, the pied crow, *Corvus albus*. Equally, adult male viduine finches, which are brood parasites of estrildine finches, vocally mimic host adult calls. It seems that a large proportion of this mimicry in both fledgling and adult brood parasites is a consequence of vocalizations being learnt in the fosterer's nest. Adult female brood parasites must locate the nests of specific host birds in order to lay their eggs. In the light of the importance of mimicked sounds at other parts of the life cycle, it is not unreasonable to assume that part of this recognition of suitable hosts by female birds is auditory recognition.

Establishment behaviour

Whether behaviour or random external processes bring associates into contact, subsequent specific behaviour to capitalize on this event must usually take place. Such a response may be termed establishment behaviour. This behaviour is only clearly seen where associates live in actual contact with one another. It is most often an attachment to, or penetration of, one partner by the other. Specific activating stimuli are often implicated in the induction of these activities.

For associating animals that use a non-behavioural transfer strategy, fortuitous contacts with their partners are the time when suddenly, appropriate establishment behaviour must become dominant. When the eggs of the rat acanthocephalan, *Moniliformis dubius*, are ingested by a cockroach, the insect's gut activates establishment behaviour. Edmonds[111] has defined *in vitro* the stimuli which probably effect their activation *in vivo*. Salt solutions with a molarity greater than 0.2 and a pH higher than 7.5 stimulated the hatching behaviour of the acanthor larvae within the eggs. CO_2 has an enhancing effect on the process and enables acanthors to hatch at lower pH values. The acanthors actively release themselves from their egg envelopes by a rhythmic cutting action of their rostellar hooks. It is possible that a chitinase secretion which they produce during hatching softens the chitinous components that exist in some of the envelopes. Once free, the larvae move in the lumen of the midgut by the combined action of their rostellar apparatus and body spines. The same movement cycles (Fig. 4.10) enable penetration of the peritrophic membrane and midgut cells of the cockroach to take place. In this way acanthors enter the cockroach haemocoele, where further development to the cystacanth larval stage occurs.

Much experimental work has been carried out on the establishment behaviour displayed by disease-causing larval helminths that directly infect man or other vertebrates. Most attention has centred upon schistosome cercariae and infective L_3 larval nematodes.

Schistosome cercariae penetrate vertebrate skin to cause the disease schistosomiasis (see also page 130). The fork-tailed swimming larvae burrow into the skin when it is wetted with cercaria-infected water. Initial penetration is almost always in depressions and discontinuities of the skin surface; skin wrinkles and hair shaft exits being specially favoured entry points (see Fig. 5.19). Attachment to the skin is first made with the ventral sucker. After this, adhesion is maintained by the oral sucker so that the anterior end of the cercaria is directed towards the skin surface. Several gland ducts open around the oral sucker, and gland contents are forced out in this region during skin penetration. Entry is made between the flattened cells of the stratum corneum by muscular contractions of the cercarial body. Early on in this process the cercarial tail detaches and the cercaria becomes a schistosomulum. Backward-pointing body spines aid the onward progress of the tailless cercaria as it moves down through the stratum granulosum and

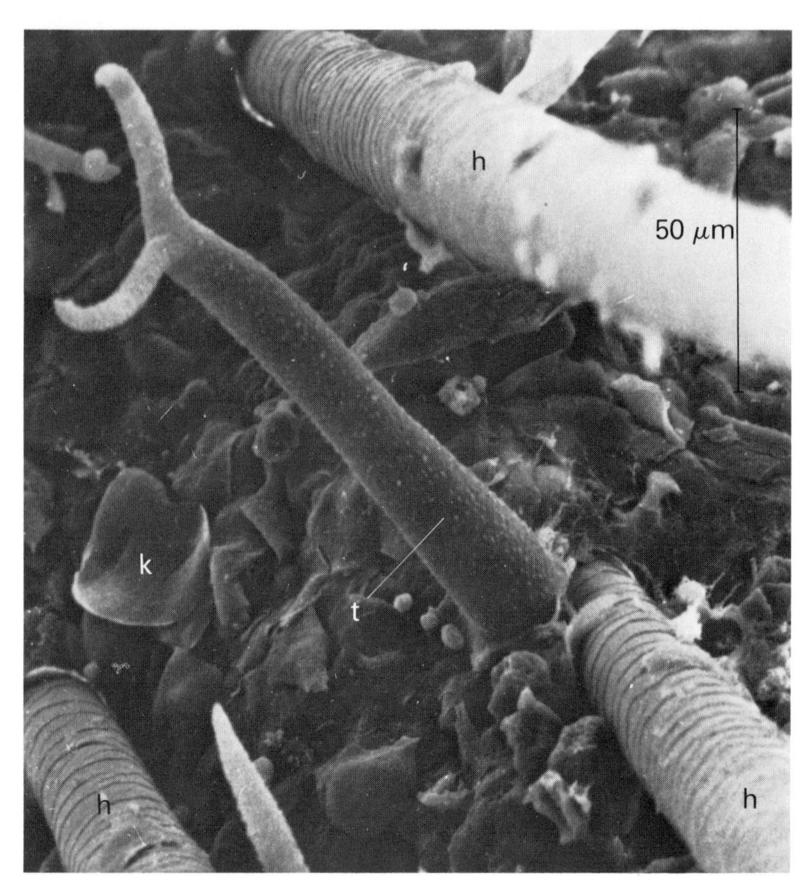

Fig. 5.19 Cercaria of *Schistosoma mansoni* penetrating the tailskin of a 10 day old mouse. Only the forked tail (t) of the cercaria still shows above the skin surface. The head has already penetrated the skin, entering alongside a hair shaft (h). k, keratinized skin cell. (Scanning electron micrograph supplied by Dr. H. D. Blankespoor, Museum of Zoology, University of Michigan, U.S.A.)

basement lamella to the dermal tissues. Many potent lytic enzymes have been demonstrated in the glandular secretions of the schistosome cercariae. Those with hyaluronidase activity probably have a weakening effect on the matrix macromolecules of collagenous connective tissue through which the schistosomula must push.

Within minutes of skin penetration, profound physiological changes occur in the schistosomula. These might be expected in an animal which has entered vertebrate soft tissues after living in, and being adapted to, a freshwater environment. One vital change is in osmotic tolerance. Schistosomula can no longer live in low osmotic pressure surroundings; if returned to freshwater, they die.

Establishment behaviour for schistosome cercariae is thus a complex of coordinated movements, secretion release and physiological alterations. Recent work has shown that in both mammal and bird schistosomes this pattern of activation is dependent on the trigger of host skin lipids. The importance of these substances has been elegantly demonstrated by Stirewalt.[341] When she removed the lipids from a rat skin, cercariae would no longer penetrate it. Once the lipid fraction was replaced, the skin regained its stimulatory properties. *In vitro* investigations have also shed light on this stimulation. MacInnis[217] has shown that dilute agar blocks impregnated with test molecules can be used as a model for host skin in cercarial penetration studies. With appropriate stimulatory molecules present, schistosome cercariae will attempt to penetrate the agar. Using this technique, other workers[315] have demonstrated that human skin lipids will induce penetration by human schistosome cercariae. Cercariae were not, however, stimulated by purified skin sterols or cholesterol; instead, they responded to the free fatty-acid fraction from human forearm skin lipid. Further separation techniques showed that the most potent stimulating molecules were to be found among the unsaturated fatty acids. It is interesting that in the tests with human skin lipids, cercarial mortality increased with increased levels of penetration response. This heightened death rate was probably a function of the physiological changes that accompany the cercaria-schistosomulum transformation. In the test system used, transformed cercariae could not survive in the low osmotic pressure conditions of the test saline.

The cercariae of the bird schistosome, *Austrobilharzia terrigalensis*, have been shown to penetrate gelatin membranes only in the presence of cholesterol.[74] This is a quite different triggering molecule from those effective for human schistosomes. The finding suggests that the schistosomes in their evolutionary utilization of different vertebrates as hosts, have become responsive to a variety of chemosensory stimuli in the different vertebrate taxa.

Lee,[191] using a combination of electron microscopy and direct observation, has given a detailed account of the establishment behaviour of the infective larvae of the rat nematode, *Nippostrongylus brasiliensis*. In these experiments, the larvae were watched as they penetrated the skins of rats, and also, in some self-sacrificing experiments, that of the author!

The stimuli that induce skin penetration in such larval nematodes have not been clearly defined. *Nippostrongylus* larvae will move up heat gradients,[268] and CO_2 is said to activate and attract some hookworm larvae.[306] It is possible that the establishment behaviour of such larvae is simply a movement up the thermal gradient that must exist through the thickness of the skin of warm-blooded vertebrates. However, surface chemical stimuli have not been rigorously searched for and until the possibility of their presence has been experimentally eliminated, they must still be regarded as possible triggers for penetration.

Maintenance behaviour

For at least one of the partners in an animal association there is an obligate or facultative need to maintain the relationship. Various aspects of the biology of associates can be understood in the light of this imperative, and behavioural specializations play a part in an integrated adaptation for what might be called relationship homeostasis. Of special interest are those behavioural components that are only biologically meaningful in the context of associations. Male and female parasitic gut nematodes are attracted heterosexually towards one another by chemical clues.[303] This is obviously adaptive behaviour on the nematodes' part, but exactly comparable behaviour occurs in free-living nematodes.[70] Hence, this reproductive behaviour by an associating animal is unlikely to be specifically adaptive in terms of the relationship itself. In other instances, however, such specific adaptation can be inferred. When behaviour of this type occurs in an established relationship it may be termed maintenance behaviour.

Maintenance behaviour by parasites

The maintenance behaviour of endoparasites is intrinsically difficult to investigate because it is often very dependent on particular features of a continuing association between host and parasite. When this association is disrupted by experimental manipulation, much of the integrated maintenance behaviour is similarly disrupted.

In vitro culture of endoparasites opens up one route of attack on this problem, and some interesting facts have emerged from studies using this technique. The importance of lateral physical constraint by the gut wall in successful cestode copulation behaviour became apparent from culture studies. In the case of the pseudophyllidean, *Schistocephalus solidus*,[326] successful insemination, in culture, was only achieved when the worms were grown in an artificial gut of cellulose tubing which provided gentle lateral compression. *In vitro* maintenance of acanthocephalans has revealed aspects of locomotory behaviour that must be important in site changes and mate-finding movements. Hammond[134] has shown that *Acanthocephalus ranae*, when detached from the intestine of its toad host, initiates activity of its hooked proboscis. This invagination-evagination behaviour is remarkably regular, the proboscis evaginating every 60–70 seconds. The regularity suggests that some form of nervous pacemaker activity, triggered by the absence of positive attachment stimuli impinging on the receptor organ at the tip of the proboscis. Activity ceases when the proboscis, by its movements, re-embeds itself.

Croll and Smith[85] have recently developed an experimental technique which would appear to be potentially fruitful for the examination of the undisturbed maintenance behaviour of endoparasites. They have directly observed *Nippostrongylus* in the rat gut by surgically externalizing a portion of infected intestine. This section, still continuous with the remainder of the gut, is gently squeezed between glass plates and bathed in physiological saline (Fig. 5.20a). The red worms of *Nippostrongylus* are visible through the translucent gut wall and their behaviour can be recorded using videotape techniques (Fig. 5.20b). One might hope that with the recent flowering of fibre-optical technology in the manufacture of gastroscopes and similar equipment such apparatus might soon be used to observe gut parasites *in situ*.

Of all the types of maintenance behaviour, site selection and alterations of site by endoparasites within their hosts are those which are most susceptible to experimental observation. Crompton[88] and Holmes,[148] in two recent reviews, have examined particular types of maintenance behaviour shown by parasitic helminths, with special emphasis on these two aspects. In most cases, site or micro-habitat specificity displayed by helminths is due to active site selection by the parasite itself.

Parasites often inhabit different sites in their hosts at different stages of their own maturation and development. The dog nematode, *Spirocerca lupi*, for instance, enters its canine host as infective larvae in dung beetles that form an unusual item in the diet of some American dogs. The larvae are freed in the stomach and start off on a remarkable series of migrations.[24] They first penetrate the gastric mucosa and enter a branch of the gastric artery. Within the wall of the artery they move up its length to the wall of the thoracic portion of the dorsal aorta. Here, for a period of up to three months, they undergo development to the immature adult phase. The nematodes subsequently burrow out of the fibrous connective tissue that has developed around them, migrate out of the aorta, and move ventrally to penetrate the outer surface of the oesophagus. They burrow through the oesophageal wall into the lumen, then move back into the submucosa

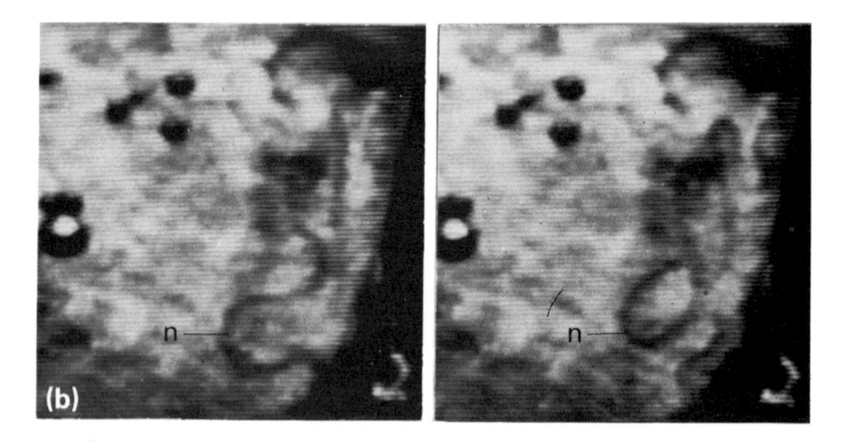

Fig. 5.20 A technique for observing living nematodes in the intact, living gut of an anaesthetized rat developed by Croll and Smith.[85] See text (page 193) for details. (**a**) The externalized loop of rat small intestine. (**b**) Two stills from a videotape monitor record of the activities of the nematode, *Nippostrongylus brasiliensis*, in the living gut of the rat using the externalized loop technique. n, *Nippostrongylus*.

or muscle layer of the oesophagus. Here, they finally cease their wanderings, complete their development and start producing eggs. These are able to escape into the oesophageal lumen along the burrow tracts through which the worms re-entered the wall from the lumen.

This nematode odyssey is an involved one, but can be regarded as a model for all developmental migrations. It is a migration in which particular larval forms appear to require the special conditions, perhaps nutritional ones, that can only be found in particular host sites. The stimuli that have been thought to bring about the directed movements are all hypothetical and very difficult to substantiate experimentally. It has been suggested, for instance, that in their movements up the wall of the gastric artery and dorsal aorta, the nematodes are responding to a pressure gradient. There is other evidence that nematodes in general can respond behaviourally to pressure. No experimental work, however, has yet been carried out to test for this sensitivity in larval *Spirocerca*.

The life cycle of the tapeworm, *Hymenolepis diminuta*, is easy to maintain in laboratories, using flour beetles and rats as intermediate and final hosts respectively. Ease of maintenance, and a relatively large size for a helminth, have led to *H. diminuta* becoming the focus for investigations from a number of behavioural standpoints. We probably know more about the behaviour of this cestode than that of any other parasitic worm. Two interrelated migration patterns have become evident in the behaviour of adult *Hymenolepis* living in the small intestine of laboratory rats.

The worms, like *Spirocerca*, have a long-term migratory behaviour linked to their maturation within the final host. This has been termed ontogenetic migration. Most workers have based their measurement of site changes on either scolex attachment points or worm biomass distributions in terms of percentage distance back along the small intestine. 16 hours after the administration of cysticercoids to a rat, young worms are distributed about a mean position of 26% along the small intestine. The worms remain in this region until day 6 of the infection, when an anterior migration starts that brings the attachment positions of mature worms to a point 13% along the intestine by day 21[352] (Fig. 5.21). During the first three weeks of the infection the worms grow rapidly, so that by day 18 they stretch, when extended, along most of the small intestine's length. Growth and anterior migration interact so that the central part of the zone occupied by the worms is displaced only slightly during development. It is about 40% of the distance along the intestine on day 10 and still at approximately 50% by day 27 (Fig. 5.22). Bråten and Hopkins[43] demonstrated experimentally that this patterned, ontogenetic migration was an

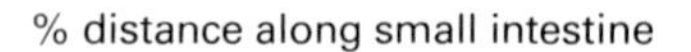

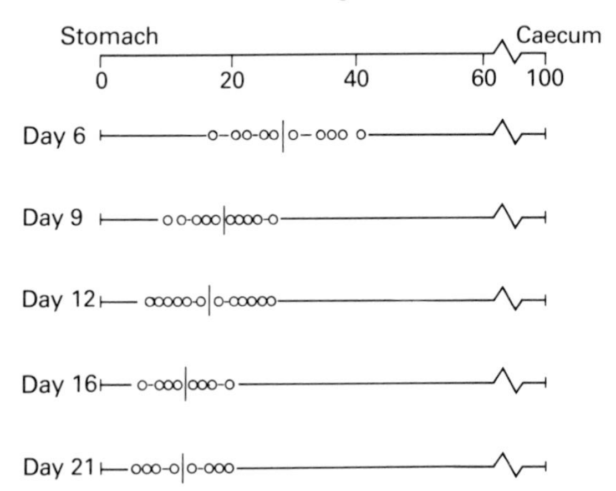

Fig. 5.21 The ontogenetic migration of the tapeworm, *Hymenolepis diminuta*, in the rat small intestine. The diagram shows the relationships between the locations of the points of attachment of individual tapeworms and days post infection. Individual attachment points are indicated by circles and the mean of these for each replicated set by a vertical bar. All the rats used had single worm infections. (From Turton.[352])

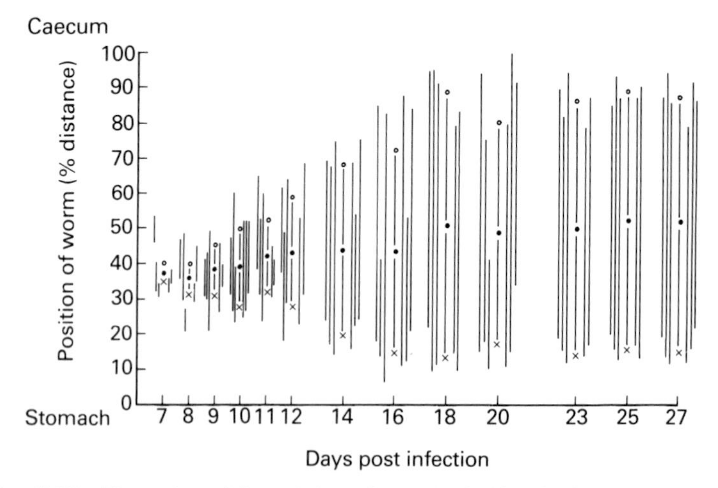

Fig. 5.22 The region of the rat's intestine occupied by single worm infections of *Hymenolepis diminuta* at various times after infection. The thin vertical lines indicate single worms, the central lines the average for each time interval. x, scolex attachment point; ●, mid-point of worm; o, posterior end of worm. (From Bråten and Hopkins.[43])

active process on the worm's part. They transplanted young worms (6.5 and 7.5 days old) into the posterior ileum of uninfected rats. Most of them migrated forwards from this aberrant position during a period of 24 hours to end up in their 'correct' position between 11% and 35% along the intestine. This response occurred against the direction of gut peristalsis. It indicated that the worms had the sensory ability to recognize an incorrect gut environment and the ability to respond in a directed way to the sensory imbalance. They also had the locomotor powers to move against gut peristalsis and, having moved, the capacity to recognize a correct gut environment and remain there. As these authors pointed out, their results produce a picture wildly at variance with the common conception of the gutless cestode as a behaviourally degenerate animal.

The anterior movement of the worms during maturation would appear to be built up from a series of active readjustments of the worms' position. This movement maintains the cestodes in a general way in the optimal micro-habitat for existence in the small intestine. It has proved difficult to analyse the mechanics of the response. Bile salt gradients do not seem to be the marker for longitudinal polarity in the gut, as posteriorly transplanted worms will still migrate forwards to their correct position in a gut with a ligatured bile duct. Gut peristalsis could be the marker, although results when opium is used as a peristalsis inhibitor have been difficult to interpret. Certainly the non-ciliated receptors on tapeworm proglottids have a structure that is consistent with a pressure-transducing function, and they could perhaps respond to the external pressures of peristalsis (see Fig. 5.8). The principal difficulty, however, in experimentally dissecting the ontogenetic movement is its coexistence with another pattern of migration. This is a migration correlated with the feeding behaviour of the rat host. When infected rats are starved for 18 to 24 hours most of the biomass of the population of adult tapeworms is spread between the anterior end of the intestine and a point 70% to 80% along its length. With single worm infections, the mean attachment point for scolices is 15% along the intestine and the worms are elongate with lengths averaging 63% of the intestinal length. Both position of attachment and the contraction state of the worms alters during the period when a rat, having fed, has food in its stomach. While food is present there and is being released into the small intestine, the worms move towards[240] and decrease in length.[22] With natural rat feeding rhythms such migrations occur with a corresponding circadian rhythm. They can also be induced by experimental feeding after starvation. In starved rats no short-term feeding migrations can be demonstrated. These facts suggest that feeding migration is an exogenously rhythmic behaviour pattern. It is not the result of intrinsic periodicity on the worm's part but a response to rhythmic rat behaviour.

The stimuli that induce feeding migrations are a matter of controversy. Partially digested products from the rat's meal entering the small intestine could stimulate the anterior movement. It has been demonstrated, however, that both glucose and olive oil will produce significant movements, so there cannot be a unitary stimulatory food molecule. Changes in the pH of the intestinal contents may be implicated, but the changes in this parameter following a meal are very complex and are influenced by the worms themselves as well as by gastric emptying. Crompton[88] has suggested that the cestode's response may be due to the effects of one or more of the hormones that integrate digestive events and gut motility during and after feeding activity. The gastro-ileal reflex, gastrin and other digestive hormones produce a multiplicity of effects in the gut, any of which might be used as stimuli to trigger feeding migration.

The purpose of the feeding migration in *Hymenolepis* is believed to be one of optimizing opportunities for absorbing organic molecules produced by host digestion. The nutrient relationship between tapeworm and rat is one of competition for nutrient molecules between the worm's microtriches and the rat's mucosal microvilli. By moving their absorptive surfaces forwards after a rat meal, the cestodes appear to be obtaining some advantage in this competition.

Maintenance behaviour in symbiotic associations

Maintenance behaviour in symbiotic associations is often difficult to recognize. In examples where microsymbiotes live inside macrosymbiotes, association homeostasis is usually carried on at a physiological or biochemical level of organization. The interactions between a cnidarian and its algal symbiotes do not often enter the realm of behaviour. Some general responses however, can be construed as maintenance behaviour on the part of the macrosymbiote.

Both the anthozoan anemone, *Condylactis*, and the scyphomedusan, *Cassiopeia*, contain intracellular zooxanthellae, usually classified as the dinoflagellate, *Symbiodinium microadriaticum*. Both the cnidarians, when harbouring the symbiotic algae, show phototactic behaviour which is thought to maintain the microsymbiotes in the optimal light conditions for photosynthesis. In the medusa, the alga-containing tissues are on the subumbrella surface. The jellyfish spend much of their time lying on the bottom in shallow water with the subumbrella surface uppermost. *Condylactis*, with symbiotes in their tentacles, actively seek medium intensity light conditions. When, however, the symbiotes are removed by 24 days in darkness, the anemones show no such phototactic behaviour.[384]

Overt maintenance behaviour is often observed in small symbiotic animals living in association with larger, relatively sessile ones. One such tropical marine example that has received much attention in this respect is the symbiosis between clownfish, *Amphiprion* and *Premnas*, and a variety of sea anemones, e.g. *Physobrachia*.[121] The brightly marked clownfish (Fig. 5.23) have an extremely intimate relationship with the anemones. The integration of the symbiosis is made possible by specific swimming and rubbing behaviour on the fish's part which

Fig. 5.23 A clown fish, *Amphiprion percula*, resting on the tentacles of its anemone partner. The fish is orange in colour, warningly marked with black and white stripes.

transfers mucus from the anemone's tentacles onto its skin surface. This appears to provide an immunity against the anemone's cnidocysts.[228] During the night the fish reside amongst the tentacles or even push their way through the cnidarian's pharynx into its enteron. In either of these locations the anemone's nematocysts provide a daunting barrier to clownfish predators. By day, the fish is a planktonic feeder above its partner, but continually makes sorties back to the anemone. For many years it was imagined that the anemone was a passive and unrewarded partner in such relationships with all the benefits of associative life accruing to the clownfish. In fact, recent underwater observations have shown the association to be even-handedly symbiotic. Fricke[120] has demonstrated that a particular common clownfish, *A. bicinctus*, will chase away predatory fish, *Chaetodon* spp. that will feed on the tentacles of aposymbiotic anemones without their clownfish partners.

Only rarely have such examples of enhanced protection from predators in symbiotic associations been subjected to close experimental scrutiny. One instance, that has undergone this sort of analysis, is the relationship between the hermit crab, *Pagurus pollicaris*, and the anemone, *Calliactis tricolor*, that lives attached to the crab's molluscan shell.[224] A serious predator of the hermit crab in the Gulf of Mexico is the oxystomatid crab, *Calappa flammea*. One cheliped of *Calappa* is modified for shearing open gastropod shells starting from the aperture. The crab uses this instrument for predating both gastropods and hermit crabs that live in empty gastropod shells. In experiments with hermit crabs in *Polinices* shells, a striking difference was noted in the behaviour of *Calappa* which attempted to feed on hermit crabs with and without attached anemones. Undecorated shells were consistently and quickly opened up and the exposed hermit crabs eaten. When anemones were present, however, manipulation by *Calappa* caused the anemone to contract and extrude its cnidocyst-laden acontial threads. When these touched the mouth parts of the predatory crab, it immediately dropped the hermit crab and the anemone. In the experimental conditions used, the presence of the anemone prevented successful predation in 91% of the trials. Non-associating anemones have acontia and extrude them when disturbed. In associations with hermit crabs, however, this behaviour pattern takes on a new significance that has great importance for the continuation of the hermit crab-anemone relationship.

It is interesting that in British waters the hermit crab, *Eupagurus prideauxi*, has an almost obligate relationship with the anemone, *Adamsia palliata*. The anemone is clasped around the shell opening so that no free shell edge is visible. The slightest disturbance of the anemone induces extrusion of the stinging acontia. The position and response of *Adamsia* prompt the hypothesis that this relationship is able to protect the hermit crab against a predator that attacks by opening the shell from the aperture end. Ross[295] has already shown that the anemone does not protect the crab from attack by an octopus in the case of *E. prideauxi*, although it does for other hermit crabs.

THE MODIFICATION OF HOST BEHAVIOUR BY PARASITES[150]

The infective phases of most parasites experience enormous mortality during the period during which successful encounters with the next host are possible. Such a host, which often must belong to a very constrained set of species, constitutes a micro-habitat that is extremely elusive in space and time. The high parasite mortalities confer evolutionary selective advantages on any aspect of the parasite's biology that enhances the possibility of making a successful contact with the appropriate host. One subtle facet of the resulting multifactorial adaptiveness involves parasites modifying the behaviour of their hosts in an advantageous manner. In almost all known cases it is an infected intermediate host that behaves unusually, making its predation by a final host more certain.

Various apparent strategies have been adopted by different parasites in this respect. They involve inducing in the host either reduced stamina, increased conspicuousness, disorientation or novel behaviour. Most of these types of perturbation can be regarded as adaptive aspects of a general induced debility of the intermediate host, but novel behaviour is best understood as the positive induction of a new behaviour pattern. The mechanisms which generate these alterations are rarely understood. Reduced nutrient availability for the host, toxin production and the mechanical or pharmacological interference with host nervous systems may all be implicated.[33,61,127,150,195]

6

The Ecology and Evolution of Intimate Associations

INTRODUCTION

The ecological and evolutionary aspects of symbiosis and parasitism are inextricably intermeshed. All modern branches of ecology have genetic and evolutionary components and the ecology of associating organisms is no exception. An obvious area of overlap is within the field of niche biology. The analysis of the niches of associating organisms can be regarded as an exercise of ecological or evolutionary significance. To take a specific example, in the colon of the tortoise, *Testudo graeca*, there coexist ten or more species of a single nematode genus, *Tachygonetria*.[307] How is one to set about explaining or examining this parasite 'species flock'? On the one hand, one is presented with an extremely tight packing of niches. An apparently homogeneous environment is supporting a congeneric species diversity that is hard to explain. The ecologically-minded experimenter will try to distinguish the important resource gradients (see page 215) along which the parasites are distributed. He will then attempt to demonstrate the degrees and types of niche overlap which occur between the species. In other words, he will try to describe niche segregation or specialization. Such attempts will always lead to a consideration of competition between the worms. On the other hand, it is quite possible to regard the *Tachygonetria*-tortoise system as an evolutionary problem. Couched in evolutionary terms the question will relate to the identity of the isolating mechanisms which have led to such extensive speciation in a sexually reproducing genus in a single environment. The two approaches are really one because the specializations that make for niche segregation between the species are likely to be identical or related to the mechanisms of evolutionary isolation.

Because of this basic unity of subject matter, ecological and evolutionary topics will be discussed together in this chapter. Broad areas of ecological information will be considered in turn and evolutionary issues incorporated into the discussion where relevant. Before attempting this exercise though, some cautionary asides are in order.

The main thrust of experimental and theoretical ecological investigation over the last few decades has concerned free-living organisms. Where interactions between organisms have been studied they have usually been of the predator-prey type. Only in the past ten to fifteen years have reliable, quantitative results or theoretical insights begun to emerge from investigations of intimate organism associations. Within this infant area of research, host-parasite and host-parasitoid interactions have received most of the attention. Quantitative ecological data on symbiotic and commensal partnerships are depressingly sparse although theoretical treatments have been initiated (see page 240).

The difficulties involved in achieving worthwhile ecological findings with intimate associates are immense and intrinsic to the systems under study. Niche dimensions of associates are likely to be parts, or attributes of other organisms and hence very troublesome to analyse. Population experiments and an understanding of bionomic strategies must usually take account of the multiplicity of developmental compartments that exist in most parasite life-cycles. This multiplicity makes attempts to measure rate parameters (births, deaths, immigrations and emigrations) occuring within and between population compartments an extensive and difficult task. What is more, population estimations usually involve the destructive sampling of hosts. Studies on energy flow between parasite and host are vitiated by the intimacy of the relationship. Separating the sinks and sources of energy can be very hard to achieve without disrupting the very processes which are under study.

All these experimental pitfalls will, in the end, be filled in and overcome by the discovery of host-parasite and symbiotic systems that can be manipulated with relative ease in the laboratory. Although field studies have a long and honourable history in this discipline, the impossibility of altering single experimental variables in the field severely constrains the horizons of such work. Much valuable progress in the general field of experimental ecology has stemmed from the utilization of appropriate laboratory models or analogues of large-scale ecological processes. Park's classic investigations[267] of inter- and intraspecific competition depended on the use of flour beetle, *Tribolium*, models. *Daphnia*, the freshwater cladoceran, was a similar pioneering laboratory analogue in the field of growth and energetics.[319] In a similar way, within the area of host-parasite interactions a number of model systems seem to offer exciting experimental possibilities. *Fasciola* and *Schistosoma* can be maintained in laboratory conditions and obviously have great applied significance because of their positions as the causative agents of clinically and economically important diseases. *Transversotrema*, an unusual ectoparasitic digenean of fish, appears to offer the bonus of population studies on a fluke whose life cycle is self-sustaining in easily maintained laboratory conditions.[10,11,375] Studies can be carried out without destructive sampling of hosts because the externally located parasites can be counted directly. Host-parasitoid interactions between insects represent in some respects an intermediate situation between predator-prey and host parasite associations. Here, the work on a number of insect parasitoids, summarized recently by Varley *et al*,[361] has yielded an impressive quantity of information and a synoptic theoretical framework exists within which this information can be considered.

DISTRIBUTIONS

The spatial distribution of a parasite or symbiote is best regarded as a hierarchy of distributions. At the most global level of the hierarchy some parasites and symbiotes show restricted geographical ranges related to such determining factors as host availability, large scale climatic patterns and geographical barriers. *Loa loa*, for instance, is a filarial worm that infects the subcutaneous and connective tissues of man. It is only found in equatorial Africa, being ecologically restricted to tropical rain-forest macro-habitats in this zone. *Loa loa* must presumably have evolved in this African ecosystem and it has not extended its range out of this area. The boundaries of the appropriate vegetation zones serve as a geographic barrier, and host availability and climatic factors also play a part in moulding the distribution. The mangrove flies, *Chrysops*, which transmit the infective L_3 larvae of the worm to man only live in the hot and humid conditions of the forests. The *Loa loa* example also demonstrates that particular parasites show distributional restrictions to particular macro-habitats within geographical areas. Usually such restrictions result from the habitat preferences of hosts to which a parasite is developmentally committed.

Within a macro-habitat, symbiotes and parasites demonstrate at least three other types of patterned distribution (Table 6.1). They have

Table 6.1 Distribution hierarchy of parasites.

Level	Type	Comments
1	Global/geographic distribution	With restrictions to particular macro-habitats within that range
2	Between-host species distribution	The expression of patterns of host specificity
3	Within-host species distribution	Frequency distribution of parasite numbers within host individuals
4	Within-host individual distribution	Patterns of micro-habitat utilization within a host individual

non-random distributions among the potential associate species within that habitat, between the individual members of the preferred associate species and, within an individual of such a species, they demonstrate preferences for locations on or in that partner. All these three facets of distribution are of special importance in the ecology of associating organisms. They arise from a central aspect of such relationships, namely the developmental and metabolic commitment of one organism to life with other species. They will be discussed in relation to host-parasite examples.

Host specificity[149,172]

The fact that an individual species of parasite inhabits a restricted range of host species generates the between-host distribution. The distribution is the physical manifestation of the phenomenon termed host specificity. Some parasites show extreme specificity by having an obligate need for a single species of host. Adult pork tapeworms, *Taenia solium*, will only develop in man. The larval development stages of the digenean, *Transversotrema patialense*, have only been described from a single type of snail, *Melanoides tuberculata*. This intense specialism is, however, only one extreme of a broad spectrum of possible host specificities. Most parasites can maintain some sort of stable relationship with a range of different host species. This range can be one of organisms of close phylogenetic affinity or of dissimilar organisms that share some crucial behavioural or ecological attribute. Examples of the first type of host grouping are legion. Adult liver flukes, *Fasciola hepatica*, provide an instance of the second type. These worms can develop and reproduce in a wide range of herbivores and omnivores including cows, sheep, rabbits and man. The important ecological link between these mammals is their feeding behaviour. They can all graze vegetation in wet places where *Fasciola* metacercarial cysts might be expected to be found. 'Grazing' in the case of man in Britain usually means eating watercress in a salad! There can be a wide disparity between the host specificities displayed by the different stages in the life cycle of one parasitic species. *Transversotrema*, with its tenacious dependence on a single snail species for larval development, will establish as an adult on a range of over ten fish species from several families.[376] *Polymorphus minutus*, a palaeacanthocephalan, carries on its larval development in a small congeneric group of amphipods of which *Gammarus pulex* is the most important. Adult *Polymorphus*, however, have been described in natural infections from 86 different bird species and one mammal, the musk rat.[86]

For most parasites it is unlikely that reproductive fitness is the same in different host species. It is difficult to assess field records of the occurrence of a parasite in a particular host in the absence of information on its reproductive condition and output there. A number of studies suggest[142,194] that successful location of a host and the ability to become established within it are not necessarily further correlated with an ability to use that host as a base for reproductive operations. Indeed, individual parasites in hosts which cannot support reproduction of the parasite are an absolute population loss in parasite population dynamics.

Such findings make it likely that many host records for parasites are accidental ones in the sense that the individual parasites concerned are unlikely to contribute much to the gene pool of succeeding generations. On the other hand, host redundancy, that is, the ability of a parasite species to use several independent, parallel host pathways to complete its life cycle, probably has evolutionary value. Two extreme evolutionary and population dynamic strategies appear feasible. One aims for high host specificity and extreme specialization for a narrow band of host types. The other strategy involves the 'insurance policy' of a wide host range. Obviously intermediate strategies will also exist. In the absence of other influences, it might be expected that high specificity would be correlated with ecological situations where contacts between parasite and preferred host are highly probable. In circumstances where this predictability is not present, parasites might be expected to employ generalist rather than specialist host strategies.

This aspect of the life-cycle pattern of a parasite has correspondences with the idea of an *r–K* continuum of bionomic strategies which has proved such a useful conceptual framework for systematizing information about the adaptive strategies of free-living organisms.[275,332] The parameters r and K refer to the intrinsic rate of natural increase of a population, and the carrying capacity of the population's habitat, respectively. The r–K continuum traverses a range of different biological strategies. At one extreme are organisms whose mode of life is dominated by their ability to rapidly colonize new, unutilized resources. These types are said to be r-selected. This type of selection favours rapid development, early reproduction and small body size. The population size is very variable in time and is usually well below the saturation level the habitat. At the other extreme are K-selected organisms. They divert far less of their energetic resources to reproduction. Instead, they are highly competitive and specialists in terms of food sources, micro-habitat utilization and other niche characteristics. Such selection favours slower development, later reproductive maturity and larger body size. K-selected populations tend to be fairly constant in size with time and exist at or near the carrying capacity of the environment. Most 'real-world' organisms exist at intermediate locations on the continuum.

In this context, it should be possible to discuss levels of host specificity in relation to r- and K-selection. High specificity could be regarded as tending toward the K extreme, very low specificity as a tendency in the r direction. It would be intriguing, therefore, to discover if levels of specificity could actually be correlated with, for instance, experimentally determined high or low values of parasitic r. Unfortunately we have no good estimates of this crucial parameter for most parasitic species. One might predict, though, that for life cycles including only one host, low host specificity would be associated with high r values and *vice versa*.

Little work has been done on the balance of generalist and specialist types of host specificity within a single widespread taxon of parasites. A recent analysis by Dritschilo *et al.*,[106] however, argues very plausibly that for ectoparasitic mites living on North American mice (cricetids), the number of mite species using a particular mouse species is related to the geographical distributional area of the mammal. Their arguments related the parasite-host relationship to the species equilibrium concept used in island biogeography.[101,212] The latter relationship is usually empirically represented in the form:

$$S = kA^z$$

where S is the equilibrium number of species that can ultimately exist on an island of area A. The dimensionless parameter, z can be thought of as an index of the rate at which new species are added as the island's area increases; it is in fact the slope of the regression line on a log S versus log A plot. The S value for islands is related to the distance from a species pool (usually the mainland) which provides emigrants for island colonization. In the mouse-mite situation, the area A is the geographical distribution range of a mammal host species, and the species pool is the sum total of all North American mite species which are potential parasites of the mice. The species number to host range area relationship for mite parasites on cricetids is of the form $S = kA^z$. The fit is particularly good for data on the host genus *Peromyscus* (the deer mouse) whose parasitic mite fauna has been thoroughly studied (see Fig. 6.1).

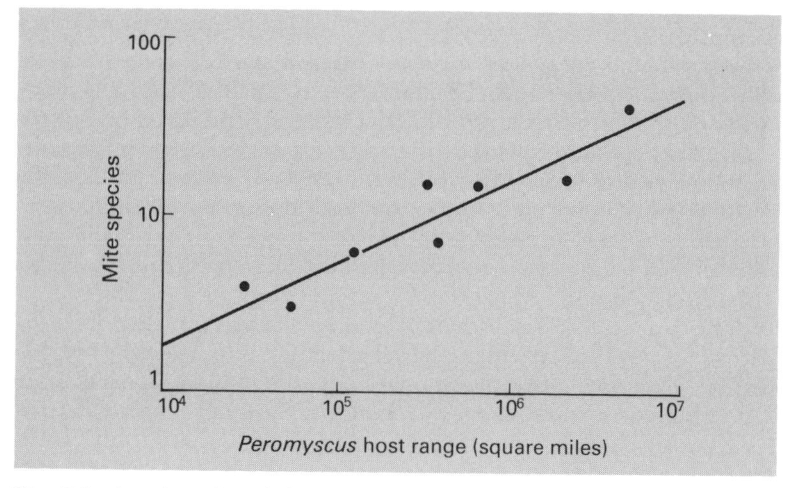

Fig. 6.1 Log-log plot of the number of mite species recorded from North American *Peromyscus* hosts against the estimated distributional ranges of the host mammals. (From Dritschilo *et al.*[106])

In the mite-mouse distribution it is obvious that there is no direct analogue of the distance of the island (host range) from the species pool (total parasitic fauna). Instead, for generalist mites, the probability of colonization of new mouse species will be related to the possibilities of close contacts of different mice between which mite transfers can take place. These probabilities will be linked to the extent of the distribution overlaps of mouse species ranges and these will increase with increasing host range areas. If the mite species on all of the mammal species considered in this study are ranked in order of increasing host specificity, the balance of generalist and specialist strategies among the mites is revealed (Fig. 6.2). Overall, the mean number of host species utilized by a single mite species is approximately eight. 25 of the 101 mite species examined were restricted to a single host species, but one 'ultrageneralist' mite was discovered on no less than 120 different host species.

The mouse-mite findings are an excellent illustration of the difficulties inherent in the interpretation of field-derived infection data. Kuris and Blaustein[182] have analysed the same data as Dritschilo *et al.* and come to quite disparate conclusions. They point out that the geographical distributional area of a North American mouse species is positively correlated with the quantity of scientific investigation performed upon it. The more widespread a mouse is, the more

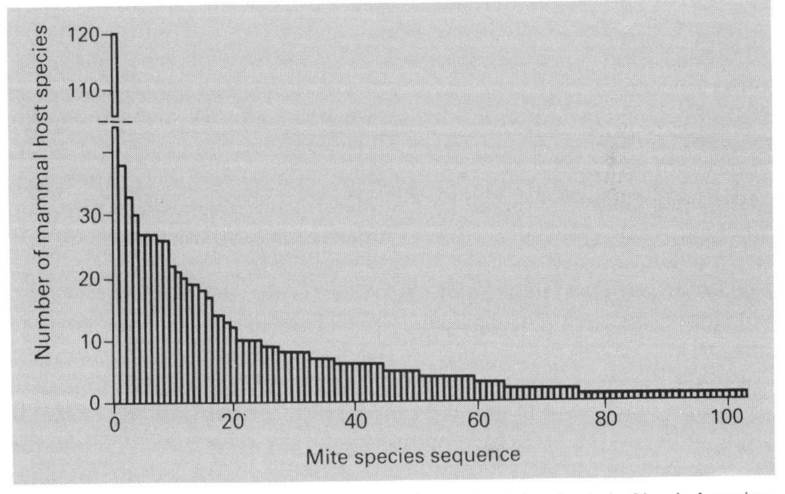

Fig. 6.2 Host specificities of mites found on cricetid rodents in North America. The ordinate is the total number of mammal species utilized by a mite species including non-cricetid mammals. The abscissa ranks the 101 mite species in descending order of host species utilization. (From Dritschilo *et al.*[106])

scientists will use it. Kuris and Blaustein found that there is, in fact, a very good positive correlation between the log of the number of papers published on a mouse over the past four years and the log of the number of mite species found on it. They suggest that the number of mite species found per rodent is a function, not of distributional area, but of intensity of study. There at present the matter rests. In general, for resolution of the problem we must await far more extensive analyses of the host specificity patterns within parasitic taxa.

A distribution of parasites of one species between different host species will provide some degree of genetic isolation between sub-populations in the alternative hosts. In many situations such isolation is given little chance to provide the impetus for parasite speciation because there is much genetic interchange between the sub-populations. This shuttling of gene types is most easily achieved when free-living, mobile infective phases are present in the life cycle.

Host specificity, with whatever precision it operates, is caused in a multifactorial manner. Specificity-generating mechanisms may operate at any point in the concatenation of behavioural and physiological events which enable a parasite to establish itself and reproduce successfully in a host.

In the case of parasites with directly infective transmission stages, specificity can be maintained by their own discriminatory powers. Kearn[168] has shown such an ability for the free-swimming, ciliated oncomiracidium larva of *Entobdella soleae* (see page 188). Discriminations based on excystment or hatching in the presence of particular physico-chemical conditions are also shown by various encysted or shelled larval forms. These factors, however, are rarely able to determine tight specificities. They simply ensure that emergence does not occur in completely inappropriate conditions. Thus, in the case of the eggs of *Entobdella soleae*, a chemosensory activation stimulus operates in addition to the primary pattern of hatching based on the natural light-dark cycle (see page 176). Fully-developed eggs of *E. soleae* can, in fact, be persuaded to hatch in darkness or light if they are in the presence of fish mucus. This stimulus is a very unspecific one. Mucus from many different fish, including both teleosts and elasmobranchs, will induce the effect. It appears to be due to urea or ammonia in the secretion.[171] Thus the circadian light-induced pattern and general chemical activation of hatching cannot themselves explain host specificity in this parasite. These influences simply release swimming oncomiracidia in situations where they are likely to be able to use their direct chemosensory discriminatory ability to attach to appropriate host fish.

In addition to the above mechanisms, there is a broad area of host specificity which is an expression of host-parasite interactions occurring after the parasite has located a potential host.

Physiological conditions may form a barrier to establishment. There are, for instance, many acanthocephalans that live in the guts of freshwater and marine teleost fish. In contrast, it appears as though no authentic records exist of elasmobranchs acting as acanthocephalan hosts.[379] It is possible that the high concentrations of urea which are present in the body fluids and gut of these fish are responsible for this selective barrier. Most acanthocephalans with aquatic life cycles lack a differentiated osmoregulatory system so perhaps they have been unable to adapt to the high urea levels as the flame cell-possessing cestodes have.

Interactions between parasites and the defensive responses of their hosts contribute very extensively to the induction of host specificity phenomena (see Chapter 5). For many parasites, the ability to evade the host's defensive measures effectively does not extent over a large range of host species. The limited powers of evasion directly determine specificity. Miracidial larvae of *Schistosoma mansoni* from Puerto Rico can penetrate snails of the host species, *Biomphalaria glabrata*, from either their home area or Brazil. Hence no specificity is operating at this early phase of establishment in the mollusc host. Two days after the penetration of Brazilian *Biomphalaria*, however, the developing larvae are destroyed in a mixed cellular defensive response involving encapsulation and phagocytosis. In the 'home' snails there is no response of this sort and larval development continues.

Distributions within a host species

In two seminarial papers in 1971, Crofton[81–2] considered the within-host distribution of parasites from a general viewpoint. The distribution involved is not a geographical one but a frequency distribution of numbers of parasite individuals among the individuals of the host population. This type of distribution has a distinctive ecological reality in intimate associations of organisms as defined in this book. It is a distribution in which the sampling unit is a single member of the host species. Crofton concluded that it was diagnostically characteristic of parasitism to produce an overdispersed (aggregated) distribution of parasites within the host population. Overdispersion implies the regular presence of host individuals with high parasite numbers that would only have vanishingly small probabilities of occurring if the parasites were distributed at random through the host population.

Along with his understanding of this distribution Crofton postulated that there was a level of parasitization, characteristic of each parasite interaction, which could kill the host. This was termed the lethal level. Its effect would be to remove a few host individuals from the right hand tail of the overdispersed distribution. Their deaths would remove a relatively large number of parasites from the parasite population. Crofton suggested that the invariably observed fact that parasites have a higher reproductive potential than their hosts, should be regarded as a counteractive influence to parasite depletion by the lethal level effect.

The vast majority of host-parasite relationships which have been studied quantitatively demonstrate the overdispersed distribution predicted by Crofton. Such distributions are characterized by a variance/mean ($s^2/\overline{x}$) ratio which is significantly greater than unity.

Mathematically there are several types of overdispersed distributions, but the one which most workers have fitted to actual data from populations is termed the negative binomial distribution. This is a contagious ('clumped') distribution defined by two parameters, the arithmetic mean ($\bar{x}$) and a positive exponent (k) which measures inversely the degree of aggregation of parasites within the host population. Fig. 6.3 demonstrates the goodness-of-fit of theoretical negative binomial distributions to data obtained from host-parasite systems as different as microfilariae in American gnats,[309] and larval ticks on small mammals in southern England.[283]

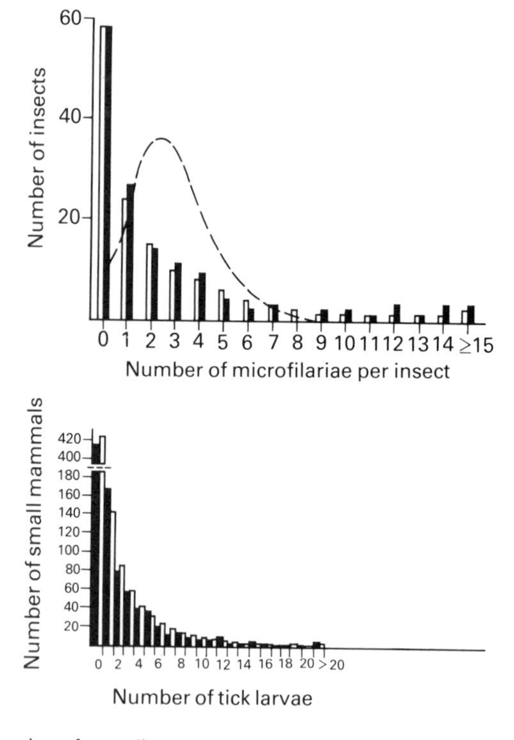

Fig. 6.3 Examples of overdispersed parasite populations within host populations. In both cases solid bars indicate observed frequencies and the open bars the best negative binomial distribution fit to the data. (**a**) The distribution of microfilariae of the filarial nematode, *Chandlerella quiscali*, in a population of the gnat, *Culicoides crepuscularis*. The dashed line indicates the best Poisson distribution fit to the data. (Based on data in Schmid and Robinson.[309]) (**b**) The distribution of larvae of the tick, *Ixodes trianguliceps*, in a mixed population of small mammals in which the shrew, *Sorex minutus*, was the quantitatively dominant host. (Based on data in Randolph.[283])

An overdispersed distribution implies that the chances of infection are not randomly assorted between the members of the host population. This unevenness is termed heterogeneity. What generates the heterogeneity of successful infections that must exist in many instances to produce the observed parasite aggregations in host populations? Within such populations, differences in susceptibility to parasites are likely to exist between different sexual, developmental, size and age categories of host. Size and developmental rates will often be diversified in a habitat showing local variability, and differences in the behaviour of individual hosts can predispose to infections or conversely make them less probable. In fact, almost any morphological, physiological or behavioural lack of homogeneity in the host population will lead to heterogeneous chances of infection and generate an overdispersed distribution.

It is quite possible, though, that within any limited class of host individuals the chance of infection will be a random variable, producing a Poisson distribution ($s^2/\overline{x} = 1$) of parasites within that class. When, however, all such subpopulations are brought together to find the total pattern of distribution, an addition (or compounding) of many separate Poisson distributions will give rise to an overdispersed distribution.

As these contagious distributions can be generated by so many different mechanisms, it is quite obvious that the fact that a parasite shows such a distribution predicts nothing very specific about its biology. In particular cases, the mechanisms producing overdispersion must be sought experimentally. In the microfilariae (*Chandlerella*)–gnat (*Culicoides*) instance, described above,[309] it seems likely that one such mechanism may be the temporal changes in microfilarial density that occur in the final host's blood. The final host of *Chandlerella quiscali* is the bird, *Quiscalus quiscula* (the purple grackle). It has been shown that peripheral blood counts of microfilariae in infected grackles peak nocturnally. Microfilarial numbers in samples taken from an individual host at the same time on successive nights vary enormously. Even with a gnat biting schedule well synchronized with the peaks in nematode numbers, the high variability of the latter will help to generate overdispersion.

Anderson[5] has discussed possible heterogeneity-producing processes in the relationship between the bream, *Abramus brama*, and its cestode parasite, *Caryophyllaeus laticeps*. The bream is a bottom-feeding fish that becomes infected with this helminth when it ingests tubificid annelids containing larval *Caryophyllaeus*. Anderson showed that the infected annelids were spatially contagiously distributed over the bottom of his stocked fishpond study area. If the feeding behaviour of individual fish were independent of others in a shoal, a single fish on finding a clump of infected tubificid worms would be likely to ingest several in a short time period. The number ingested would be significantly larger than those ingested by its companions. Over a time period, such events would be expected to produce heterogeneity by a contagious input of infective stages. If, on the other hand, it is considered that the number of larvae ingested by a fish in a short time interval is a randomly dispersed variable, each fish will eventually achieve an adult parasite population by a non-contagious infection process. Variability between the feeding behaviour of individual fish though, will produce different equilibrium population sizes. The total fish population will have a parasite distribution which could be overdispersed by a compounding of a series of essentially random processes.

Overdispersed within-host distributions preponderate in host-parasite systems but occasionally other distributions occur. In laboratory conditions, over short intervals, Poisson distributions can be achieved. Also, a few cases[7] have been uncovered in which underdispersed (regular) distributions exist. Such distributions, with variance/mean ratios significantly less than unity, demonstrate a narrow range of parasite number classes than would be expected if infections were random events.

Site selection within a host[88,148]

It has long been recognized that specimens of individual species of parasites are not distributed at random throughout the bodies of their hosts. Instead, they occupy species-characteristic locations or sites that have been termed micro-habitats. This name emphasizes the contrast between the external macro-habitat within which the host exists and the host-maintained habitat of the parasite. In most examples, which have been examined with care, the micro-habitat specificity is due mainly to an active behavioural discrimination on the part of the parasite. Chapter 5 (pages 193–198) has described some of the complex migrations which parasites undergo within their hosts. These migrations are often equally site-specific, taking parasites from one circumscribed micro-habitat to another.

The microhabitat-defined spatial distribution of a parasite, which is the most particular of its hierarchy of distributions, is amenable to direct observational analysis. With appropriate precautions and standardization of techniques[88] it is relatively easy to describe the pattern of parasite locations within a host. As a result, much descriptive detail exists on this topic and many examples are described in two recent reviews.[88,148] The most striking illustrations of this type of distribution are the instances where parasites have become profoundly altered morphologically for attachment to extremely restricted host micro-habitats. At this structural level, Williams *et al.*[379] have provided descriptions of the methods of attachment of some gut-inhabiting tapeworms of elasmobranch rays. A consideration of two such cestodes which both utilize the gut of *Raja radiata* will demonstrate the morphological modifications that can occur in relation to micro-habitat specificity (see Fig. 1.1b).

Pseudanthobothrium hanseni and *Phyllobothrium piriei* are both found anchored to the villi of the spiral valve region of the intestine in *Raja*. Adults of the former species are found predominantly on tiers 1, 2 and 3 of the helical valve and are further restricted in location by showing preference for the posterior-facing surfaces of these tiers. Adult *P. piriei*, in contrast, are restricted to tiers 5, 6 and 7. Villous morphology is different in the anterior (1, 2, 3) and posterior (5, 6, 7) tiers. The anterior zone bears long villi, those posteriorly placed are short. Related to this difference one finds correlated specializations of the cestode attachment organs (scolices). That of *Pseudanthobothrium* has four, deep, cup-shaped suckers (bothria) which fit neatly over the tops of the long villi. It also possesses a long, apical proboscis tipped with a sucker which probes downwards between the villi bases for attachment. In contrast, the lobes of the scolex of *Phyllobothrium* are secondarily dissected so they look like muscular holly leaves. Each crenulation of the lobe edge fits round a separate short villus. In this way, in a zone of short villi, this cestode laterally extends its attachment area for firm anchorage. The two types of scolex specialization would appear to have irrevocably restricted these two tapeworms to their respective micro-habitats.

Further descriptions of micro-habitat specificity will be included in the following section.

NICHE BIOLOGY

Before embarking on a discussion of the niche biology of parasites it will probably be useful to summarize some of the salient aspects of modern niche theory.[19,78,94,276]

The niche of an organism can be regarded as its range in ecological rather than geographical space. For a particular species, the niche is quantified by a description of the segments of environmental resources which the species utilizes. If each resource is regarded as an axis describing a continuous variable, a species will utilize a particular portion of that resource gradient (Fig. 6.4). If there are p such axes the niche will be a characteristic closed region of p-dimensional space (Fig.

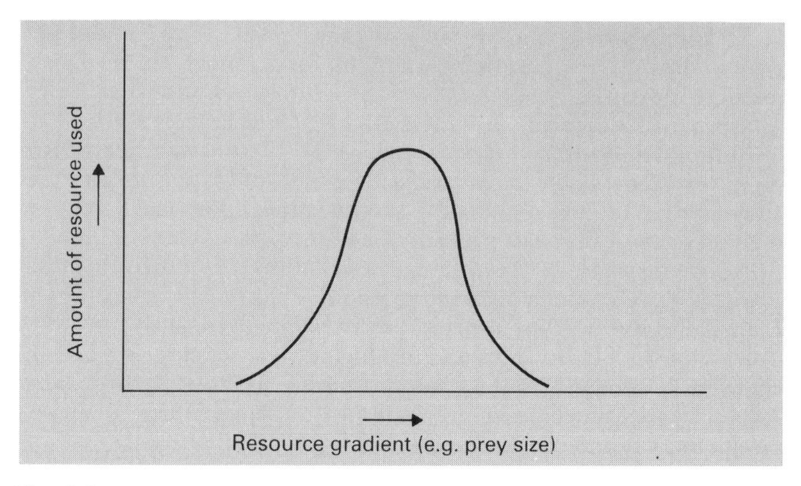

Fig. 6.4 Diagram of a resource utilization curve, demonstrating the differential use of a resource over (in this case) a range of sizes.

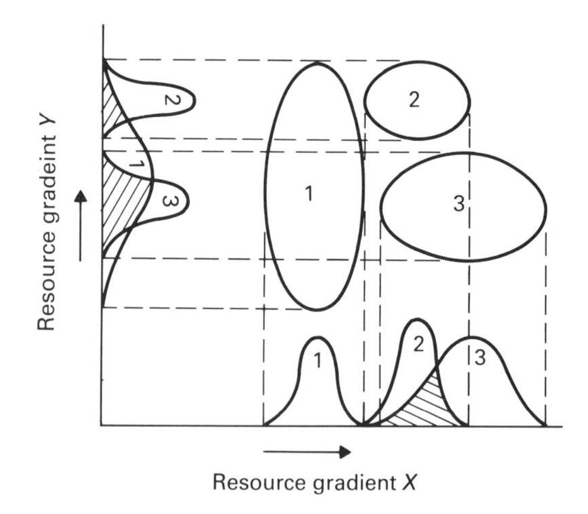

Fig. 6.5 The niches of three species in a two-dimensional space defined by the resource gradients *X* and *Y*. The resource utilizations of *X* and *Y* by species 1, 2 and 3 are shown individually as two sets of bell-shaped curves. In both gradients extensive utilization overlaps occur (hatched zones). In *Y*, species 1 overlaps the utilization range of both 2 and 3. In *X*, species 2 and 3 overlap. When two-dimensional descriptions of niches (tinted ellipses) are considered, however, all overlap is eliminated.

6.5). With three axes the niche can be conceived as a three-dimensional volume of utilizable resource values. With four or more axes the niche volume (hypervolume) is not directly conceivable and has to be treated mathematically. Luckily, as Levins[196] has pointed out, in most circumstances the only axes (niche dimensions) which need to be considered are those important resource variables which serve to separate the competing species under study. In most systems of free-living animals three or four niche dimensions are sufficient to explain species segregations, in the sense that they will provide a description of the niche, which although incomplete, eliminates most niche overlap between species.

Along such primary resource gradients one finds a serial array of species with utilization curves which only partly overlap (Fig. 6.6).

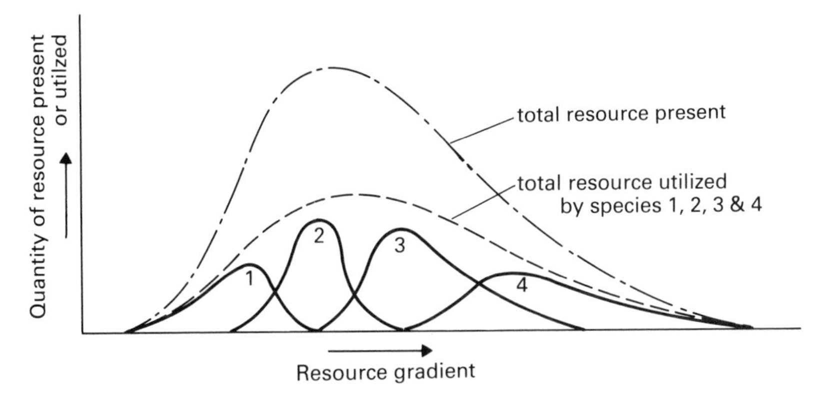

Fig. 6.6 A segregated array of resource utilization curves. Utilization curves are shown for species 1–4 along with the total supply of the resource in the environment and the total utilization of the resource by the four competing species. Utilization overlaps are shown as hatched areas.

These arrays of species are said to be segregated from each other or to show niche displacement or segregation. The utilization curve of each species on a particular gradient has a characteristic spread of values, or niche breadth, which can be mathematically indicated by the variance of those values. Packing of niches onto resource gradients almost always involves overlaps of utilization curves. Such niche overlaps, as they have come to be known, probably correspond in some way to the intensity of competition that exists for that resource between two species with adjacent niches. It must be remembered, though, that competition can occur in at least two different ways, namely interference competition and exploitation competition.[276] The former involves direct interactions like interspecific conflicts over territoriality. The latter describes situations where species are jointly utilizing the same finite resource such as a particular prey species. Different types of competition might well be related to niche overlaps in quite separate ways.

A generalization about competition called 'Gause's Principle' has often been an albatross round the neck of niche biology. Restated in modern terminology Gause's principle (or the competitive exclusion principle) may be stated thus: 'two populations (species) cannot long coexist if they compete for a vital resource, limitation of which is the direct and only factor limiting both populations'.[94] Put like this, in relation to single-resource competition, the postulate is uncontroversial. It has, however, been applied out of this context, in situations of more complicated, multi-resource competition, and has given rise to much unnecessary academic dissention. Perhaps Cody[78] has best resolved the misunderstandings in the following terms:

'Competing species do not completely exclude each other, but show considerable overlap in resource use. Moreover, there are some combinations of resource distribution and density which actually favour convergence of utilization curves. Thus, the questions we should be asking as a consequence of Gause's results should not centre on why the 'principle' is being 'violated', but rather on how much overlap of resource use is tolerable? How does this overlap vary with resource predictability and density? To what extent are species distributions and abundances predictable from a knowledge of resources types and productivity?'

Niche biology of parasites[148]

Most descriptions of the niches of parasites have concerned themselves with endoparasitic helminths, although some studies on the site preferences and feeding behaviour of ectoparasitic helminths and arthropods and blood-dwelling protozoans are also relevant.

Closely related or ecologically equivalent species of parasites which coexist in a host might be expected to interact in one of two ways. Considering pairs of species, the two types can either compete in ways which lead to perhaps the complete exclusion of one, or they can interact competitively so as to segregate their niches. As has been explained above, these two possibilities are not independent of one another. Time scales are an important consideration here. In the life span of individual parasites it is possible that 'interactive segregation'[148] will occur. That is, the actual niches of the competing species will alter their conformations, when the species are present together, so as to reduce niche overlap to Cody's 'tolerable' amount. On an evolutionary time scale, if the two species meet often in the same host micro-environment, it might be expected that there would be selective advantage in these short term responses becoming genetically 'fixed'. In this situation, niche breadth would be adjusted even in the temporary absence of the competitor. In practice, it is often difficult to separate the influences of these two types of segregation. In most situations where parasites are competing, it is likely that they are operating together.

The mutability of niches described above has necessitated a particular terminology which differentiates between the niche in the absence of all competitors and other enemies (the fundamental niche) and the actual niche in a particular competitive milieu (the realized niche).

What are the resource gradients that we must use to understand the niche segregations of parasitic species? It is likely that time, nutrient sources and micro-habitat are among the crucial niche dimensions. It is also possible that the total pool of potential host species can be regarded as a resource 'axis'.

Time as a parasite resource gradient

It is possible to treat time as a resource in communities of parasites, in that parasites of different species can utilize very similar micro-habitats or other resources in a particular host in a relatively non-overlapping pattern along a time axis. As Cody[78] has pointed out, time is conceptually a different type of resource from food or micro-habitat. It is completely exhausted at a uniform rate and yet is perfectly renewable at the same rate.

It does appear that parasites can utilize otherwise similar niches at different time slots in a seasonal displacement pattern. The nematodes, *Cucullanus heterochrous* and *C. minutus*, develop in the intestine of the flounder, *Platichthys flesus*, at different times of the year.[218] The congeneric nematodes have similar life cycles which include larval worms of both species moulting twice in the host's gut wall and then entering the gut lumen from March to September with a peak between May and July. Once in the gut lumen *C. minutus* matures quickly, producing eggs in the summer and dying in the autumn. The adults of *C. heterochrous* develop much more slowly, maturing in the autumn and not dying until the late winter or summer of the following year. This difference in maturation period has been interpreted as a partial temporal segregation of the two species.[148] It must be pointed out, however, that the linear distribution of the two nematodes along the intestine differs somewhat (see Fig. 6.7b). Also, the linear

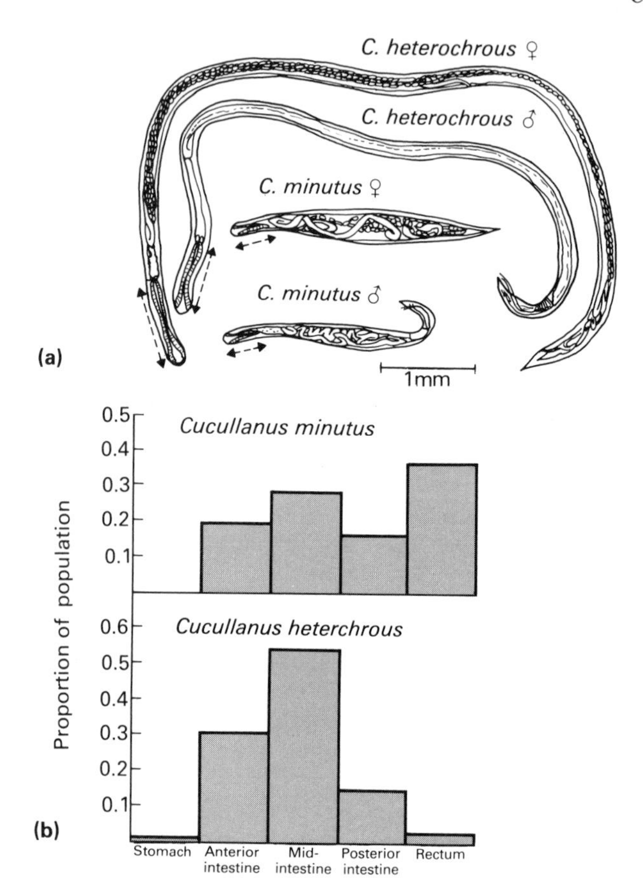

Fig. 6.7 The competing nematodes, *Cucullanus minutus* and *C. heterochrous*, from the gut of the flounder, *Platichthys flesus*. (**a**) The morphologies of the two species demonstrating the difference in the size of the muscular pharynx (←-→). (**b**) The longitudinal distribution of the two species along the flounder gut. (Both redrawn from MacKenzie and Gibson.[218])

dimensions of their anterior feeding structures differ in the approximate ratio 1 : 1.9 (Fig. 6.7a). These facts suggest that in both the micro-habitat and nutrient resource gradients the two nematodes might show significantly different utilization curves.

Nutrient sources as a parasite resource gradient

It has been commonly observed in free living animals that in a continuous resource gradient of food types (say sizes of seed grains or insects), niche displacement by apportionment of different food types occurs. This is often correlated with morphological specializations of feeding structures such as the beaks of birds. Hutchinson has suggested that in a segregated array of competing species, differences of at least 10% in linear dimensions are found between adjacent species in the array.[155] Little careful work has been carried out on this aspect of helminth niches. This is an unfortunate gap in our knowledge because there is every reason to believe that among, for instance, gut-dwelling helminths, great specialization for the acquisition of particular food types will have occurred in different parasite species. This prediction stems from the general premise of niche theory that in a predictable superabundance of food it is a tenable, indeed favoured, strategy to become highly specialized for a particular narrow category of nutrients. In other words, this aspect of niche breadth is reduced and niche packing density can increase.

Cannon[58] has looked at two closely related and cohabiting digeneans of the perch gut, *Crepidostomum* and *Bunodera*. They utilize the anterior portion of the gut of the host concurrently. The two species probably feed on both gut wall material and gut contents with their muscular oral suckers. Oral sucker diameters of the two digeneans do indeed differ by 10% and a differential food utilization might be inferred from this discrepancy. In another study, the ectoparasitic mites of small mammals have been divided into three different 'guilds' based on the utilization of different nutrient sources namely blood, tissue fluids and hair.[106]

A more extensive series of studies relate to the *Tachygonetria* – tortoise system[273,307–8] which has already been mentioned (see page 202). The system provides an exceptionally clear example of both nutrient source and micro-habitat operating as crucial niche parameters. It will be assessed in this light in the next section.

Feeding structures can be analysed at a morphological level in bulk-feeding parasites like digeneans in order to provide clues about differential nutrient sources. To what extent though, are tapeworms and acanthocephalans distributed along nutrient-based resource gradients? Differences in the utilization of nutrients between pairs of these gutless endoparasites would have to involve differences in the patterns of active transport systems in their outer surfaces. It is, unfortunately, very difficult to gain enough quantitative and comparative information on these systems to come to firm conclusions about interspecific differences. One pioneering study, however, by Uglem and Beck[354] has tried to obtain such data for a pair of competing congeneric acanthocephalans. The host-parasite system which they utilized consisted of the two species of *Neoechinorhynchus*, *N. cristatus* and *N. crassus*, which as adults inhabit characteristic regions of the intestine of the large-scale sucker, *Catostomus macrocheilus*. The surface biochemical system studied was the aminopeptidase (APase) enzyme activity that is the first phase of a spatially linked digestive-absorptive process for peptides at the surface of acanthocephalans (see chapter 3, page 103). Uglem and Beck, in a long series of experiments, tested the specificity characteristics of the APase activity of the two worms. They assayed worm homogenates for APase activity against a series of ten different B-naphthylamine derivatives of amino acids to obtain an 'activity spectrum' for each worm's APase activity. It was found that the two species possessed APase activities with distinctively different activity spectra with respect to the naphthylamines. *N. cristatus* had a broader profile of APase specificity than *N. crassus* and had quantitatively higher levels of activity. So it seems possible that helminths that absorb nutrients directly across their body walls could segregate on the basis of a differential utilization of nutrients.

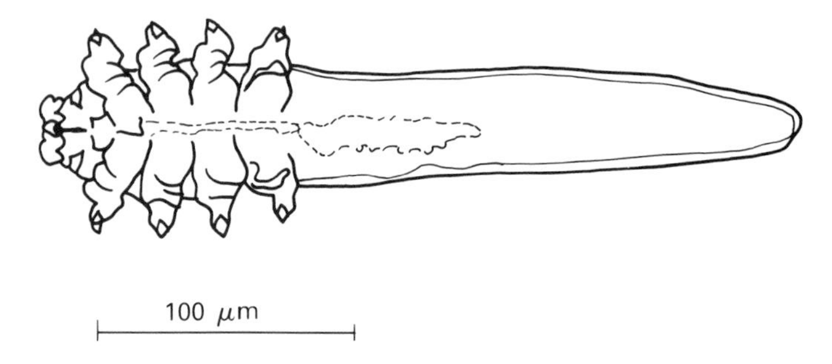

Fig. 6.8 The follicle mite, *Demodex folliculorum*. A female from a human hair follicle.

The micro-habitat resource

Spatial separations between parasites in a host seem to be an extremely important part of niche segregation. This generalization can be illustrated by examples from a number of parasitic groups.

Until 1972[100] it was assumed that the interestingly specialized parasitic mite, *Demodex*, which inhabits human skin consisted of a single species, *D. folliculorum*. These extraordinary arachnids have become worm-like in shape (Fig. 6.8). Their general micro-habitat is human facial skin especially that of the nose and forehead. The long-observed polymorphism of the mites has been been explained by the description of two coexisting species, *D. folliculorum* and *D. brevis* which have beautifully segregated micro-habitat preferences. The larger form, *D. folliculorum*, lives in the actual follicles of simple hairs above the level of the sebaceous glands. Individuals of *D. brevis*, the smaller species, mostly inhabit the acini of sebaceous glands linked to vellus hairs. The two species have different feeding habits. *D. folliculorum* consumes follicular epithelial cells, and *D. brevis* apparently eats sebaceous gland cells.

Schad[307] has made an extremely detailed analysis of *Tachygonetria* micro-habitat utilization in the elongate colon of the tortoise. Some of his distributional results are summarized in Fig. 6.9. Special attention

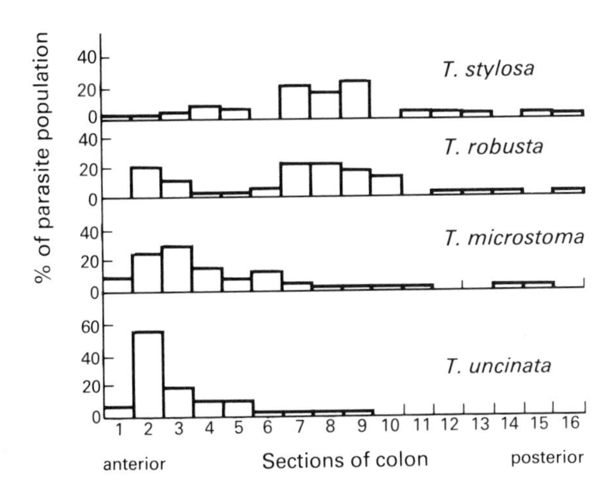

Fig. 6.9 The *Tachygonetria* – tortoise system. The longitudinal distribution of four *Tachygonetria* species in the colon of the tortoise, *Testudo graeca*. The four species illustrated are those that show a strong affinity for the paramucosal lumen zone. (Redrawn from data in Schad.[307])

was paid to the mature females of eight morphologically distinguishable species in the genus. Longitudinal distribution among sixteen longitudinally defined colon regions was noted as was radial distribution from the mucosal surface to the central lumen. The latter aspect of the spatial distribution was assessed in an ingenious manner. Portions of infected colon were dropped into liquid air. While still frozen, with the worms fixed immobile in their original positions, central lumen subsamples were chopped out with a cork borer to be compared with a mucosal annulus. If only longitudinal distributions were considered, there was considerable spatial overlap between the ranges of the eight species. However, four species (*T. uncinata*, *T. microstoma*, *T. robusta*, and *T. stylosa*) showed radial distributions demonstrating a strong aggregation in the mucosal zone. The remaining four (*T. numidica*, *T. macrolaimus*, *T. dentata*, and *T. conica*) are broadly distributed between mucosal and central luminal zones. With this second spatial axis also considered, overlap between species was almost completely abolished. Thus, *T. uncinata* and *T. numidica*, which have very similar longitudinal distributions at the anterior end of the colon, are well segregated radially. One problematic species-pair, however, remains: *T. robusta* and *T. stylosa* have similar distributions both longitudinally and radially, and are therefore not segregated spatially to any significant extent. The two species can be easily segregated when nutrient sources are taken into account. *T. robusta* is an indiscriminate feeder on colonic contents, *T. stylosa* is a specialist bacterial feeder. Petter[273] has made a detailed analysis of the morphological differences between this interesting pair of nematodes. She showed that in mature females the muscular oesophagus, a crucial segment of the feeding apparatus, was very different in size in the two species. In *Tachygonetria* (*Mehdiella*) *stylosa* it had a length of 1.3 mm whereas in *T. robusta* it was 1.9 mm long; once again apparently justifying Hutchinson's[155] generalization about trophic organ size (see page 221).

Overall, Schad's results appear to demonstrate that even within this densely packed species flock, three niche dimensions are sufficient to describe meaningful segregations.

The host species resource

It is difficult to consider a range of potential host species as a resource gradient. Such an assemblage of species cannot be regarded as a linear axis of a continuous variable. It must, however, be thought of as a resource that is competed for by parasites. A single example of fish parasites will illustrate the principle.

Populations of four different species of chimaerid ratfish, *Hydrolagus* spp., for instance, each contain two different species of the cestodarian genus *Gyrocotyle*.[357–8] These eight parasite species have divided up this particular section of the chimaerid host resource in a partially overlapping fashion among themselves. Apportionment, in fact, goes to greater lengths. In each parasite pair of species, one (the larger of the pair) is dominant at the individual fish level. So, in *Hydrolagus collei*, the large *G. fimbriata* is competitively superior to *G. parvispinosa*.[316] Of 415 ratfish examined, only three had mixed infections. In three different ratfish populations of *H. collei*, unispecific *G. fimbriata* infections represented between 70% and 90%

of the infected fish. Thus, at a population level the two parasites coexist, whereas at the level of most fish individuals the large parasitic species can competitively exclude its smaller cousin.

Competitive exclusion[133,148]

The *Gyrocotyle*–ratfish example of exclusion cited above is an instance of a fairly familiar phenomenon among parasites sharing a host species. Evidence for competitive exclusion is normally based upon field data which show that mixed infections of A and B in a shared host species are less common than the incidences of A and B individually would suggest on a random assortment basis. Holmes[148] has collected a number of references to such competitive interactions and they almost all relate to pairs of species which are close ecological analogues of one another.

It is quite conceivable that some competitive relationships of this sort involve the defensive responses of the host. So, when the gills of a carp are infected with the monogenean, *Dactylogyrus vastator*, a cellular reaction of the gill epithelium is induced with increased mucus production and epithelial hyperplasia. This defensive reaction makes infected gills unsuitable substrates for several other closely related *Dactylogyrus* species, and eventually for the initiator of the response itself.[263] In a similar way, Heyneman,[140] working with infections in white mice, showed that the tapeworm, *Hymenolepis nana*, produced an immune response in the host which was more effective against the closely related *Hymenolepis diminuta* than against itself.

One extreme form of competitive exclusion between parasites actually involves interspecific predation of one parasite on another. Several species of echinostome digeneans have snail-inhabiting rediae that can kill and eat the larval developmental stages of other digeneans in the same snail.[27] This ability is so marked in some echinostomes that they have been seriously considered as biological control agents which would remove the larvae of dangerous digeneans like schistosomes from a snail population.

Interactive and selective segregations

One lesson that has been learnt from extensive studies on the niche biology of free-living animals like birds is that changes in niche position and breadth can occur in response to an altered competitive regime[78] or changing environmental conditions. This suggests that for some animals there may be few compelling physiological or genetic checks which constrain a species to particular sections of resource gradients. As Darlington has put it,[94] '... niches are made by the organisms that occupy them; their limits are largely determined by competing organisms, and the whole complex ecological structure that results is flexible and capable of evolving both in detail and as a whole'. Using another terminology, one could say that the fundamental niche is often more voluminous than would be inferred from consideration of any instantaneously realized niche.

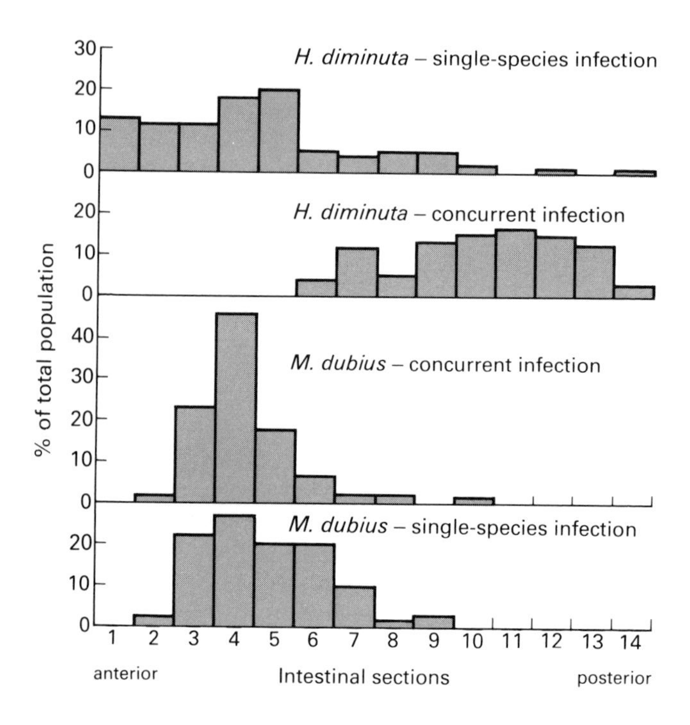

Fig. 6.10 Longitudinal distributions of the cestode, *Hymenolepis diminuta*, and the acanthocephalan, *Moniliformis dubius*, in the intestine of the rat. The effects on the distributions of single-species and concurrent infections. (Redrawn from data in Holmes.[147])

In parasitic examples these alterations in niche boundaries due to varying patterns of competition or environmental variables can be short-term behavioural ones or long-term genetically fixed changes. The former category is termed interactive site segregation, the latter selective site segregation. Only a few convincing instances of the behaviourally mediated change have been described, although it seems unlikely that the phenomenon is, in reality, rare. Holmes[147] showed that interactive segregation probably occurs in concurrent infections of the cestode, *Hymenolepis diminuta*, and the acanthocephalan, *Moniliformis dubius*, in the rat. When present together, both helminths are restricted to parts of their normal ranges in the non-competitive situation. The tapeworm alters its distribution more than the acanthocephalan does (Fig. 6.10). In a similar way, Chappell[69] has noted that when the tapeworm, *Proteocephalus filicollis*, occurs together with the acanthocephalan, *Neoechinorhynchus rutili*, in the stickleback alimentary canal, the spatial distributions of the parasites are quite different from those in single-species infections. Specifically, in concurrent infections, both adults and larval forms of *P. filicollis* attached more anteriorly whereas specimens of *N. rutili* attached posteriorly in the rectum. In single-species infections both parasites demonstrated a wider dispersion throughout the gut (Table 6.2). This

Table 6.2 Numbers of *Proteocephalus filicollis* and *Neoechinorhynchus rutili* attached in different stickleback gut zones in single species and concurrent infections. (Based on data in Chappell.[69])

Species	Type of infection	Number of worms		
		Ant. intestine	Post. intestine	Rectum
P. filicollis (adults)	Single species	91	77	0
	Concurrent	20	4	0
P. filicollis (plerocercoids)	Single species	74	138	0
	Concurrent	18	10	0
N. rutili	Single species	0	23	19
	Concurrent	0	11	26

pattern of a broadening of the utilizable range of micro-habitats when competitive pressure is released has been commonly observed for free-living systems[28,91,130] from taxa as different as freshwater triclad flatworms, Appalachian salamanders and the Bermudan birds.

The understanding of the niche biology of parasites is poised at an exciting stage. Despite the sparseness of the evidence it is obvious that many of the more subtle static and dynamic features of the niches of free-living animals can be expected to reveal themselves soon in parasitic examples. Rapid advances are occurring now, and the impetus of continuing results from free-living systems should make the study of the niche biology of associating organisms a stimulating and active area.

THE POPULATION DYNAMICS OF PARASITES[8]

If the study of the niche biology of parasites is in its infancy, the quantitative investigation of host-parasite population dynamics is in a neonatal state, when compared with the analogous studies which have been carried out on free-living systems. As was mentioned above (page 203), the many separate yet intercommunicating population compartments in the life history of most parasites preclude any

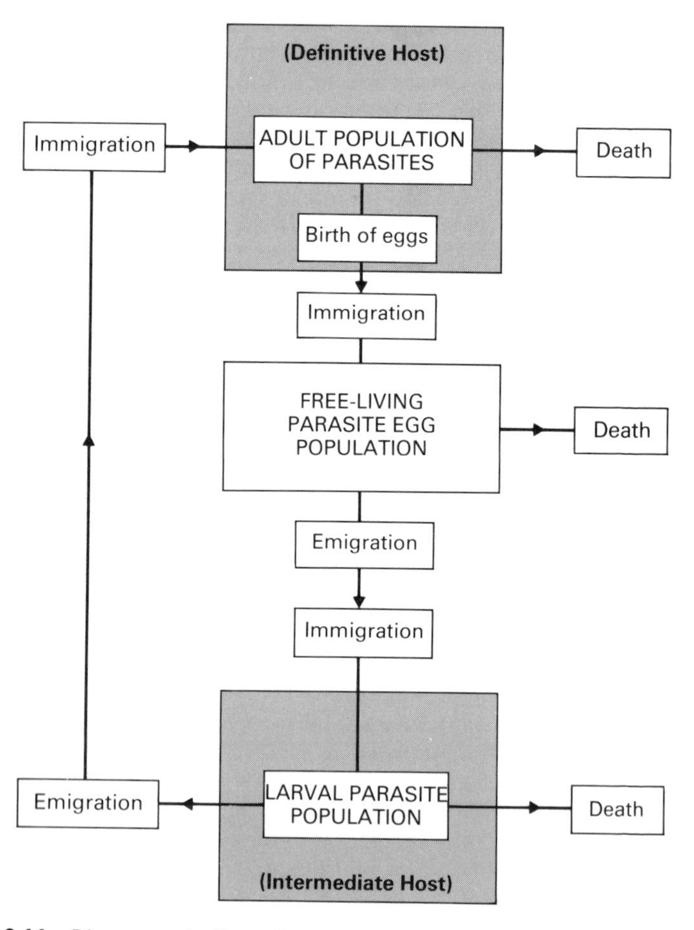

Fig. 6.11 Diagrammatic flow chart of the indirect life cycle of the cestode, *Caryophyllaeus laticeps* (see text). This representation displays the subpopulations of the parasite and some of the population parameters that control the flow of parasites through the cycle. (A simplified version of a diagram in Anderson.[8])

straightforward and simple-minded analysis of the quantitative way in which parasites flow through the life cycle. Happily, however, Anderson[8] has recently assembled a conspectus of modern findings and principles in this difficult area. Much of what follows in this section leans heavily on his synoptic account.

Figure 6.11 illustrates the subpopulations and population processes that exist in the indirect life cycle of the cestode, *Caryophyllaeus laticeps*. This helminth utilizes bream, *Abramis brama*, as a final (definitive) host and various tubificid annelids as intermediate hosts. Three parasitic subpopulations exist (adults, eggs and larvae), connected by three discrete immigration processes and one birth process. Each of the subpopulations is subject to a separate death process. The processes taken together, that is, births, deaths, immigrations and emigrations, are the absolute determinants of the parasite population dynamics. The rate parameters of each of these processes will, of course, be influenced by a wide variety of environmental and competitive factors and this makes any rigorous understanding of their natures immensely difficult. The establishment of these parameters and their functional relationships with other factors, however, is the *sine qua non* for meaningful quantitative results in the field of parasite populations dynamics. Figure 6.11 is, it will be noted, an incomplete map of the population dynamics of the *Caryophyllaeus*–bream-tubificid interactions. No account has been taken of the population processes of the two hosts or of the predator-prey interaction which links them and constitutes an obligatory link in the life cycle. For the purpose of this chapter attention will, however, be concentrated mainly on the population dynamics of parasites alone.

The rate parameters of the parasite's population processes can only be determined with any ease in the controllable conditions of the laboratory experiment. With the necessarily limited numbers of experimental hosts and parasites that can be maintained for such studies, random variability in the process can significantly affect the outcomes of experiments. Because of this problem, mathematical treatments of the data which take into account such variability (stochastic formulations) must supplement the more normal methods which assume infinitely large populations of animals (deterministic formulations).[8] This requirement has the depressing effect of moving the quantitative formulations into areas beyond the capabilities of the average biologist.

Parasite subpopulations

The size of many adult and larval parasitic subpopulations within hosts is determined by the combination of immigration (infection) and death processes. This type of causality produces quite different population dynamics from those occurring in free-living systems controlled by constant birth and death rates. The immigration-death process is relevant whenever the parasites concerned do not multiply themselves within the host under consideration. They may produce eggs or larvae but these do not mature in the host. Instead, they are passed out in an emigration process. In these conditions, dynamics of the following form will apply[8]:

$$\frac{dN_t}{dt} = \lambda - \mu N_t$$

Where dN_t/dt is the rate of change of the number of parasites (N_t) in the system at time t, λ is the immigration rate of new individuals and μ is the death rate of individuals in the subpopulation considered. In an interaction with characteristic values of immigration and death rates, it is inevitable that an equilibrium population N^* will be achieved such that $N^* = \lambda/\mu$. This relationship has been experimentally tested using the *Transversotrema* model system.[11] By maintaining a constant immigration rate of approximately two new adults per fish per week and directly estimating the rise of the adult fluke population an equilibrium value of the adult population was determined (Fig. 6.12). The mean death rate for the stable age distribution of flukes, which is obtained once the equilibrium population size is achieved, was determined independently. This death rate was estimated indirectly by observing the survivorship of synchronously established populations of 14 adult parasites on individual fish. The experimentally observed equilibrium population size correspond closely to that predicted by the $N^* = \lambda/\mu$ relationship.

Some parasites, of course, do multiply within a host. Larval digeneans in molluscs and larval cestodes in vertebrates do so. In addition, all parasitic protozoans and bacteria have multiplicative phases within hosts. If this extra population process is playing a part in the determination of population size, different population growth characteristics are the result such that[8]:

$$\frac{dN_t}{dt} = \lambda + (\alpha - \mu)N_t$$

where α is a constant birth rate per parasite per unit of time. In these circumstances, the equilibrium value of the population size ($\hat{N}$) is given by

$$\hat{N} = \lambda / -(\alpha - \mu)$$

The behaviour of this model depends on the arithmetic relationships

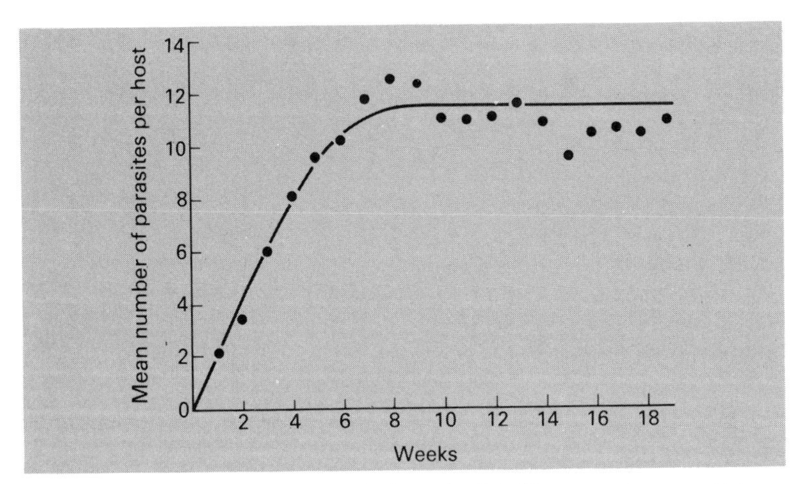

Fig. 6.12 The growth of a population of adult *Transversotrema patialense* (denoted as the mean number of parasites per fish) subject to a constant weekly immigration rate of 2.14 parasites per host. (●) represent experimentally observed points and (———) prediction of an age-structured immigration-death model which takes account of the age-dependent nature of the death rate of adults. (Based on a figure in Anderson *et al.*[11]) This paper also gives a fuller account of the immigration-death model.)

between μ, the death rate and α, the birth rate. If μ is greater than α the term, $-(\alpha-\mu)$, is positive and a plateau equilibrium population results. If α is greater than μ, the term is negative and no equilibrium value is possible. Instead, the system demonstrates exponential growth. In practice, of course, unconstrained exponential increase cannot occur in a parasitic subpopulation. The finite carrying capacity of an individual host means that eventually either such growth will kill the host and hence all the parasites, or increasing parasite densities will reduce the population growth rate.

Death rates

If μ, the death rate, were constant for a parasite population, the latter would decline exponentially in the absence of any new immigrants (see Fig. 6.13). This behaviour, however, is rarely encountered because many factors make the death rate non-constant.

Age-dependent death rates are common and several examples have been noted in which μ increases exponentially with time (see Fig. 6.14). This type of age dependence may be due to the exhaustion of finite nutrient reserves in a non-feeding but active larval stage, immune processes, which after a refractory period operate with

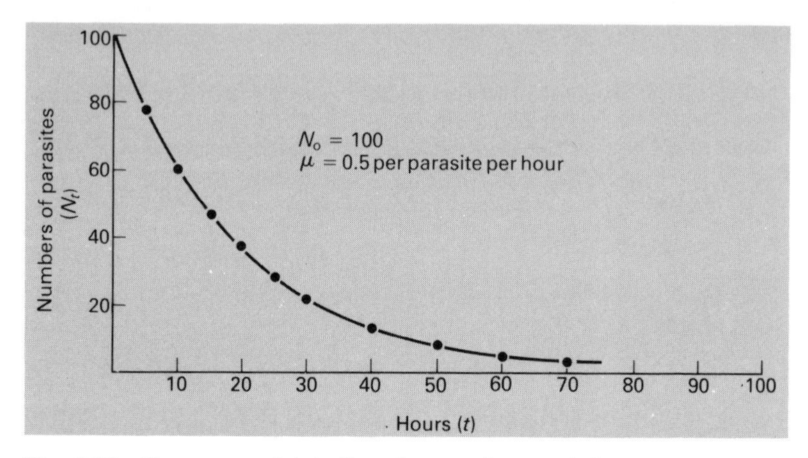

Fig. 6.13 The exponential decline of a parasite population experiencing no immigrations and possessing a constant death rate per parasite per unit of time. In such circumstances, the population decline is determined by

$$N_t = N_0 \exp(-\mu t)$$

where N_t is the number of parasites at time t, N_o is the initial number and μ is the death rate.

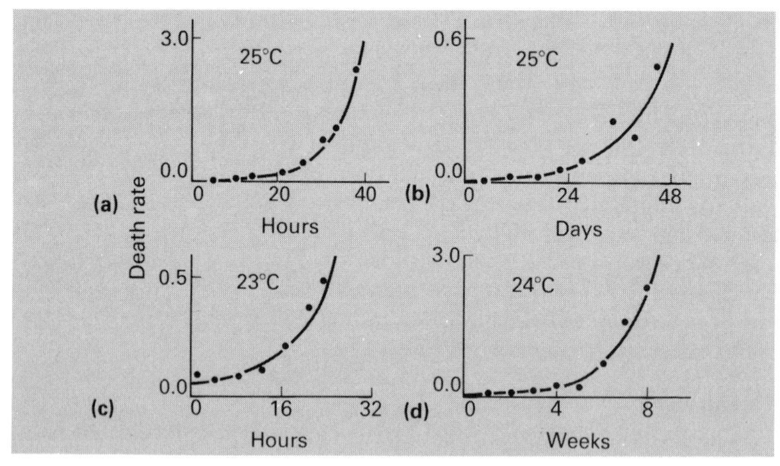

Fig. 6.14 A selection of parasite instantaneous death rates which increase exponentially with increasing parasite age. In each case, (●) denotes experimental points and (———) the best exponential fit to the data. (**a**) Cercariae of *Transversotrema patialense*, death rate per 4 hours. (From Anderson and Whitfield.[10]) (**b**) L_3 larvae of *Bunostomum trignocephalum*, death rate per day. (From Narain, B. (1965). *Parasitology*, **55**, 551–8.) (**c**) Miracidia of *Schistosomatium douthitti*, death rate per hour. (From Oliver, J. H. and Short, R. B. (1956). *Expl Parasit.*, **5**, 238–49.) (**d**) Adult *T. patialense*, death rate per week. (From Anderson *et al.*[11]) (redrawn from Anderson.[8])

increasing potency, or several other possible mechanisms including the portmanteau concept of 'simple' senescence. One effect of an age-dependent death process of this type can be gauged from Fig. 6.15 which is an example of the type of survival resulting from an exponentially increasing μ with age. Although this figure represents survivorship it also represents the stable age structure at the equilibrium population size given a constant immigration rate. From this it can be seen that, demographically, populations of this sort will have a preponderance of young individuals.

Temperature-dependent death rates have been described, especially from parasites in poikilothermic hosts. The death rate of adult *Caryophyllaeus* in bream, for instance, rises exponentially with rising temperature[6] (Fig. 6.16). This means that with ambient temperatures in a water body fluctuating cyclically over 12 months due to climatic changes, the death rate will also be a periodic function of time. This relationship can produce cyclical seasonal oscillations in the

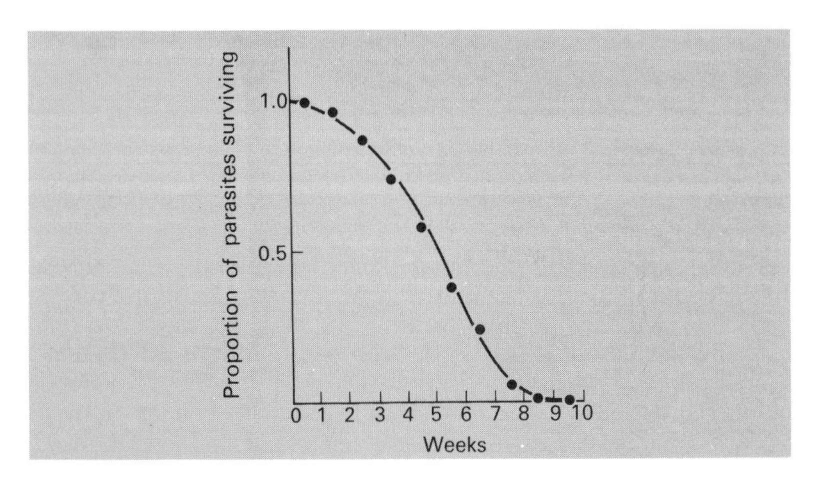

Fig. 6.15 The survival characteristics of a population of adult *Transversotrema patialense* on the fish final host, *Brachydanio*, at 24 °C. The initial parasite density was 14 parasites per fish. (●) represents experimental points, and (———) the expected proportions predicted by an age-dependent survival model which assumes an exponentially increasing instantaneous death rate. The model is of the form:

$$P_t = \exp\left[\frac{a}{b}\left(1 - \exp(bt)\right)\right]$$

where P_t is the proportion surviving at time t and a and b are the constants in the exponential equation for the death rate: $\mu(t) = a \exp(bt)$ (Adapted from a figure in Anderson *et al.*[11])

population size. The population processes can be treated formally thus:

$$\frac{dN_t}{dt} = \lambda - \mu(t)N_t$$

where (t) is a death rate which is a periodic function of time.

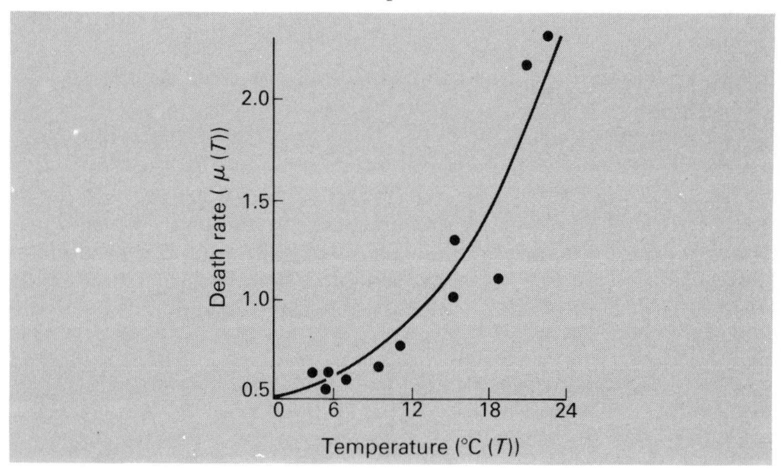

Fig. 6.16 The relationship between water temperature (T) and the death rate of adult *Caryophyllaeus laticeps* in the gut of the bream. (●) are experimental points, (———) is the best fit exponential model of the form $\mu(T) = a \exp(bT) + c$ where *a, b* and *c* are constants. (From Anderson.[6])

Similarly, almost any other significant environmental parameter can affect the death rate; outside hosts, temperature, salinity and pH are important. On and within them, host defensive responses can markedly increase death rates. It will usually be the case that the overall resultant death rate of a particular parasitic subpopulation is influenced by a series of factors which may not all be independent of one another.

Immigration (infection) rates

The infective population processes of parasites can be of several types. Table 6.3 categorizes some of these. In every type, except for that of the chance contact of a passive, free-living infective state with a host, a behavioural interaction between host and parasite or between two hosts is implicated in the parasite immigration process. This being the case, a quantification of the influences that determine a particular immigration rate is an onerous experimental task. The simplest possible relationship would be one where the rate of infection was directly proportional to the population density of the parasitic infective

Table 6.3 Some parasite infection processes.

Transmission stage of parasite	Mode of transmission to host	Examples
1 Passive, free-living infective stage	Chance contact	Rust spores (fungus) settling on plant host
2 Passive, free-living infective stage	Parasite actively predated by host	Ingestion of *Ascaris* (nematode) eggs by man.
3 Active, free-living infective stage	Parasite actively predated by host	Coracidium larva of *Diphyllobothrium* (cestode) is ingested by copepods.
4 Active, free-living infective stage	Parasite penetrates or attaches to host	(i) Schistosome (digenean) cercariae penetrate mammalian skin (ii) Transversotrematid (digenean) cercariae attach to fish surface
5 Passive or active larval stage in blood-feeding vector	Vector mediates infection during feeding bout on final host	L_3 larvae of *Loa* (nematode) transferred to human blood by the fly *Chrysops*
6 Passive or active larval stage in blood of final host	Vectors mediates infection during feeding bout on final host	L_1 microfilariae of *Loa* acquired by *Chrysops* from human blood
7 Passive or active larval stage in parent host	Within-parent parasite transfer to offspring	Human, trans-placental congenital *Syphilis* (spirochaete)
8 Passive infective stage in or on intermediate host	Final host ingests intermediate host and parasite simultaneously	Ducks feeding on *Gammarus* containing *Polymorphus* (acanthocephalan) cystacanths

stage. It is probable that this happy state of affairs only occurs under carefully controlled laboratory conditions (see Fig. 6.17, for instance). In the real world the size, age and behaviour of individuals in the target host population will affect infection success as will the age structure of the population of infective stages.

Transversotrema cercariae have an age-dependent infection rate when assessed in standardized infection conditions against size-matched fish (see Fig. 6.18). Infection rates are not simply or directly dependent on either cercarial survival or total activity levels, but can be related to both if it is assumed that all three indices are related to the exponential depletion of a finite nutrient reserve, glycogen.[10]

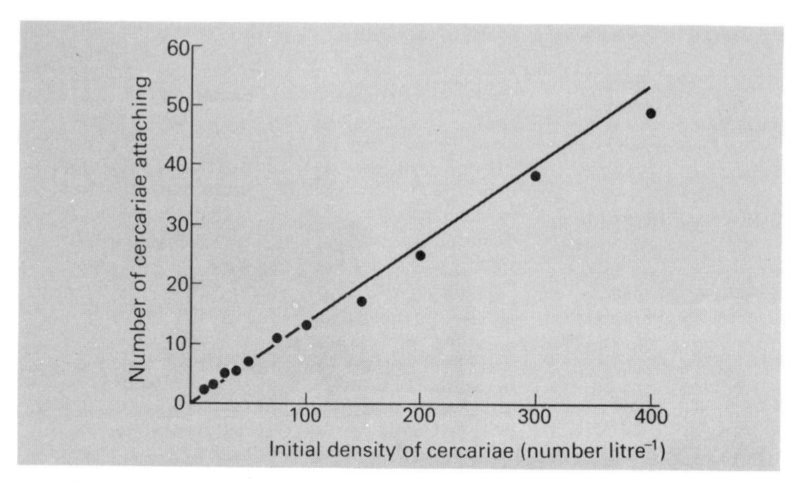

Fig. 6.17 *Transversotrema patialense.* The relationship between initial cercarial density and the number of cercariae attaching to five fish. In the experiments five fish were placed in one litre of water with different initial numbers of cercariae. After 5 minutes the fish were removed and the total number of parasites that had attached to the five fish were counted. (●) indicate experimental points and (———) the best straight line to fit to the data. (Anderson, Whitfield and Dobson, unpublished results.)

The size and age structure of free-living populations of active infective stages like cercariae will depend upon patterns of larval release from the antecedent host. The release of cercariae from snails has been shown to be influenced by a multiplicity of factors and some of these are demonstrated in Fig. 6.19.

Birth rates

The production of new individuals by a reproductive process in a parasitic life cycle can either directly add new members to the population that produced them, or the new individuals can emigrate and enter the next subpopulation in the life cycle. Examples of the former situation include the multiplicative stages of parasitic protozoa, proliferative larval tapeworms like hydatid cysts, and monogeneans like *Gyrodactylus* which viviparously produce juvenile worms. The second category is extremely common and includes all cases where helminths produce eggs or larvae which leave the host by a variety of routes. Microfilariae leave via the blood, the eggs of *Schistosoma haematobium* by a urine route, those of *Fasciola* in the faeces. In the bizarre case of *Ollulanus*, a stomach nematode of the cat, eggs leave the host in the cat's vomit!

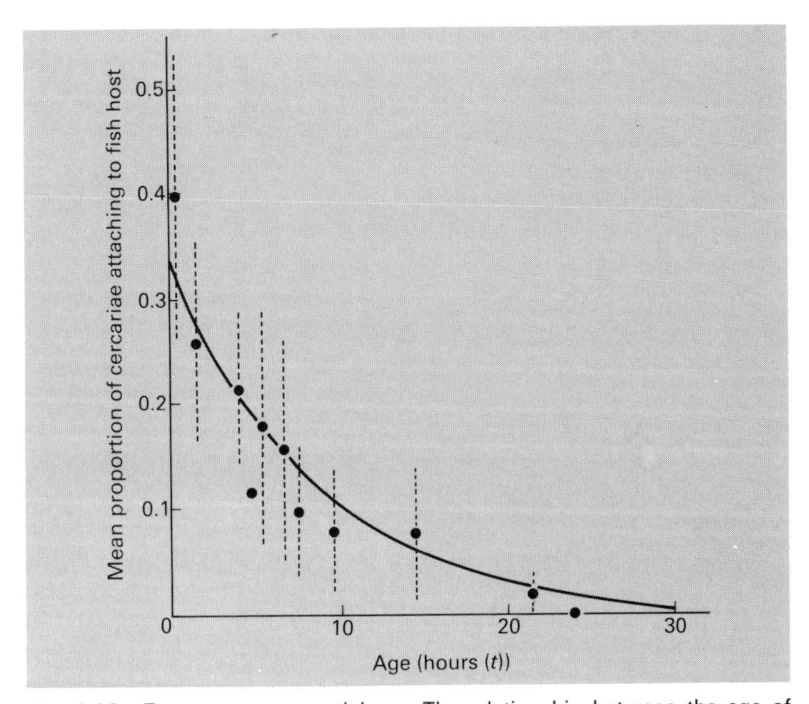

Fig. 6.18 *Transversotrema patialense.* The relationship between the age of cercariae and the mean proportion of a group of similarly aged cercariae which attach to a fish in a unit period of time (30 minutes). (●) indicate experimental point with 95% confidence limits and (——) the best exponential fit to the data of the form:
proportion attaching = $x \exp(-yt)$ where $x = 0{\cdot}332$ and $y = 0{\cdot}114$. (From Anderson *et al.*[11])

An ectoparasitic example from fish will give some conception of the complexity of the determining factors which control birth processes.

Adult *Transversotrema* on the surface of the teleost fish, *Brachydanio rerio*, have a maximum life span of about 9.5 weeks at 24°C.[11] If the egg output per worm per day is plotted against the time course of a synchronously applied infection of 14 worms, Fig. 6.20 results. A complex pattern of-age dependence in egg output is immediately apparent. It will be seen that after a refractory (prepatent) period of about two or three days egg production begins with an increasing rate. By about three weeks post infection a plateau of egg production of about 18 eggs per worm per week has been reached which then tapers off in the second half of the life span.[11] At present, it

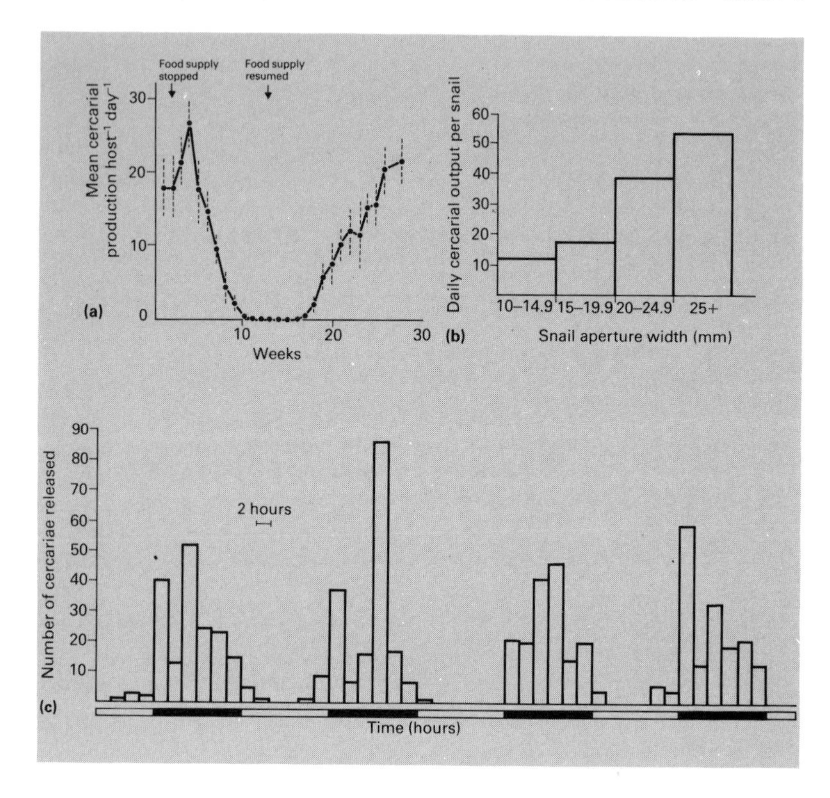

Fig. 6.19 Factors influencing the release of cercariae from snails. (**a**) Snail food levels. The influence of host starvation on the daily cercarial production (*Transversotrema patialense*) by individuals of *Melanoides tuberculata.* (●) are mean daily cercarial production rates, estimated over one week periods for four infected snails, with 95% confidence limits of the means. The cessation of the snails' food supply produces a slow decline in the production of cercariae until no cercariae are released. On resumption of the food supply, the output is reestablished. (From Anderson *et al.*[11]) (**b**) Snail size. The relationship between shell size of naturally infected *Melanoides* in Penang, Malaysia and the mean daily output of *Transversotrema* cercariae by the snails. (Data kindly supplied by Dr. C. Betterton, Universiti Sains, Penang, Malaysia.) (**c**) Dark-light cycles. The relationship between the number of cercariae (*Transversotrema patialense*) released from a group of five snails (*Melanoides tuberculata*) in successive 2 hour periods while subject to a LD 12:12 regime indicated by the solid (dark) and open (light) bars under the histogram. The great bulk of cercariae are released in the dark periods. (Whitfield and Anderson, unpublished results.)

is difficult to disentangle the complex of developmental factors which influence the form of this age-dependent birth rate. The prepatent phase is probably the consequence of both the time necessary for the readjustment of cercarial physiology from a freshwater-tolerant to a fish surface-tolerant state and also for the differentiation of vitelline

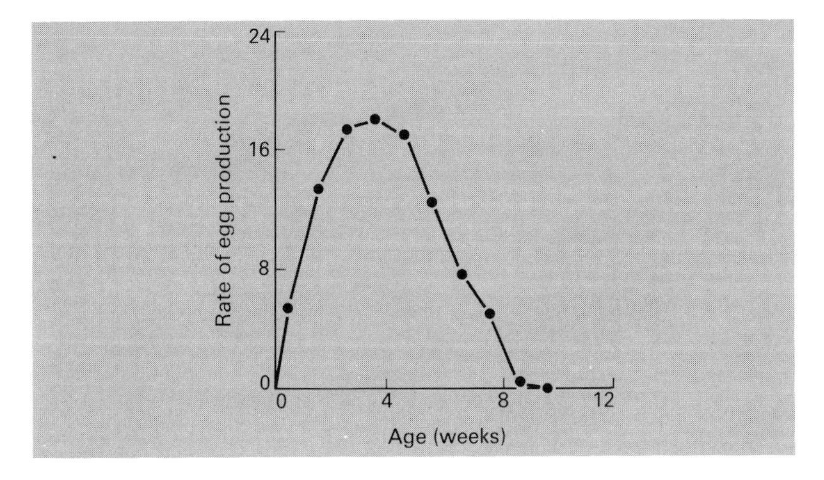

Fig. 6.20 The age-dependency of the fecundity of adult *Transversotrema patialense*. (●) represent the mean rates of egg production per parasite per week. (Redrawn from Mills, C. A. (1977). *Ph.D. dissertation.* University of London.

cells. Oocytes and sperm are already present in the free-swimming cercaria. The period of rising egg-release rate is concurrent with an increase in fluke size and the number of differentiated vitelline cells. Causal factors that contribute to the plateau and decline phases are not at present understood.

The regulation of parasite and host populations[8,41]

Populations of parasitic organisms are regulated. Most of the host-parasite relationships that have been observed carefully in recent historical time both persist and do not show indefinite increases in parasite numbers. These simply-made observations demonstrate that regulation must be operating. If it were not, parasite populations would be frequently observed to increase indefinitely with the eventual extinction of the host species or to become extinct themselves.

It is an accepted tenet of modern ecology that two distinctly different types of influence can affect the sizes of populations. One type, the so-called density-independent influences, include factors such as climatic changes. They operate outside the context of the population itself, imposing an arbitary, external effect on population size. The other, converse, type is termed a density-dependent influence. It is an influence on a population process which is in some way a function of the population size itself. Among free-living animals it is often true that density-independent influences determine the directions of perturbations in population size, whereas particular

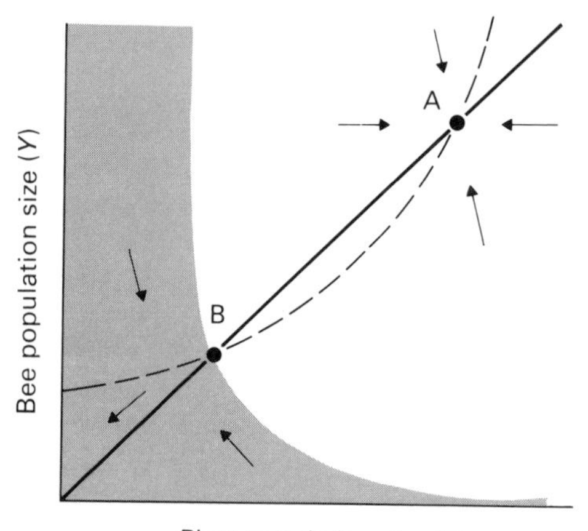

Fig. 6.21 A summary of the stability properties of a model for a mutually obligatory symbiosis between a bee and a plant population. The model was produced and discussed by May.[233] The bee population is limited by a carrying capacity proportional to the size of the plant population. As this is so, possible equilibrium values of Y can be plotted as (———) where $dY/dt = 0$. If the plant population density is too low, bees may have difficulty in finding more than one plant and mediating necessary cross-fertilization. Plant reproduction may thus fall below replacement levels. Equally, there will be a maximum permissible plant population size determined by environmental factors such as available space which are independent of the bee population size. Taking these two influences together, the possible equilibrium values of X, the plant population size, follow a curvilinear path (— — —) where $dX/dt = 0$.

The directions in which plant and bee population trajectories move among the possible concomitant values of X and Y are shown by the arrows. All pairs of population values originating *outside* the tinted area are attracted to the stable point A. The other intersection of $dY/dt = 0$ and $dX/dt = 0$, that is point B, is unstable. All population trajectories originating *inside* the tinted area are inexorably drawn to the origin. In other words both species become extinct. Taken alone, this model suggests that mutually obligatory symbioses of this type will only demonstrate long-term persistence in relatively unchanging environments. In fluctuating environments, the boundary into the tinted danger zone would eventually be crossed and extinction would occur.

types of density-dependent influences provide an important source of restoring regulatory forces that bring about a population's ability to counteract perturbations. It has been argued[12] that in special circumstances density-independent regulation can also occur. Some biologists have used the language of cybernetics in this area and called density-dependent regulatory forces negative feedback because this is often the nature of their effect.

How can these ecological concepts, developed for free-living systems, be applied to intimate associations? Both Bradley[40–1] and Anderson[8] have recently reviewed this question with relation to host-parasite systems and the following discussion mainly reflects their ideas. May[233] has also considered the theoretical aspects of the stability properties of models for symbiotic associations (for example see Fig. 6.21).

Bradley[40] has suggested that the regulation of parasitic populations can occur in three different ways. The mechanisms that he proposes can act independently or together. His Type 1 regulation involves complex density-independent regulation due to the combined, out of phase effects of a large number of separate external influences such as temperature and humidity. Type 2 regulation is regarded as regulation brought about by the increased likelihood of host deaths in those individuals carrying unusually large numbers of parasites, a similar concept to Crofton's[81–2] lethal level. Bradley has pointed out that host individuals which exhibit so-called sterile immunity are approximately equivalent in a regulatory sense to a host that dies due to a large parasitic burden. Bradley's Type 3 regulatory mode is linked with density-dependent control of within-host parasite numbers.

To take an example whose mechanism has already been described in some detail (see pages 130–5), adult schistosomes of man certainly are subject to this form of regulation. Early schistosome colonizers of a host can evade immune attack by 'disguising' themselves with host antigens. All subsequent penetrating cercariae, however, are likely to be killed by the host's powerful immune response. Thus, the interaction of host defences and parasitic counter-measures produces a profound constraint on adult parasite numbers whatever the external numbers of cercariae.

Looked at from a wider perspective, it is obvious that Bradley's type 2 and type 3 regulation modes are density-dependent regulatory processes whereas type 1 represents a low grade form of density-independent regulation. Theoretical studies[8] have shown that direct or indirect parasite life cycles in which no density-dependent influences are present must necessarily be unstable. If this mathematical conclusion can be extrapolated to the real world, it suggests that density-dependent influences must be operating somewhere in the Bradley type I regulatory mode.

The concomitant immunity phenomenon in schistosome infections provides an example of a density-dependent regulation which is largely produced by the host. Intraspecific competitive effects, however, can also generate such processes and they may operate on any population process of the parasites. Fig. 6.22 brings together several examples of density-dependent alterations of population processes

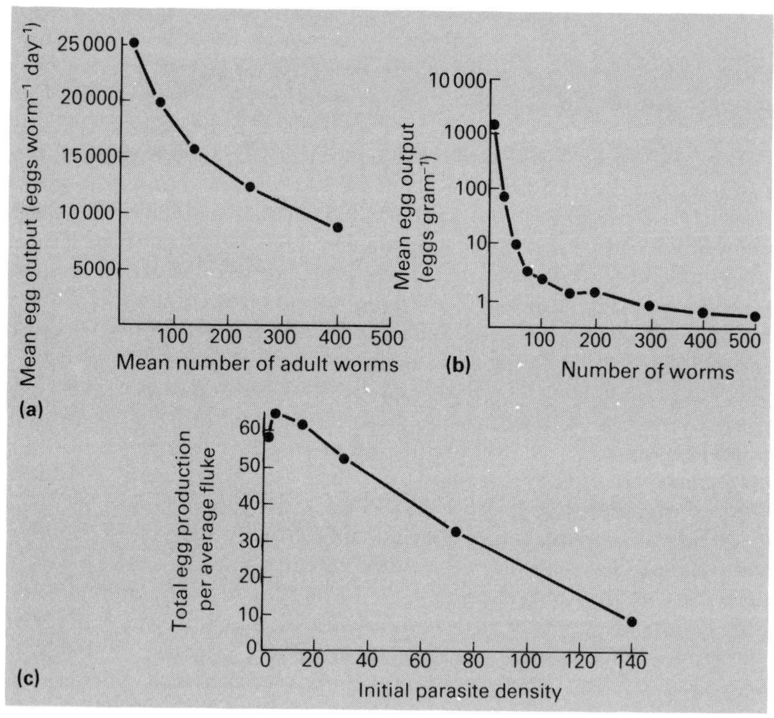

Fig. 6.22 Effects of intraspecific influences on parasite population parameters. (**a**) Relationship between mean daily egg output per adult worm of *Fasciola hepatica* and the mean number of worms present in the definitive host (sheep) 13–19 weeks post infection. (Based on data in Boray, J. C. (1969). *Adv. Parasit.*, **7**, 96–210.) (**b**) Relationship between mean egg output per gram of *Hymenolepis nana* and the number of worms per mouse. Infections initiated with cysticercoids. (Redrawn from Ghazal, A. M. and Avery, R. A. (1974). *Parasitology*, **69**, 403–16.) (**c**) *Transversotrema patialense.* The relationship between the initial density of flukes (parasites per fish) and the total lifetime egg production of an average worm in the resulting infection. (From Mills, C. A. (1977). *Ph.D. dissertation.* University of London.)

which have been quantitatively characterized, and which must have some regulatory properties.

Anderson[9] has recently produced a theoretical analysis of the regulatory influence of parasites on host population growth. This study indicated that overdispersion of parasite numbers per host, density dependent parasite mortality or reproduction and parasite-induced host mortality that increases more than linearly with parasite load can all stabilize host population growth. Other phenomena such as parasitic castration of hosts, parasite reproduction within a host which directly increases the parasite burden, and time delays in the larval transmission stages were shown to have destabilizing effects. Anderson has suggested that the population rate parameter values that one observes in real world host-parasite interactions are not haphazard. He feels that only sets of values which mean that destabilizing influences are more than offset by stabilizing ones will actually exist and persist.

At a level of consideration which, at present, is more philosophical than experimentally tractable, is the question of the intrinsic stability or otherwise of complex parasitic life cycles. Posing the problem more specificially, we can compare *Entobdella* and a strigeid digenean like *Cotylurus*. The former parasite has only one host and its larvae are directly infective to that same host type. The digenean example has a final host and at least two intermediate hosts. Both of these must be traversed developmentally before the definitive host species can be re-engaged. Does the complexity exemplified by the digenean system beget stability or does it destabilize? Verbal, qualitative arguments can be proposed to support either conclusion. Complexity could, on the one hand, provide multiple possibilities for density-dependent effects and hence extra regulatory power. On the other hand, the linking of development to the necessary continued presence of three other types of animal might seem a dangerously vulnerable strategy in the face of the known extinction rates of species. The mathematical treatment of this dichotomy of possibilities has been discussed by May.[234] He considered that there is no underlying, theoretical reason why complexity should imply stability. This, although an interesting and important conclusion, does not help us explain the exuberant superabundance of complex parasitic life cycles that persist today.

The overwhelming applied significance of parasite population regulation is self-evident. Hundreds of millions of people today are infected with debilitating, crippling and killing parasitic diseases. Hundreds of millions of pounds have been spent on attempts to control or eradicate these diseases. Many of the relative failures in the field of control probably stem from a failure to realise that many, if not most, of these diseases are highly regulated. If this is so, attempts to break the cycle of parasitic transmission by attacks on vectors or critical points in the life cycle are liable to be of doubtful benefit unless the nature of the regulatory processes is understood. It is in the very nature of density-dependent regulation, for instance, to show great resilience in the face of violent attempts to perturb the parasite population size. In this light, Bradley[41] has argued strongly that the acquisition of information on regulation must be considered a vital component of the development of control measures.

EVOLUTIONARY ASPECTS OF ASSOCIATIONS

Many evolutionary processes have been implied or explicitly described above in connection with parasite distributions, niche biology and population dynamics. Some specifically evolutionary topics, however, remain to be discussed separately. It is encouraging that in one of these topics symbiotic relationships rather than parasitic ones are the experimental subjects of choice. Hard evidence is a rare commodity in discussions on evolutionary relationships and all such exercises must be undertaken with the knowledge that most speculations will be extremely difficult to test. Read[284] summarized some of these difficulties when he wrote (speaking of symbiosis in the general sense outlined on page 7): 'To treat the evolution of symbiosis is a very difficult task. The symbiote and its host have undergone evolution, the former with varying degrees of linkage to host evolution. . . . Any discussion of the evolution must be highly speculative'.

The evolution of associative ways of life

Intimate associations of organisms are of ubiquitous occurrence in the living world. The wide and discontinuous distribution of these relationships among the different taxa of plants, animals and micro-organisms suggest that the associations have evolved independently and on numerous occasions.

Within any particular taxon it is reasonable to assume that associative modes of existence have arisen from the free-living ways of life of pre-existing members of the group. In groups like the flatworms, nematodes, fungi, algae and insects, associating and non-associating types both still exist today. In the light of such coexistence one can at least make informed speculations about the evolution of the ways of life of the associating types. Other groups like viruses, acanthocephalans and mesozoans are entirely associative in their habits. Ideas about the origins of these taxa and the selective pressures which shaped their evolution as associating organisms must be regarded as almost unverifiable conjectures.

In some groups of animals it is possible to make patterns of possible evolutionary descent which link actual modes of life that are employed today. These patterns which link free-living existences to associating ones can be regarded as models of the changes which have occurred on an evolutionary time scale.

Among crustaceans, insects, arachnids and molluscs it is likely that one much-used evolutionary pathway has been free-living animal-ectocommensal-ectoparasite-endoparasite. Among the mites, for instance, one can find examples of free living forms, and species which belong to a so-called 'phoretic guild' of mites. These latter types do not feed on their small mammal 'hosts' but simply hang onto them as a dispersal mechanism. Ectoparasitic mites of the same small mammals feed on hair and skin externally and as the culmination of this sequence we have endoparasitic forms like *Demodex* that live inside their hosts.

The above 'transdermal' route to endoparasitism seems, however, to have been less common than what might be called an ingestive route. So many modern-day endoparasites gain entry to a host as eggs or larvae when the host eats them, that it is difficult not to conclude that this route has evolutionary significance. One may imagine that in the evolutionary past a variety of free-living organisms were ingested by animals and were not necessarily killed in the process. Endosymbiosis or endoparasitism could arise in this way if the colonizing individuals had the physiological ability to withstand internal host conditions. Equally, parasites or symbiotes already living in association with an invertebrate might gain entry to the vertebrate gut by being ingested along with their partner.

Co-evolution of hosts and parasites

Several groups of parasites have segregated themselves from their phylogenetic neighbours by virtue of their early restriction to a particular line of host evolutionary development. Many examples of this phenomenon have been documented and they have usually been taken to imply a long-standing evolutionary coexistence of host and parasite. In other words, if a homogeneous and discrete group of parasites is only found in a similarly discrete taxon of hosts, and within these hosts it has had time to show adaptive radiation up to the generic or familial level, the period of coexistence must have been a long one. So, for instance, adults of the two, highly specialized, tapeworm families (Tetrarhynchida and Tetraphyllida) are only found in elasmobranch fish (Fig. 6.23).

One thorough-going analysis of the co-evolution of hosts and

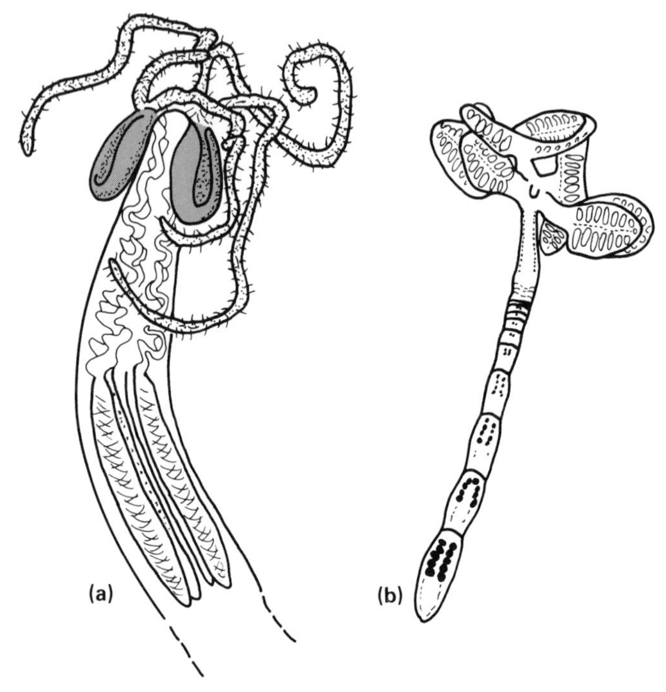

Fig. 6.23 (**a**) A tetrarhynchid cestode. The anterior end of *Christianella trygonis-brucconis* from the spiral valve of the sting ray, *Trygon brucco.* Note the four spiny, retractable proboscides. (Redrawn from Joyeux and Baer.[165]) (**b**) A tetraphyllid cestode. A whole worm of *Rhinebothrium hawaiiensis,* a form from the gut of Hawaiian sting rays. (Redrawn from Cornford, E. M. (1974). *J. Parasit.,* **60**, 942–8.)

parasites has been provided for the ectoparasitic insects of birds by Rothschild and Clay.[297] In discussing the possibilities of using patterns of parasitic speciation to provide clues about evolutionary relationships between hosts, these workers have carefully thought about many of the possible misinterpretations that can occur in this field. A consideration of the feather lice of birds illustrates the general concepts involved.

The ancestors of the birds differentiated in a distinctive way from other groups of reptiles about 120×10^6 years ago. At that period, in the Jurassic, forms like *Archeopteryx* were alive. Within another $50–60 \times 10^6$ years an immense adaptive radiation of birds had occurred and most of the modern orders were established. Today there are about 8600 species of birds and over the past 120×10^6 years many tens of thousands of additional bird species must have arisen and become extinct. There is good reason to believe that the evolution of the feather lice has lasted almost as long as that of birds (*Archeopteryx* fossils in lithographic limestone show perfectly normal-looking feathers). On the other hand, it appears that the pace of feather lice evolution has been slower than that of their more adaptable hosts. As a consequence of this differential speciation rate and the complete commitment of feather lice to life on bird feathers (see Chapter 3, page 84), related bird hosts now often harbour closely related mallophagan feather lice. So, for instance, the head-inhabiting feather lice of all the waders, a bird grouping containing many families, all belong to a single genus.

Some workers have suggested that this type of relationship can be used to disclose evolutionary affinities between birds when other independent evidence is equivocal. Rothschild and Clay[297] described specific examples of such attempts. One concerned the flamingoes. These water birds are parasitized by three genera of feather lice, *Anatoecus*, *Anaticola* and *Trinoton*, which have otherwise only been recorded from the anseriform birds, that is, ducks, geese and swans. This correspondence in parasitic fauna perhaps suggests a close evolutionary connection between the Anseriformes and the flamingoes. Such a suggestion, however, runs contrary to some taxonomic opinion which considers that the flamingoes should be allied with the storks and herons. These birds, though, have no feather lice genera which they share with either flamingoes or the Anseriformes. Evidence of this type must be used with a knowledge of the possible misunderstandings that are attendant upon its interpretation, but if these pitfalls are recognized it seems that a knowledge of host-parasite co-evolution can be a useful, ancillary taxonomic tool. It is a tool, however, that must be used with caution and in the context of other, independent evidence about lineages.

Symbiosis and the origin of eukaryotic cells

The study of organisms in intimate associations or symbiology, to use the more succinct term used by Read,[284] is intrinsically multidisciplinary. This being so, it is instructive and almost inevitable that the last topic to be discussed in this book should involve an amalgam of many conceptual backgrounds and investigatory techniques. The topic is the serial endosymbiosis theory of the origin of eukaryotic organelles[229–30,302] and the arguments of its promulgators and critics.

Simply stated, the theory suggests that three classes of eukaryotic organelles, namely mitochondria, cilia/flagella and photosynthetic plastids, have originated by hereditary endosymbiosis. Adherents to the theory and its detractors have used biochemical, cell biological, ultrastructural, palaeontological, ecological, genetic and evolutionary evidence mixed with a stimulating dash of pure speculation. At a symposium held at Bristol University, England, in September 1974, the proceedings of which have been published,[162,229,282] advocates from both sides of the controversy stated their views and reviewed the evidence.

The version of the theory proposed by Margulis[229] suggests that all eukaryotic organisms had anaerobic, fermentative, heterotrophic, prokaryotic ancestors. In the early Precambrian, adaptive radiation among prokaryotes would have led to the evolution of a range of different cell types. Among those that might have developed are mycoplasm-like fermenters that utilized anaerobic glycolysis, mobile spirochaetes, photosynthetic blue-green algae and aerobic bacteria that could oxidize small organic acids by the tricarboxylic acid (Krebs) cycle.

The first phase of serial endosymbiosis would involve the invasion of the cytoplasm of an anaerobic fermenter by an aerobic, Krebs-cycle-using bacterium. The genetic and functional integration of such endosymbiotes is supposed to have led to these aerobes becoming mitochondria-like organelles. Resulting from this integration of two organism-types would be a mitochondrion-containing cell that was ancestral to all other eukaryotes. It has been hypothesized by some workers that the selection pressure for this symbiote acquisition was the accumulation of oxygen in the previously anaerobic primitive atmosphere due to the activity of photosynthetic blue-green algae. Without the aerobic symbiotes the new atmosphere might have been distinctly toxic to the anaerobic fermenters.

Margulis then suggests that nuclear segregation from the cytoplasm occurred autogenously, that is by synthetic processes mediated by the original 'host' cell cytoplasm. After this, a second symbiotic acquisition would occur. Surface associations of motile spirochaetes with the mitochondrion-containing cell are presumed to have developed. Modern analogues of this process are known (Chapter 2, page 26–8). After much development and integration, including the passage of spirochaete genes into the central 'host' nuclear genome, the spirochaetes became the cilia and flagella of modern cells and contributed the necessary systems for the elaboration of all microtubule-containing structures. Especially important in this respect would be centrioles and spindles round which complex chromosomal mitosis and meiosis could be organized.

The final symbiotic step was the development of a photosynthetic organelle. It is surmised to have originated from the integration of blue-green algal symbiotes into the now-complex cell system.

Many variations on this plan of the theory have been suggested by basically sympathetic workers (see for instance Taylor,[344] but as stated above it serves as a foundation for more detailed arguments. Some of the more powerful arguments in favour of the theory are:

1) During the processes by which each symbiote type was acquired as an organelle a slow process of metabolic and genetic integration must have occurred. In these processes, the symbiote changes from a largely autonomous organism to an entity dependent on information and substrates supplied by the host cell. It is possible to regard the kappa particles in the cytoplasm of *Paramecium* as modelling a 'halfway house' in this integration process. These particles are largely autonomous and are inherited independently of the nucleus. They require, however, various nuclear-coded products of the *Paramecium* for their maintenance.
2) Mitochondria contain their own genetic system. This consists of specific (mt) DNA as well as the means for its replication and probably for its protein-producing expression. The organelles contain a DNA-dependent RNA polymerase, ribosomes, mRNA, tRNA and all the other factors needed for protein synthesis. What is more, mitochondria have a sexual process, in the sense that recombinant DNA with new linkage patterns can occur when two dissimilar sets of mitochondria meet at a cell cross. All this biochemical machinery can be interpreted as being the remnants of a previously autonomous cellular organization present in an aerobic bacterial symbiote.
3) At the geological interface between the Precambrian (3–0.6×10^9 years ago) and the Phanerozoic (0.6×10^9 years ago until today) there is a major discontinuity in the microfossil record. This could be due to the explosive adaptive radiation of eukaryotes after they had acquired their new, symbiotically derived organelles.
4) Present-day bacteria, such as *Paracoccus denitrificans*,[162] possess respiratory chains, patterns of oxidative phosphorylation and cytochrome structure which are remarkably similar to those of modern eukaryotic mitochondria.

Counter-arguments to the theory have been based principally on biochemical evidence. This poses a problem because the formulation of the theory itself has developed in the main from comparative ultrastructural ideas and it is not always possible to integrate the two sets of arguments. Raff and Mahler[282] have produced the most telling

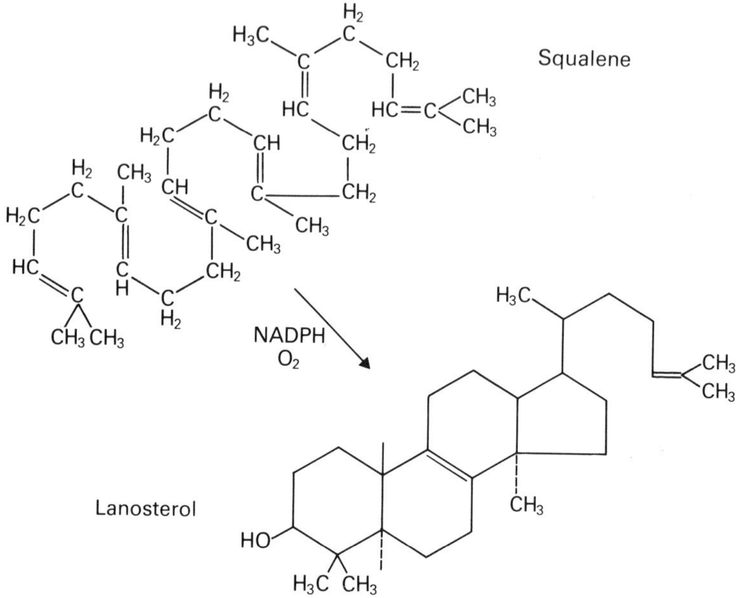

Squalene

NADPH
O_2

Lanosterol

Fig. 6.24 The cyclization of the hydrocarbon squalene to the tetracyclic steroid lanosterol. The reaction takes place under the influence of the enzyme squalene oxidocyclase and requires molecular oxygen.

arguments against the central precepts of the theory mainly in relation to the origin of the mitochondrion. They are:

1) No modern examples exist of endosymbioses in which a prokaryote is the host cell.
2) It is difficult to imagine mechanisms which would have enabled the large scale transfer of genes from symbiote to host nucleus to occur. This process is a prerequisite for the serial endosymbiosis theory.
3) The theory predicts that the background eukaryotic cytoplasm in contrast to the mitochondrion, should be of fundamentally anaerobic organization. This requirement follows from the contention that the primitive anaerobic fermenting prokaryote acquired its oxygen-utilizing systems from the aerobic 'protomitochondrion'. Unfortunately for the theory it can be shown that the background cytoplasm of most eukaryotic cells is committed to an intrinsically aerobic biochemistry. The cells contain oxygen-detoxifying enzymes. The cytoplasm itself

carries superoxide dismutases (see Chapter 4, page 124) and catalases for protection against the highly reactive superoxide radical. In fact all modern, strictly anaerobic bacteria possess neither superoxide dismutase nor catalase.[214] It is thus particularly odd that the matrix cytoplasm of eukaryotic cells contains both if the endosymbiosis theory is correct.

4) There are biosynthetic pathways in the background, non-mitochondrial cytoplasm of eukaryotes and also in prokaryotes which have an obligate requirement for molecular oxygen. The final cyclization step in the synthesis of steroids is a case in point (Fig. 6.24). No anaerobic analogue of this crucial synthesis is known from any organism. The universal presence of the aerobic pathway in eukaryotic cells is difficult to square with the requirement of the endosymbiosis theory that the 'host' cell should be intrinsically anaerobic.

5) Evidence exists which provides backing for an alternative explanation of the origin of genetic and synthetic machinery in the mitochondrion. Firstly, the various mitochondrial RNA polymerases that have been studied are quite different from both eukaryotic nuclear and bacterial polymerases. Also, there is evidence in the eukaryotic yeasts that the mitochondrial RNA polymerase is coded for on a nuclear gene. Secondly, ribosomal evolution since the Precambrian appears to have consisted of three divergent sets, cytoplasmic eukaryotic, mitochondrial eukaryotic and prokaryotic. There are no clearly closer links between eukaryotic mitochondrial ribosomes and prokaryotic ribosomes than there are between prokaryotic types and eukaryotic cytoplasmic ribosomes.

   The alternative theory of mitochondrial origin suggests that portions of the plasma membrane of a proto-eukaryote had respiratory systems built onto their inner surfaces. If these regions budded inwards to form more efficient closed systems, problems could arise because of the limited permeability of the membrane. In particular, it would be a barrier to ribosomes and ribosomal RNA. These would need to be on the inside of the closed organelle as subunits of cytochrome oxidase appear to require synthesis *in situ*. Thus, a genetic and protein-producing system would be required within the protomitochondrion. Raff and Mahler[282] suggest that the organelle was donated a stable plasmid containing the appropriate genes for ribosomal construction.

6) Many present-day prokaryotes have membrane-bound organelles. Photosynthetic bacteria have thylakoids and many

bacterial types have mesosomes. These are developed from intuckings of the outer plasma membrane. Such systems could be analogous to the untucking postulated during the autogenous method of mitochondrial production.

Other arguments, counter-arguments and rebuttals exist on both sides. The debate continues and the need to test hypotheses generates ideas for new experiments. This active area of theoretical conceptions and experimental investigations could serve as an illustrative microcosm of the whole field of modern symbiology.

References

1. ABRAM, D., CASTRO e MELO, J. and CHOU, D. (1974). Penetration of *Bdellovibrio bacteriovorus* into host cells. *J. Bact.*, **118**, 663–80.
2. AIKAWA, M., HUFF, C. G., and SPRINZ, H. (1967). Fine structure of the asexual stages of *Plasmodium elongatum*. *J. Cell Biol.*, **34**, 229–49.
3. AIKAWA, M. and STERLING, C. R. (1974). *Intracellular Parasitic Protozoa*. Academic Press, New York.
4. AIKAWA, M. and THOMPSON, P. G. (1971). Localization of acid phosphatase activity in *Plasmodium berghei* and *P. gallinaceum*: an electron microscopic observation. *J. Parasit.*, **57**, 603–10.
5. ANDERSON R. M. (1974a). Population dynamics of the cestode *Caryophyllaeus laticeps* (Pallas, 1781) in the bream (*Abramis brama* (L)). *J. Anim. Ecol.*, **43**, 305–21.
6. ANDERSON, R. M. (1974b). Mathematical models of host-helminth parasite interactions. In *Ecological Stability*, USHER, M. B. and WILLIAMSON, M. H. (eds). Chapman and Hall, London.
7. ANDERSON, R. M. (1974c). An analysis of the influence of host morphometric features on the population dynamics of *Diplozoon paradoxum*. (Nordmann, 1832). *J. Anim. Ecol.*, **43**, 873–87.
8. ANDERSON, R. M. (1976). Dynamic aspects of parasite population ecology. In *Ecological Aspects of Ecology*, KENNEDY, C. R. (ed). North Holland Publishing Co., Amsterdam.
9. ANDERSON, R. M. (1978). The regulation of host population growth by parasitic species. *Parasitology*, **76**, 119–57.
10. ANDERSON, R. M. and WHITFIELD, P. J. (1975). Survival characteristics of the free-living cercarial population of the ectoparasitic digenean, *Transversotrema patialense* (Soparkar, 1924). *Parasitology*, **70**, 295–310.
11. ANDERSON, R. M., WHITFIELD, P. J. and MILLS, C. A. (1977). An experimental study of the population dynamics of an ectoparasitic digenean *Transversotrema patialense*: the cercarial and adult stages. *J. Anim. Ecol.*, **46**, 555–80.
12. ANDREWARTHA, H. G. and BIRCH, L. C. (1954). *The Distribution and Abundance of Animals*. University of Chicago Press, Chicago.
13. ARME, C. (1975). Tapeworm – host interactions. *Symp. Soc. exp. Biol.*, **29**, 505–32.
14. ARME, C. and COATES, A. (1973). *Hymenolepis diminuta*: active transport of α-aminoisobutyric acid by cysticercoid larvae. *Int. J. Parasit.*, **3**, 553–60.
15. ARME, C. and READ, C. P. (1970). A surface enzyme in *Hymenolepis diminuta* (Cestoda). *J. Parasit.*, **56**, 514–16.
16. ASCH, H. L. and READ, C. P. (1975). Transtegumental absorption of amino acids by male *Schistosoma mansoni*. *J. Parasit.*, **61**, 378–79.
17. ASHTON, P. J. and WALMSLEY, R. D. (1976). The aquatic fern *Azolla* and its *Anabaena* symbiont. *Endeavour*, **124**, 39–43.
18. ASKEW, R. R. (1971). *Parasitic Insects*. Heinemann, London.

19. AYALA, F. J. (1972). Competition between species. *Am. Scient.*, **60**, 348–57.
20. BACH, F. H. (1975). Collaborating T and B cells. *Nature*, **256**, 88–89.
21. BAER, J. G. (1952). *Ecology of Animal Parasites*. University of Illinois Press, Urbana.
22. BAILEY, G. N. A. (1971). *Hymenolepis diminuta*: circadian rhythm in movement and body length in the rat. *Expl. Parasit.*, **29**, 285–91.
23. BAILEY, J. A., CARTER, G. A., BURDEN, R. S. and WAIN, R. L. (1975). Control of rust diseases by diterpenes from *Nicotiana glutinosa*. *Nature*, **255**, 328–30.
24. BAILEY, W. S. (1972). *Spirocerca lupi*: a continuing enquiry. *J. Parasit.*, **58**, 3–22.
25. BANG, B. G. and BANG, F. B. (1975). Cell recognition by mucus secreted by urn cell of *Sipunculus nudus*. *Nature*, **253**, 634–35.
26. BANNISTER, L. (1977). Invasion of red cells by *Plasmodium*. *Symp. Brit. Soc. Parasit.*, **15**, 27–56.
27. BASCH, P. F. (1970). Relationships of some larval strigeids and echinostomes (Trematoda): Hyperparasitism, Antagonism and 'Immunity' in the snail host. *Expl Parasit.*, **27**, 193–216.
28. BEAUCHAMP, R. S. A. and ULLYETT, P. (1932). Competitive relationships between certain species of freshwater triclads. *J. Ecol.*, **20**, 200–8.
29. BELL, E. A. (1976). 'Uncommon' amino acids in plants. *FEBS Letters*, **64**, 29–35.
30. BELL, E. A., FELLOWS, L. E. and QURESHI, M. Y. (1976). 5-hydroxy-L-tryptophan; taxonomic character and chemical defence in *Griffonia*. *Phytochemistry*, **15**, 823.
31. BELTON, C. M. and HARRIS, P. J. (1967). Fine structure of the cuticle of *Acanthatrium oregonense* (Macy). *J. Parasit.*, **53**, 715–24.
32. BENNETT, C. E. and THREADGOLD, L. T. (1975). *Fasciola hepatica*: development of tegument during migration in mouse. *Expl Parasit.*, **38**, 38–55.
33. BETHEL, W. M. and HOLMES, J. C. (1973). Altered evasive behaviour and responses to light in amphipods harboring acanthocephalan cystacanths. *J. Parasit.*, **59**, 945–56.
34. BIRD, A. F. (1975). Symbiotic relationships between nematodes and plants. *Symp. Soc. exp. Biol.*, **29**, 351–71.
35. BLITZ, N. M. and SMYTH, J. D. (1973). Tegumental ultrastructure of *Raillietina cesticillus* during the larval-adult transformation, with emphasis on the rostellum. *Int. J. Parasit.*, **3**, 561–70.
36. BLOODGOOD, R. A., MILLER, K., FITZHARRIS, T. P. and MCINTOSH, J. R. (1974). The ultrastructure of *Pyrsonympha* and its associated microorganisms. *J. Morph.*, **143**, 77–105.
37. BOHLOOL, B. B. and SCHMIDT, E. L. (1974). Lectins: a possible basis for specificity in the *Rhizobium*-legume root nodule symbiosis. *Science*, **185**, 269–71.
38. BORET, J. (1967). Contribution à la morphologie et à la biologie de *Diplozoon paradoxum* von Nordman, 1832. *Bulletin de la Société neuchâteloise des sciences naturelles*, **90**, 63–159.
39. BRADLEY, D. E. 1967). Ultrastructure of bacteriophages and bacteriocins. *Bact. Rev.*, **31**, 230–314.
40. BRADLEY, D. J. (1972). Regulation of parasite populations. A general theory of the epidemiology and control of parasitic infections. *Trans. R. Soc. trop. Med. Hyg.*, **66**, 697–708.

41. BRADLEY, D. J. (1974). Stability in host-parasite systems. In *Ecological Stability*. USHER, M. B. and WILLIAMSON, M. H. (eds). Chapman and Hall, London.
42. BRADLEY, D. J. and MCCULLOUGH, F. S. (1973). Egg output stability and the epidemiology of endemic *Schistosoma haematobium*. II An analysis of the epidemiology of endemic *S. haematobium*. *Trans. R. Soc. trop. Med. Hyg.*, **67**, 491–500.
43. BRÅTEN, T. and HOPKINS, C. A. (1969). The migration of *Hymenolepis diminuta* in the rat's intestine during normal development and following surgical transplantation. *Parasitology*, **59**, 891–905.
44. BRENNER, S. (1973). The genetics of behaviour. *Brit. med. Bull.*, **29**, 269–71.
45. BRENNER, S. (1974). The genetics of *Caenorhabditis elegans*. *Genetics Princeton*, **77**, 71–94.
46. BRESCIANI, J. and KØIE, M. (1970). On the ultrastructure of the epidermis of the adult female of *Kronborgia amphipodicola* Christensen and Kanneworff, 1964 (Turbellaria, Neorhabdocoela). *Ophelia*, **8**, 209–30.
47. BREZNAK, J. A. (1975). Symbiotic relationships between termites and their intestinal microbiota. *Symp. Soc. exp. Biol.*, **29**, 559–80.
48. BRIDGEN, P. J., CROSS, G. A. M. and BRIDGEN, J. (1976). N-terminal amino acid sequences of variant-specific surface antigens from *Trypanosoma brucei*. *Nature*, **263**, 613–14.
49. BROOKER, B. E. (1970) Desmosomes and hemidesmosomes in the flagellate *Crithidia fasciculata*. *Z. Zellforsch.*, **105**, 155–66.
50. BROOKER, B. E. (1971a). Flagellar adhesion of *Crithidia fasciculata* to millipore filters. *Protoplasma*, **72**, 19–31.
51. BROOKER, B. E. (1971b). Flagellar attachment and detachment of *Crithidia fasciculata* to the gut wall of *Anopheles gambiae*. *Protoplasma*, **73**, 191–202.
52. BROOKER, B. E. (1972). The sense organs of trematode miracidia. In *Behavioural Aspects of Parasite Transmission*, CANNING, E. U. and WRIGHT, C. A. (eds). Academic Press, London.
53. BROOKER, B. E. and FULLER, R. (1975). Adhesion of lactobacilli to the chicken crop epithelium. *J. Ultrastruct. Res.*, **52**, 21–31.
54. BRUESKE, C. H. and BERGESON, G. B. (1972). Investigation of growth hormones in xylem exudate and root tissue of tomato infected with root-knot nematode. *J. exp. Bot.*, **23**, 14–22.
55. BURNET, F. M. (1969). *Self and Non-self*. Cambridge University Press, London.
56. BUXTON, P. A. (1939). *The Louse : an account of the lice which infest man, their medical importance and control*. Edward Arnold, London.
57. BYRAM, J. E. and FISHER, F. (1973). The absorptive surface of *Moniliformis dubius* (Acanthocephala) I. Fine Structure. *Tissue and Cell*, **5**, 553–79.
58. CANNON, L. R. G. (1971). The life cycles of *Bunodera sacculata* and *B. luciopercae* (Trematoda: Allocreadidae) in Algonquin Park, Ontario. *Can. J. Zool.*, **49**, 1417–29.
59. CAPRA, J. D. and EDMUNDSON, A. B. (1977). The antibody combining site. *Scient. Am.*, **236** (1), 50–9.
60. CAPRA, J. D. and KEYHOE, J. M. (1975). Hypervariable regions, idiotypy and antibody-combining site. *Adv. Immun.*, **20**, 1–37.
61. CARNEY, W. P. (1969). Behavioural and morphological changes in carpenter ants harboring dicrocoeliid metacercariae. *Am. Midl. Nat.*, **82**, 605–11.

62. CAULLERY, M. (1952). *Parasitism and Symbiosis.* Sidgwick and Jackson Ltd., London.
63. CHANG, K-P. (1975). Haematophagous insect and haemoflagellate as hosts for prokaryotic endosymbionts. *Symp. Soc. exp. Biol.*, **29**, 407–28.
64. CHANG, K-P. and DWYER, D. M. (1976). Multiplication of a human parasite (*Leishmania donovani*) in phagolysosomes of hamster macrophages *in vitro. Science*, **193**, 678–80.
65. CHANG, K-P. and MUSGRAVE, A. J. (1973). Morphology, histochemistry and ultrastructure of mycetome and its rickettsial symbiotes in *Cimex lectularis* L. *Can. J. Microbiol.*, **19**, 1075–81.
66. CHANG, K-P. and TRAGER, W. (1974). Nutritional significance of symbiotic bacteria in two species of hemoflagellates. *Science*, **183**, 531–32.
67. CHAPMAN, D. (1969). *Introduction to Lipids.* McGraw-Hill, London.
68. CHAPMAN, H. D. and WILSON, R. A. (1970). The distribution and fine structure of the integumentary papillae of the cercaria of *Himasthla secunda* (Nicoll). *Parasitology*, **61**, 219–27.
69. CHAPPELL, L. H. (1969). Competitive exclusion between two intestinal parasites of the three-spined stickleback, *Gasterosteus aculeatus* L. *J. Parasit.*, **55**, 775–78.
70. CHENG, R. and SAMOILOFF, M. R. (1971). Sexual attraction in the free-living nematode *Panagrellus silusiae* (Cephalobidae). *Can. J. Zool.*, **49**, 1443–48.
71. CHRISTENSEN, A. M. and KANNEWORFF, B. (1964). *Kronborgia amphipodicola* gen. et sp. nov., a dioecious turbellarian parasitizing ampeliscid amphipods. *Ophelia*, **1**, 147–66.
72. CHRISTENSEN, A. M. and KANNEWORFF, B. (1965). Life history and biology of *Kronborgia amphipodicola* Christensen and Kanneworff (Turbellaria, Neorhabdocoela). *Ophelia*, **2**, 237–51.
73. CIRILLO, V. P. (1961). Sugar transport in microorganisms. *Ann. Rev. Microbiol.*, **15**, 197–218.
74. CLEGG, J. A. (1969). Skin penetration by cercariae of the bird schistosome, *Austrobilharzia terrigalensis*: the stimulatory effect of cholesterol. *Parasitology*, **59**, 973–89.
75. CLEGG, J. A. (1972). The schistosome surface in relation to parasitism. *Symp. Brit. Soc. Parasit.*, **10**, 19–40.
76. CLEGG, J. A., SMITHERS, S. R. and TERRY, R. J. (1971). Concomitant immunity and host antigens associated with schistosomiasis. *Int. J. Parasit.*, **1**, 43–49.
77. CLEVELAND, L. R. and GRIMSTONE, A. V. (1964). The fine structure of the flagellate *Mixotricha paradoxa* and its associated microorganisms. *Proc. R. Soc. B.*, **159**, 668–86.
78. CODY, M. L. (1974). *Competition and the Structure of Bird Communities.* Princeton University Press.
79. COFFEY, M. D. (1975). Obligate parasites of higher plants, particularly rust fungi. *Symp. Soc. exp. Biol.*, **29**, 297–323.
80. COX, F. E. G. (1975). Factors affecting infections of mammals with intraerythrocytic protozoa. *Symp. Soc. exp. Biol.*, **29**, 429–52.
81. CROFTON, H. D. (1971a). A quantitative approach to parasitism. *Parasitology*, **62**, 179–93.
82. CROFTON, H. D. (1971b). A model of host-parasite relationships. *Parasitology*, **63**, 343–64.
83. CROLL, N. A. (1972). Behaviour of larval nematodes. In *Behavioural Aspects of Parasite Transmission.* CANNING, E. U. and WRIGHT, C.A. (eds).

Academic Press, London.

84. CROLL, N. A. and AL-HADITHI, I. (1972). Sensory basis of activity in *Ancylostoma tubaeforme* infective larvae. *Parasitology*, **64**, 279–91

85. CROLL, N. A. and SMITH, J. M. (1977). The location of parasites within their hosts: the behaviour of *Nippostrongylus brasiliensis* in the anaesthetised rat. *Int. J. Parasit.*, **7**, 195–200.

86. CROMPTON, D. W. T. (1969). On the environment of *Polymorphus minutus* (Acanthocephala) in ducks. *Parasitology*, **59**, 19–28.

87. CROMPTON, D. W. T. (1970). *An Ecological Approach to Acanthocephalan Physiology*. Cambridge University Press, Cambridge.

88. CROMPTON, D. W. T. (1973). The sites occupied by some parasite helminths in the alimentary tract of vertebrates. *Biol. Rev.*, **47**, 27–83.

89. CROMPTON, D. W. T. and LEE, D. L. (1965). The fine structure of the body wall of *Polymorphus minutus* (Goeze, 1782) (Acanthocephala). *Parasitology*, **55**, 357–64.

90. CROSS, G. A. M. (1978). Antigenic variation in trypanosomes. *Proc. R. Soc. B*, **202**, 55–72.

91. CROWELL, K. L. (1962). Reduced interspecific competition among the birds of Bermuda. *Ecology*, **43**, 75–88.

92. CURTIS, A. S. G. (1972). Adhesive interactions between organisms. *Symp. Brit. Soc. Parasit.*, **10**, 1–21.

93. DAMIAN, R. T., GREENE, N. D. and HUBBARD, W. J. (1973). Occurrences of mouse α_2-macroglobulin antigenic determinants on *Schistosoma mansoni* adults with evidence on their nature. *J. Parasit.*, **59**, 64–73.

94. DARLINGTON, P. J. (1972). Competition, competitive repulsion and coexistence. *Proc. nat. Acad. Sci. U.S.A.*, **69**, 3151–55.

95. DAVENPORT, D. (1966). The experimental analysis of behaviour in symbioses. In *Symbiosis* Vol. I. HENRY, S. M. (ed). Academic Press, London.

96. DAVIES, C. (1976). *Development and* in vitro *Culture of Some Digenea*. Ph.D. thesis. University of London.

97. DAWES, B. and HUGHES, D. L. (1964). Fascioliasis: the invasive stages of *Fasciola hepatica* in mammalian hosts. *Adv. Parasit.*, **2**, 97–168.

98. DEAN, D. A. (1974). *Schistosoma mansoni*: adsorption of human blood group A and B antigens by schistosomula. *J. Parasit.*, **60**, 260–63.

99. DEMPSTER, J. P. (1977). The scientific basis of practical conservation: factors limiting the persistence of populations and communities of animals and plants. *Proc. R. Soc., B*, **197**, 69–76.

100. DESCH, C. and NUTTING, W. B. (1972). *Demodex folliculorum* (Simon) and *D. brevis* Akbulatova of man: redescription and reevaluation. *J. Parasit.*, **58**, 169–77.

101. DIAMOND, J. M. and MAY, R. M. (1976). Island biogeography and the design of natural reserves. In *Theoretical Ecology: principles and applications*. Blackwell Scientific Publications, Oxford.

102. DIKE, S. C. and READ, C. P. (1971). Relation of tegumentary phosphohydrolase and sugar transport in *Hymenolepis diminuta*. *J. Parasit.*, **57**, 1251–55.

103. DOBSON, C. H. (1975). Coevolution of orchids and bees. In *Coevolution of Animals and Plants*. GILBERT, L. E. and RAVEN, P. H. (eds). University of Texas Press, Austin and London.

104. DONISTHORPE, H. ST. J. K. (1927). *The Guests of British Ants: their habits and life histories*. George Routledge and Sons, London.

105. DOUGLAS, J. (1975). *Bacteriophages*. Chapman and Hall, London.

106. DRITSCHILO, W., CORNELL, H., NAFUS, D. and O'CONNOR, B. (1975). Insular biogeography: of mice and mites. *Science*, **190**, 467–69.
107. DROPKIN, V. H. (1969). Cellular responses of plants to nematode infections. *A. Rev. Phytopath.*, **7**, 101–22.
108. DUCKETT, J. G., TOTH, R. and SONI, S. L. (1975). An ultrastructural study of the *Azolla, Anabaena azollae* relationship. *New Phytol.*, **75**, 111–18.
109. EATON, R. D. P., MEEROVITCH, E. and COSTERTON, J. W. (1969). A surface-active lysosome in *Entamoeba histolytica. Trans. R. Soc. trop. Med. Hyg.*, **63**, 678–81.
110. EATON, R. D. P., MEEROVITCH, E. and COSTERTON, J. W. (1970). The functional morphology of pathogenicity in *Entamoeba histolytica. Ann. trop. Med. Parasit.*, **64**, 299–304.
111. EDMONDS, S. J. (1966). Hatching of the eggs of *Moniliformis dubius* (Archiacanthocephala). *Expl Parasit.*, **19**, 216–26.
112. EDMUNDS, M. (1966). Protective mechanisms in the Eolidacea. (Mollusca: Nudibranchia). *J. Linn. Soc.* (Zoo), **46**, 46–71.
113. EIBL-EIBESFELDT, I. (1955). Uber Symbiosen, Parasitismus und andere besondere zwischenartlicke Besiehungen tropischer Meeresfische. *Z. Tierpsychol.*, **12**, 203–19.
114. ERASMUS, D. A. (1967). The host-parasite interface of *Cyathocotyle bushiensis* (Kahn), 1962 (Trematoda: Strigeoidea). II. Electron microscope studies of the tegument. *J. Parasit.*, **53**, 703–14.
115. ERASMUS, D. A. (1972). *The Biology of Trematodes.* Edward Arnold, London.
116. FAIRLAMB, A. H., OPPERDOES, F. R. and BORST, P. (1977). New approach to screening drugs for activity against African trypanosomes. *Nature*, **265**, 270–71.
117. FEATHERSTONE, D. W. (1972). *Taenia hydatigena* IV. Ultrastructural study of the tegument. *Z. Parasit.*, **38**, 214–32.
118. FEDER, H. M. (1966). Cleaning symbiosis in the marine environment. In *Symbiosis* Vol. I. HENRY, S. M. (ed). Academic Press, London.
119. FEENY, P. (1975). Biochemical coevolution between plants and their insect herbivores. In *Coevolution of Animals and Plants.* GILBERT, L. E. and RAVEN, P. H. (eds). University of Texas Press, Austin and London.
120. FRICKE, H. W. (1974). Öko-Ethologie des monogamen Anemonen-fisches *Amphiprion bicinctus. Z. Tierpsychol.*, **36**, 429–513.
121. FRICKE, H. W. (1975). The role of behaviour in marine symbiotic animals. *Symp. Soc. exp. Biol.*, **29**, 581–94.
122. FRIEND, D. S. (1966). The fine structure of *Giardia muris. J. Cell Biol.*, **29**, 317–32.
123. GALUN, M., PARAN, N. and BEN-SHAUL, Y. (1971). Electron microscopic study of the lichen. *Dermatocarpon hepaticum* (Ach) Th.Fr. *Protoplasma*, **73**, 457–68.
124. GILLES, M. T. and WILKES, T. J. (1969). A comparison of the range of attraction of animal baits and of carbon dioxide for some West African mosquitoes. *Bull. ent. Res.*, **59**, 441–56.
125. GILLES, M. T. and WILKES, T. J. (1970). The range of attraction of single baits for some West African mosquitoes. *Bull. ent. Res.*, **60**, 225–35.
126. GOLDRING, O. L., CLEGG, J. A., SMITHERS, S. R. and TERRY, R. J. (1975). Acquisition of human blood group antigens by *Schistosoma mansoni. Clin. exp. Immun.*, **26**, 181–87.
127. GOODCHILD, C. and FRANKENBERG, D. (1962). Voluntary running in the golden hamster, *Mesocricetus auratus* (Waterhouse, 1839) infected with

Trichinella spiralis (Owen, 1835). *Trans. Am. microsc. Soc.*, **81**, 292–98.

128. GRAY, A. R. (1967). Some principles of the immunology of trypanosomes. *Bull. Wld Hlth Org.*, **37**, 177–93.

129. GRAY, A. R. and LUCKINS, A. G. (1976). Antigenic variations in salivarian trypanosomes. In *Biology of the Kinetoplastida*, Vol. 1. Academic Press, London.

130. HAIRSTON, N. G. (1951). Interspecies competition and its probable influence upon the vertical distribution of Appalachian salamanders of the genus *Plethodon*. *Ecology*, **32**, 266–74.

131. HALE, M. E. (1974). *The Biology of Lichens* (2nd Edition). Edward Arnold, London.

132. HALTON, D. W. and MORRIS, G. P. (1969). Occurrence of cholinesterase and ciliated sensory structures in a fish gill fluke *Diclidophora merlangi* (Trematoda: Monogenea). *Z. Parasit.*, **33**, 21–30.

133. HALVORSEN, O. (1976). Negative interaction among parasites. In *Ecological Aspects of Parasitology*. KENNEDY, C. R. (ed). North Holland Publishing Co., Amsterdam.

134. HAMMOND, R. A. (1966). The proboscis mechanism of *Acanthocephalus ranae*. *J. exp. Biol.*, **45**, 203–13.

135. HAMMOND, R. A. (1967). The mode of attachment within the host of *Acanthocephalus ranae* (Schrank, 1788), Lühe, 1911. *J. Helminth.*, **41**, 321–28.

136. HARTLEY, P. H. T. (1964). Feeding habits. In *A New Dictionary of Birds*. THOMSON, A. L., (ed). Nelson, London.

137. HARTZELL, A. (1967). Insect ectosymbiosis. In *Symbiosis* Vol. II, HENRY, S. M. (ed). Academic Press, New York.

138. HENRY, S. M. (ed). (1966). Foreword in *Symbiosis* Vol. I. Academic Press, New York.

139. HERTIG, M., TALIFERRO, W. H. and SCHWARTZ, B. (1937). The terms symbiosis, symbiont and symbiote. *J. Parasit.*, **23**, 326–29.

140. HEYNEMAN, D. (1962). Studies on helminth immunity II. Influence of *Hymenolepis nana* (Cestoda: Hymenolepidae) in dual infections with *H. diminuta* in white mice and rats. *Expl Parasit.*, **12**, 7–18.

141. HILL, D. J. and SMITH, D. C. (1972). Lichen physiology XII. The 'inhibition technique'. *New Phytol.*, **71**, 15–30.

142. HINE, P. M. and KENNEDY, C. R. (1974). Observations on the distribution, specificity and pathogenicity of the acanthocephalan *Pomphorhynchus laevis* (Müller). *J. Fish Biol.*, **6**, 665–79.

143. HOCKLEY, D. J. (1973). Ultrastructure of the tegument of *Schistosoma*. *Adv. Parasit.*, **11**, 233–305.

144. HOCKLEY, D. J. and MCLAREN, D. J. (1973). *Schistosoma mansoni*: changes in the outer membrane of the tegument during development from cercaria to adult worm. *Int. J. Parasit.*, **3**, 13–25.

145. HOLBERTON, D. V. (1973). Fine structure of the ventral disk apparatus and the mechanism of attachment in the flagellate *Giardia muris*. *J. Cell Sci.*, **13**, 11–41.

146. HOLBERTON, D. V. (1974). Attachment of *Giardia* – A hydrodynamic model based on flagellar activity. *J. expl Biol.*, **60**, 207–21.

147. HOLMES, J. C. (1961). Effects of concurrent infections on *Hymenolepis diminuta* (Cestoda) and *Moniliformis dubius* (Acanthocephala) I. General effects and comparison with crowding. *J. Parasit.*, **47**, 209–16.

148. HOLMES, J. C. (1973). Site selection by parasitic helminths: interspecific interactions, site segregation, and their importance to the development

of helminth communities. *Can. J. Zool.*, **51**, 333–47.
149. HOLMES, J. C. (1976). Host selection and its consequences. In *Ecological Aspects of Parasitology* KENNEDY, C. R. (ed). North Holland Publishing Co., Amsterdam.
150. HOLMES, J. C. and BETHEL, W. M. (1972). Modification of intermediate host behaviour by parasites. In *Behavioural Aspects of Parasite Transmission*, CANNING, E. U. and WRIGHT, C. A. (eds). Academic Press, London.
151. HONIGBERG, B. M. (1967). Chemistry of parasitism among some Protozoa. In *Chemical Zoology*, Vol. I, FLORKIN, M. and SCHEER, B. T. (eds). Academic Press, London.
152. HONIGBERG, B. M. (1970). Protozoa associated with termites and their role in digestion. In *Biology of Termites* Vol. 2, KRISHNA, K. and WEESNER, F. M. (eds). Academic Press, New York.
153. HOWARD, B. H. (1967). Intestinal microorganisms of ruminants and other vertebrates. In *Symbiosis* Vol. II, HENRY, S. M. (ed). Academic Press, New York.
154. HUDSON, K. M., BYNER, C., FREEMAN, J. and TERRY, R. J. (1976). Immunodepression, high IgM levels and evasion of the immune response in murine trypanosomiasis. *Nature*, **264**, 256–8.
155. HUTCHINSON, G. E. (1959). Homage to Santa Rosalia or why are there so many animals? *Am. Nat.*, **93**, 145–59.
156. ITO, S. (1969). Structure and function of the glycocalyx. *Fed. Proc.*, **28**, 12–29.
157. JAHNS, H. M. (1973). Anatomy, morphology and development. In *The Lichens*, AHMADJIAN, V. and HALE, M. E. (eds). Academic Press, London.
158. JENNINGS, D. H. and LEE, D. L. (eds). (1975). Symbiosis *Symp. Soc. exp. Biol.*, **29**.
159. JENNINGS, J. B. (1971). Parasitism and commensalism in the Turbellaria. *Adv. Parasit.*, **9**, 1–27.
160. JHA, R. K. and SMYTH, J. D. (1969). *Echinococcus granulosus*: ultrastructure of microtriches. *Expl Parasit.*, **25**, 232–44.
161. JHA, R. K. and SMYTH, J. D. (1971). Ultrastructure of the rostellar tegument of *Echinococcus granulosus* with special reference to the biogenesis of mitochondria. *Int. J. Parasit.*, **1**, 169–77.
162. JOHN, P. and WHATLEY, F. R. (1975). *Paracoccus denitrificans* : a present-day bacterium resembling the hypothetical free-living ancestor of the mitochondrion. *Symp. Soc. exp. Biol.*, **29**, 39–40.
163. JONES, M. G. K. and NORTHCOTE, D. H. (1972). Nematode-induced syncytium – a multinucleate transfer cell. *J. Cell Sci.*, **10**, 789–809.
164. JONES, T. C., YEH, S. and HIRSCH, J. G. (1972). The interaction between *Toxoplasma gondii* and mammalian cells. I. Mechanism of entry and intracellular fate of the parasite. *J. exp. Med.*, **136**, 1157–72.
165. JOYEAUX, C. and BAER, J. G. (1936). Cestodes, Vol. 30. In *Faune de France*. Lechavalier et Fils, Paris.
166. KARAKASHIAN, M. (1975). Symbiosis in *Paramecium bursaria*. *Symp. Soc. exp. Biol.*, **29**, 145–74.
167. KARPLUS, I., TSURNAMAL, M. and SZLEP, M. (1972). Associative behaviour of the fish *Cryptocentrus cryptocentrus* (Gobiidae) and the pistol shrimp *Alphaeus djiboutensis* (Alpheidae) in artificial burrows. *Marine Biology*, **15**, 95–104.
168. KEARN, G. C. (1967). Experiments in host finding and host specificity in the monogenean skin parasite *Entobdella soleae*. *Parasitology*, **57**, 585–605.

169. KEARN, G. C. (1971). The physiology and behaviour of the monogenean skin parasite *Entobdella soleae* in relation to its host (*Solea solea*). In *Ecology and Physiology of Parasites*. FALLIS, A. M. (ed). University of Toronto Press, Toronto.
170. KEARN, G. C. (1973). An endogenous circadian hatching rhythm in the monogenean skin parasite, *Entobdella soleae*, and its relationship to the activity rhythm of the host (*Solea solea*). *Parasitology*, **66**, 101–22.
171. KEARN, G. C. (1976). Body surfaces of fishes. In *Ecological Aspects of Parasitology*. KENNEDY, C. R. (ed). North Holland Publishing Co., Amsterdam.
172. KENNEDY, C. R. (1975). *Ecological Animal Parasitology*. Blackwell Scientific Publications, Oxford.
173. KERSHAW, K. A. and MILLBANK, J. W. (1970). Nitrogen metabolism in lichens. II. The partition of cephalodial-fixed nitrogen between the mycobionts and phycobiont of *Peltigera aphthosa*. *New Phytol.*, **69**, 75–79.
174. KHALIL, L. F. (1961). On the capture and destruction of miracidia by *Chaetogaster limnaei* (Oligochaeta). *J. Helminth.*, **35**, 269–74.
175. KIES, L. (1974). Elektronmikroscopische Untersuchungen an *Paulinella chromatophora* Lauteborn, einer Thekamöbe mit blau-grun Endosymbionten (Cyanellen). *Protoplasma*, **80**, 69–89.
176. KOCH, A. (1967). Insects and their endosymbionts. In *Symbiosis*, Vol. II, HENRY, S. M. (ed). Academic Press, New York.
177. KØIE, M. (1971a). On the histochemistry and ultrastructure of the daughter sporocyst of *Cercaria baccini* Lebour 1911. *Ophelia*, **9**, 145–63.
178. KØIE, M. (1971b). On the histochemistry and ultrastructure of the redia of *Neophasis lageniformis* (Lebour, 1910) (Trematoda, Acanthocolpidae). *Ophelia*, **9**, 113–43.
179. KØIE, M. (1973). The host-parasite interface and associated structures of the cercaria and adult *Neophasis lageniformis* (Lebour, 1910). *Ophelia*, **12**, 205–19.
180. KRUPA, P. L. and BOGITSH, B. J. (1972). Ultrastructural phosphohydrolase activities in *Schistosoma mansoni* sporocysts and cercariae. *J. Parasit.*, **58**, 495–514.
181. KULLENBERG, B. (1952). Recherches sur la biologie florale des *Ophrys*. *Bulletin de la Société d'histoire naturelle de l'Afrique du Nord*, **43**, 53–62.
182. KURIS, A. M. and BLAUSTEIN, A. R. (1977). Ectoparasitic mites on rodents: application of the island biogeography theory? *Science*, **195**, 596–8.
183. LACKIE, J. M. (1975). The host specificity of *Moniliformis dubius* (Acanthocephala), a parasite of cockroaches. *Int. J. Parasit.*, **5**, 301–7.
184. LACKIE, J. M. and ROTHERHAM, S. (1972). Observations on the envelope surrounding *Moniliformis dubius* (Acanthocephala) in the intermediate host *Periplaneta americana*. *Parasitology*, **65**, 303–8.
185. LAGUNOFF, D. and CURRAN, D. E. (1972). Role of bristle-coated membrane in the uptake of ferritin by rat macrophages. *Expl Cell Res.*, **75**, 337–46.
186. LANGE, P. T. (1966). Bacterial symbiosis with plants. In *Symbiosis*, Vol. I, HENRY, S. M. (ed). Academic Press, New York.
187. LARSCH, J. E. and WEATHERLY, N. F. (1975). Cell-mediated immunity against certain parasitic worms. *Adv. Parasit.*, **13**, 183–222.
188. LAWRENCE, J. D. (1973). The ingestion of red blood cells by *Schistosoma mansoni*. *J. Parasit.*, **59**, 60–3.
189. LEE, D. L. (1966). The structure and composition of the helminth cuticle.

Adv. Parasit., **4**, 187–254.

190. LEE, D. L. (1970). The ultrastructure of the cuticle of adult female *Mermis nigrescens* (Nematoda). *J. Zool.*, **161**, 513–18.
191. LEE, D. L. (1972). Penetration of mammalian skin by the infective larva of *Nippostrongylus brasiliensis*. *Parasitology*, **65**, 499–505.
192. LEE, D. L. (1972). The structure of the helminth cuticle. *Adv. Parasit.*, **10**, 345–77.
193. LEE, D. L., LONG, P. L., MILLARD, B. J. and BRADLEY, J. (1969). The fine structure and method of feeding of the tissue parasitizing stages of *Histomonas meleagridis*. *Parasitology*, **59**, 171–84.
194. LEONG, T. S. (1975). *Metazoan Parasites of Fishes of Cold Lake, Alberta : a community analysis*. Ph.D. Thesis, University of Alberta, Edmonton.
195. LESTER, R. J. G. (1971). The influence of *Schistocephalus* plerocercoids on the respiration of *Gasterosteus* and a possible resulting effect on the behaviour of the fish. *Can. J. Zool.*, **49**, 361–66.
196. LEVINS, R. (1968). *Evolution in Changing Environments*. Princeton University Press.
197. LLEWELLYN, J. (1954). Observations on the food and gut pigment of the Polyopisthocotylea (Trematoda: Monogenea). *Parasitology*, **44**, 428–37.
198. LLEWELLYN, J. (1965). The evolution of parasitic Platyhelminthes. *Symp. Brit. Soc. Parasit.*, **3**, 47–78.
199. LOM, J. (1972). On the structure of the extruded microsporidian polar filament. *Z. Parasit.*, **38**, 200–13.
200. LUMSDEN, R. D. (1966). Cytological studies on the absorptive surface of cestodes. I. The fine structure of the strobilar integument. *Z. Parasit.*, **27**, 355–82.
201. LUMSDEN, R. D. (1972). Cytological studies on the absorptive surfaces of cestodes. VI. Cytochemical evaluation of electrostatic charge. *J. Parasit.*, **58**, 229–34.
202. LUMSDEN, R. D. (1974). Relationship of extrinsic polysaccharides to the tegument glycocalyx of cestodes. *J. Parasit.*, **60**, 374–75.
203. LUMSDEN, R. D., OAKS, J. A. and ALWORTH, W. L. (1970). Cytological studies on the absorptive surfaces of cestodes. IV. Localization and cytochemical properties of membrane-fixed cation binding sites. *J. Parasit.*, **56**, 736–47.
204. LUMSDEN, R. D., OAKS, J. A. and MUELLER, J. F. (1974). Brush border development in the tegument of the tapeworm *Spirometra mansonoides*. *J. Parasit.*, **60**, 209–26.
205. LYON, F. and WOOD, R. K. S. (1976). The hypersensitive reaction and other responses of bean leaves to bacteria. *Ann. Bot.*, **40**, 479–91.
206. LYONS, K. M. (1969a). Sense organs in monogenean skin parasites ending in a typical cilium. *Parasitology*, **59**, 611–23.
207. LYONS, K. M. (1969b). Compound sensillae in monogenean skin parasites. *Parasitology*, **59**, 625–36.
208. LYONS, K. M. (1972). Sense organs in monogeneans. In *Behavioural Aspects of Parasite Transmission*, CANNING, E. U. and WRIGHT, C. A. (eds). Academic Press, London.
209. LYONS, K. M. (1973a). The epidermis and sense organs of the Monogenea and some related groups. *Adv. Parasit.*, **11**, 193–232.
210. LYONS, K. M. (1973b). Epidermal fine structure and development in the oncomiracidium larva of *Entobdella soleae* (Monogenea). *Parasitology*, **66**, 321–33.

211. LYONS, K. M. (1973c). Scanning and transmission electron microscope studies on the sensory sucker papillae on the fish parasite *Entobdella soleae* (Monogenea). *Z. Zellforsch.*, **137**, 471–80.
212. MACARTHUR, R. H. and WILSON, E. O. (1967). *The Theory of Island Biogeography*. Princeton University Press, Princeton, New Jersey.
213. MCCAUL, T. F. (1976). *A Fine Structural and Cytochemical Investigation into Pathogenicity of* Entamoeba histolytica *Strains Using Cell Line Monolayers*. Ph.D. thesis. Faculty of Medicine, University of London.
214. MCCORD, J. M., KEELE, B. B. Jr. and FRIDOVICH, I. (1971). An enzyme-based theory of obligate anaerobiasis: the physiological function of superoxide dismutase. *Proc. natn. Acad. Sci., U.S.A.*, **68**, 1024–27.
215. MCCULLOUGH, F. S. and BRADLEY, D. J. (1973). Egg output stability and the epidemiology of *Schistosoma haematobium* I. Variation and stability in *S. haematobium* egg counts. *Trans. R. Soc. trop. Med. Hyg.*, **67**, 475–90.
216. MCDEVITT, H. O. and BODMER, W. F. (1974). HL-A, immune response genes and disease. *Lancet*, Vol. **1**, 1269–75.
217. MACINNIS, A. J. (1969). Identification of chemicals triggering cercarial penetration responses of *Schistosoma mansoni*. *Nature*, **224**, 1221–22.
218. MACKENZIE, K. and GIBSON, D. (1970). Ecological studies of some parasites of plaice, *Pleuronectes platessa* (L) and flounder *Platichthys flesus* (L). *Symp. Brit. Soc. Parasit.*, **8**, 1–47.
219. MCLAREN, D. J. (1974). The anterior glands of adult *Necator americanus* (Nematoda: Strongyloidea): I. Ultrastructural studies. *Int. J. Parasit.*, **4**, 25–37.
220. MCLAREN, D. J. (1976). Sense organs and their secretions. In *The Organisation of Nematodes*, CROLL, N. A. (ed). Academic Press, London.
221. MCLAREN, D. J., BURT, J. S. and OGILVIE, B. M. (1974). The anterior glands of *Necator americanus* (Nematoda: Strongyloidea): II. Cytochemical and functional studies. *Int. J. Parasit.*, **4**, 39–46.
222. MCLAREN, D. J., CLEGG, J. A. and SMITHERS, S. R. (1975). Acquisition of host antigens by young *Schistosoma mansoni* in mice: correlation with failure to bind antibody *in vitro*. *Parasitology*, **70**, 67–75.
223. MCLAUGHLIN, J. J. A. and ZAHL, P. A. (1966). Endozoic algae. In *Symbiosis*, Vol. I, HENRY, S. M. (ed). Academic Press, New York.
224. MCLEAN, R. B. and MARISCAL, R. N. (1973). Protection of a hermit crab by its symbiotic sea anemone *Calliactis tricolor*. *Experientia*, **29**, 128–30.
225. MCVICAR, A. H. (1972). The ultrastructure of the parasite host interface of three tetraphyllidean tapeworms of the elasmobranch *Raja naevus*. *Parasitology*, **65**, 77–88.
226. MANNING, M. J. and TURNER, R. J. (1976). *Comparative Immunobiology*. Blackie, Glasgow.
227. MANSOUR, T. E. (1959). Studies on the carbohydrate metabolism of the liver fluke, *Fasciola hepatica*. *Biochim. biophys. Acta*, **34**, 456–64.
228. MARISCAL, R. M. (1971). Experimental studies on the protection of the anemone fishes from the sea anemones. In *Aspects of the Biology of Symbiosis*, CHENG, T. C. (ed). University Park Press, Baltimore.
229. MARGULIS, L. (1975). Symbiotic theory of the origin of eukaryotic organelles: criteria for proof. *Symp. Soc. exp. Biol.*, **29**, 21–38.
230. MARGULIS, L. (1976). A Review: genetic and evolutionary consequences of symbiosis. *Expl Parasit.*, **39**, 277–349.
231. MARSHALL, J. P. (1977). *Aspects of the Biology of* Pomphorhynchus laevis *(Acanthocephala) in its intermediate host* Gammarus pulex. M.Phil. Thesis. April 1977. Department of Biological Sciences, Portsmouth

Polytechnic.

232. MATRICON-GONDRAN, M. (1971). Étude ultrastructurale des récepteurs sensoriels tégumentaires de quelques Trématodes Digénétiques larvaires. *Z. Parasit.*, **35**, 318–33.

233. MAY, R. M. (1976a). Models for two interacting populations. In *Theoretical Ecology: principles and applications*, MAY, R. M. (ed). Blackwell Scientific Publications, Oxford.

234. MAY, R. M. (1976b). Patterns in multi-species communities. In *Theoretical Ecology: principles and applications*, MAY, R. M. (ed). Blackwell Scientific Publications, Oxford.

235. MEAD-BRIGGS, A. R. and RUDGE, A. J. B. (1960). Breeding of the rabbit flea *Spillopsyllus cuniculi* (Dale); requirement of a 'factor' from a pregnant rabbit for ovarian maturation. *Nature*, **187**, 1136–37.

236. MEEUSE, B. J. D. (1961). *The Story of Pollination*. Ronald Press, New York.

237. MELLANBY, K. (1943). *Scabies*. Oxford University Press, London.

238. MELLANBY, K. (1944). The development of symptoms, parasitic infection and immunity in human scabies. *Parasitology*, **35**, 197.

239. MERCER, E. H. and NICHOLAS, W. L. (1967). The ultrastructure of the capsule of the larval stages of *Moniliformis dubius* (Acanthocephala) in the cockroach *Periplaneta americana*. *Parasitology*, **57**, 169–74.

240. METTRICK, D. F. (1971). Effect of host dietary constituents on intestinal pH and the migrational behaviour of the rat tapeworm, *Hymenolepis diminuta*. *Can. J. Zool.*, **49**, 1513–25.

241. MILLBANK, J. W. and KERSHAW, K. A. (1973). Nitrogen metabolism. In *The Lichens* AHMADJIAN, V. and HALE, M. E. (eds). Academic Press, New York.

242. MIRELMAN, D., GALUN, E., SHARON, N. and LOTAN, R. (1975). Inhibition of fungal growth by wheat germ agglutinin. *Nature*, **256**, 414–16.

243. MORRIS, G. P. and THREADGOLD, L. T. (1967). A presumed sensory structure associated with the tegument of *Schistosoma mansoni*. *J. Parasit.*, **53**, 537–39.

244. MORSETH, D. J. (1966). The fine structure of the tegument of adult *Echinococcus granulosus*, *Taenia hydratigena* and *Taenia pisiformis*. *J. Parasit.*, **52**, 1074–85.

245. MOSER, J. C. (1964). Inquiline roach responds to trail-marking substance of leaf-cutting ants. *Science*, **143**, 1048–49.

246. MUNDIM, M. H., ROITMAN, I., HERMANS, M. A. and KITAJIMA, E. W. (1974). Simple nutrition of *Crithidia deanei*, a reduviid trypanosomatid with an endosymbiont. *J. Protozool.*, **21**, 518–21.

247. MUNDY, P. J. (1973). Vocal mimicry of their hosts by nestlings of the great spotted cuckoo and striped crested cuckoo. *Ibis*, **115**, 602–4.

248. MURRAY, P. K., URQUHART, G. M., MURRAY, M. and JENNINGS, F. W. (1973). The responses of mice infected with *T. brucei* to the administration of sheep erythrocytes. *Trans. R. Soc. trop. Med: Hyg.*, **67**, 267.

249. MUSCATINE, L. and CERNICHIARI, E. (1969). Assimilation of photosynthetic products of zooxanthellae by reef coral. *Biol. Bull.*, **137**, 506–23.

250. MUSCATINE, L., COOK, C. B., PARDY, R. L. and POOL, R. R. (1975). Uptake, recognition and maintenance of symbiotic *Chlorella* by *Hydra viridis*. *Symp. Soc. exp. Biol.*, **29**, 175–204.

251. MUSCATINE, L. and LENHOFF, H. M. (1965). Symbiosis of hydra and algae II. Effects of limited food and starvation on growth of the symbiotic and

aposymbiotic hydra. *Biol. Bull.*, **129**, 316–28.

252. MUSGRAVE, A. J. (1964). Insect mycetomes. *Can. Entomol.*, **96**, 377–89.
253. NACHTIGALL, W. (1974). *Biological Mechanisms of Attachment: the comparative morphology and bioengineering of organs for linkage, suction and adhesive.* Springer Verlag, Berlin.
254. NEWTON, B. A. (1956). A synthetic growth medium for the trypanosomid flagellate *Strigomonas (Herpetomonas) oncopelti. Nature*, **177**, 279–80.
255. NICHOLAS, W. L. (1972). The fine structure of the cuticle of *Heterotylenchus. Nematologia*, **19**, 138–40.
256. NIERENBERG, L. (1972). The mechanism for the maintenance of species integrity in sympatrically occurring equitant oncidiums in the Caribbean. *Am. Orchid Soc. Bull.*, **41**, 873–82.
257. NOLLEN, P. M. (1968). Uptake and incorporation of glucose, tyrosine, leucine and thymidine by adult *Philophthalmus megalurus* (Cort, 1914) (Trematoda), as determined by autoradiography. *J. Parasit.*, **54**, 295–304.
258. NUTTMAN, C. J. (1971). The fine structure of ciliated nerve endings in the cercaria of *Schistosoma mansoni. J. Parasit.*, **57**, 855–59.
259. OAKS, J. A. and LUMSDEN, R. D. (1971). Cytological studies on the absorptive surfaces of cestodes V. Incorporation of carbohydrate-containing macromolecules into tegument membranes. *J. Parasit.*, **57**, 1256–68.
260. OWEN, G. (1972). Lysosomes, peroxisomes and bivalves. *Science Progress, Oxford*, **60**, 299–318.
261. PALING, J. E. (1969). The manner of infection of trout gills by the monogenean parasite, *Discocotyle sagittata. J. Zool., Lond.*, **151**, 293–309.
262. PAN, C-T. (1963). Generalized and focal tissue responses in the snail *Australorbis glabrata* infected with *Schistosoma mansoni. Ann. N.Y. Acad. Sci.*, **113**, 475–85.
263. PAPERNA, I. (1964). Competitive exclusion of *Dactylogyrus extensus* by *D. vastator* (Trematoda: Monogenea) on the gills of reared carp. *J. Parasit.*, **50**, 94–98.
264. PAPPAS, P. W. and READ, C. P. (1972a). The absorption of pyridoxine and riboflavin by *Hymenolepis diminuta. J. Parasit.*, **58**, 417–21.
265. PAPPAS, P. W. and READ, C. P. (1972b). Trypsin inactivation by intact *Hymenolepis diminuta. J. Parasit.*, **58**, 864–71.
266. PAPPAS, P. W. and READ, C. P. (1974). Relation of nucleoside transport and surface phosphohydrolase activity in *Hymenolepis diminuta. J. Parasit.*, **60**, 447–52.
267. PARK, T. (1962). Beetles, competition and populations. *Science*, **138**, 1369–75.
268. PARKER, J. C. and HALEY, A. J. (1960). Phototactic and thermotactic responses of the filariform larvae of the rat nematode *Nippostrongylus muris. Expl Parasit.*, **9**, 92–97.
269. PATERSON, W. D., ALLEN, B. R. and BEVERIDGE, G. W. (1973). Norwegian scabies during immunosuppressive therapy. *Br. Med. J.*, **3**, 211–12.
270. PAULSON, R. E. and WEBSTER, J. M. (1972). Ultrastructure of the hypersensitive reaction in roots of tomato, *Lycopersicon esculentum* L. to infection by the root-knot nematode, *Meloidogyne incognita. Physiol. Pl. Path.*, **2**, 227–34.
271. PEARSON, I. G. (1963). Use of the chromium radioisotope ^{51}Cr to estimate blood loss through ingestion of *Fasciola hepatica. Expl Parasit.*, **13**,

186–93.

272. PETERSON, P. A., RASK, L., SEGE, K., KLARESKOG, L., ANUNDI, H. and ÖSTBERG, L. (1975). Evolutionary relationship between immunoglobulins and transplantation antigens. *Proc. Natn. Acad. Sci., U.S.A.*, **72**, 1612.

273. PETTER, A. J. (1962). Redescription et analyse critique de quelques éspeces d'oxyures de la tortue greque (*Testudo graeca* L.). Diversité des structures cephaliques (II). *Annls Parasit. hum. comp.*, **37**, 140–52.

274. PEVELING, E. (1973). Fine structure. In *The Lichens*, AHMADJIAN, V. and HALE, M. E. (eds). Academic Press, London.

275. PIANKA, E. R. (1970). On r- and K- selection. *Am. Nat.*, **104**, 592–97.

276. PIANKA, E. R. (1976). Competition and niche theory. In *Theoretical Ecology: principles and applications*, MAY, R. M. (ed). Blackwell Scientific Publications, Oxford.

277. PLATZER, E. G. and ROBERTS, L. S. (1969). Developmental physiology of cestodes. V. Effects of vitamin deficient diets and host coprophagy prevention on development of *Hymenolepis diminuta*. *J. Parasit.*, **55**, 1143–52.

278. POINAR, G. O. and HESS, R. J. (1972). Food uptake by the insect-parasitic nematode, *Sphaerularia bombi* (Tylenchida) *J. Nemat.*, **4**, 270–77.

279. POLJAK, R. J. (1975). Three dimensional structure, function and genetic control of immunoglobulins. *Nature*, **256**, 373–76.

280. POSTGATE, J. (1975). *Rhizobium* as a free living nitrogen fixer. *Nature*, **256**, 363.

281. POTTS, G. W. (1973). Cleaning symbiosis among British fish with special reference to *Crenilabrus melops* (Labridae). *J. mar. biol. Ass. U.K.*, **53**, 1–10.

282. RAFF, R. A. and MAHLER, H. R. (1975). The symbiont that never was: an enquiry into the evolutionary origin of the mitochrondion. *Symp. Soc. exp. Biol.*, **29**, 41–92.

283. RANDOLPH, S. E. (1975). Patterns of distribution of the tick *Ixodes trianguliceps* Birula on its hosts. *J. An. Ecol.*, **44**, 451–74.

284. READ, C. P. (1970). *Parasitism and Symbiology*. Ronald Press Co., New York.

285. READ, C. P. (1973). Contact digestion in tapeworms. *J. Parasit.*, **59**, 672–77.

286. READ, C. P., ROTHMAN, A. H. and SIMMONS, J. E. (1963). Studies on membrane transport with special reference to parasite-host integration. *Ann. N.Y. Acad. Sci.*, **113**, 154–205.

287. READ, D. P., FEENY, P. P. and ROOT, R. B. (1970). Habitat selection by the aphid parasite *Diaeretiella rapae* (Hymenoptera: Braconidae) and hyperparasite *Charips brassicae* (Hymenoptera: Cynipidae). *Can. Entomol.*, **102**, 1567–78.

288. REES, G. (1966). Light and electron microscopical studies of the redia of *Parorchis acanthus* Nicoll. *Parasitology*, **56**, 589–602.

289. RICHARDSON, J. S., THOMAS, K. A., RUBIN, B. H. and RICHARDSON, D. C. (1975). Crystal structure of bovine Cu, Zn superoxide dismutase at 3Å resolution: chain tracing and metal ligands. *Proc. Natn. Acad. Sci., U.S.A.*, **72**, 1349.

290. RIDING, I. L. (1970). Microvilli on the outside of a nematode. *Nature*, **226**, 179–80.

291. ROBERTS, W. L., SPEER, C. A. and HAMMOND, D. M. (1971). Penetration of *Eimeria larimerensis* sporozoites into cultured cells as observed with the

light and electron microscopes. *J. Parasit.*, **57**, 615–25.

292. ROCHE, M. and TORRES, C. M. (1960). A method for *in vitro* study of hookworm activity. *Expl Parasit.*, **9**, 250–56.

293. ROITT, I. M. (1974). *Essential Immunology*. (2nd Edition). Blackwell, London.

294. ROSE, N. R., MILGROM, F. and VAN OSS, C. J. (1973). *Principles of Immunology*. Macmillan, London.

295. ROSS, D. M. (1971). Protection of hermit crabs (*Dardanus spp*) from octopus by commensal sea anemones (*Calliactis spp*) *Nature*, **230**, 401–2.

296. ROTHERHAM, S. and CROMPTON, D. W. T. (1972). Observations on the early relationship between *Moniliformis dubius* (Acanthocephala) and the haemocytes of the intermediate host *Periplaneta americana*. *Parasitology*, **64**, 15–21.

297. ROTHSCHILD, M. and CLAY, T. (1952). *Fleas, Flukes and Cuckoos : a study of bird parasites*. Collins, London.

298. ROTHSCHILD, M. and FORD, B. (1966). Hormones of the vertebrate host controlling ovarian regression and copulation of the rabbit flea. *Nature*, **211**, 261–66.

299. ROTHSCHILD, M. and FORD, B. (1969). Does a pheromone-like factor from the nestling rabbit stimulate impregnation and maturation in the rabbit flea. *Nature*, **221**, 1169–70.

300. RUIZ, J. M. (1951). Nota sôbre a cercariofagia de um oligochaeta do gênero *Chaetogaster* v. Baer, 1827. *Anais da Faculdade de farmácia e odontologia Universidade de Sao Paulo*, **9**, 51–56.

301. RUTHERFORD, T. A. and WEBSTER, J. M. (1974). Transcuticular uptake of glucose by the entomophilic nematode *Mermis nigrescens*. *J. Parasit.*, **60**, 804–8.

302. SAGAN, L. (1967). On the origin of mitosing cells. *J. theor. Biol.*, **14**, 225–75.

303. SALM, R. W. and FRIED, B. (1973). Heterosexual chemical attraction in *Camallanus sp.* (Nematoda) in the absence of worm mediated tactile behaviour. *J. Parasit.*, **59**, 434–36.

304. SALT, G. (1970). *The Cellular Defense Reactions of Insects*. Cambridge University Press, Cambridge.

305. SANGER, F., AIR, G. M., BARRELL, B. G., BROWN, N. L., COULSON, A. R., FIDDES, J. C., HUTCHINSON III, C. A., SLOCOMBE, P. M. and SMITH, M. (1977). Nucleotide sequence of bacteriophage Φx174 DNA. *Nature*, **265**, 687–95.

306. SASA, M., SHIRASAKA, R., TANAKA, H., MIURA, A., YAMAMOTO, H. and KATHIRA, K. (1960). Observations on the behaviour of infective larvae of hookworms and related nematode parasites, with notes on the effect of carbon dioxide in the breath as a stimulant. *Jap. J. exp. Med.*, **30**, 433–47.

307. SCHAD, G. A. (1963). Niche diversification in a parasite species flock. *Nature*, **198**, 404–6.

308. SCHAD, G. A. (1963). The ecology of co-occurring congeneric pinworms in the tortoise *Testudo graeca*. *Proc. XVI Int. Cong. Zool.*, **1**, 223–24.

309. SCHMID, W. D. and ROBINSON, E. J. (1972). The pattern of a host-parasite distribution. *J. Parasit.*, **57**, 907–10.

310. SCHREVEL, J. and VIVIER, E. (1966). Étude de l'ultrastructure et du rôle de la region antérieure de gregarines parasites d'annelids polychaetes. *Parasitologica*, **2**, 17–28.

311. SEEL, D. C. (1973). Egg laying in the cuckoo. *Br. Birds*, **66**, 528–35.
312. SHARON, N. and LIS, H. (1972). Lectins: cell-agglutinating and sugar specific proteins. *Science*, **177**, 949–53.
313. SHEFFIELD, H. G., GARNHAM, P. C. C. and SHIROSIMI, T. (1971). The fine structure of the sporozoite of *Lankasteria culicis*. *J. Protozool.*, **18**, 98–105.
314. SHER, A., SMITHERS, S. R. and MACKENZIE, P. (1975). Passive transfer of acquired resistance to *Schistosoma mansoni* in laboratory mice. *Parasitology*, **70**, 347–57.
315. SHIFF, C. J., CMELIK, S. H. W., LEY, H. E. and KRIEL, R. L. (1972). The influence of human skin lipids on the cercarial penetration response of *Schistosoma haematobium* and *Schistosoma mansoni*. *J. Parasit.*, **58**, 476–80.
316. SIMMONS, J. E. and LAURIE, J. S. (1972). A study of *Gyrocotyle* in the San Juan Archipelago, Puget Sound, U.S.A., with observations on the host *Hydrolagus collei* (Lay and Bennett). *Int. J. Parasit.*, **2**, 59–77.
317. SKAER, R. J. (1965). The origin and continual replacement of epidermal cells in the planarian *Polycelis tenuis* (Ijima). *J. Embryol. exp. Morph.*, **13**, 129–39.
318. SLEIGH, M. (1973). *The Biology of Protozoa*. Edward Arnold, London.
319. SLOBODKIN, L. B. (1961). *Growth and Regulation of Animal Populations*. Holt, Rinehart and Winston, New York.
320. SMITH, D. C. (1973). *Symbiosis of Algae with Invertebrates*. Oxford University Press, Oxford.
321. SMITH, D. C. (1973). *The Lichen Symbiosis*. Oxford University Press, Oxford.
322. SMITH, D. C. (1975). Symbiosis and the biology of lichenized fungi. *Symp. Soc. exp. Biol.*, **29**, 373–405.
323. SMITH, J. H., REYNOLDS, E. S. and VON LICHTENBERG, F. (1969). The integument of *Schistosoma mansoni*. *Am. J. trop. Med. Hyg.*, **18**, 28–49.
324. SMITHERS, S. R. and TERRY, R. J. (1965). Naturally acquired resistance to experimental infections of *Schistosoma mansoni* in the rhesus monkey (*Macaca mulatta*). *Parasitology*, **55**, 701–10.
325. SMITHERS, S. R., TERRY, R. J. and HOCKLEY, D. J. (1969). Host antigens in schistosomiasis. *Proc. R. Soc., B*, **171**, 483–94.
326. SMYTH, J. D. (1959). Maturation of larval pseudophyllidean cestodes and strigeid trematodes under axenic conditions, the significance of nutritional levels in platyhelminth development. *Ann. N.Y. Acad.Sci.*, **77**, 102–25.
327. SMYTH, J. D. (1969). *The Physiology of Cestodes*. Oliver and Boyd, London.
328. SMYTH, J. D. (1972). Changes in the digestive-absorptive surface of cestodes during larval/adult differentiation. *Symp. Brit. Soc. Parasit.*, **10**, 41–70.
329. SMYTH, J. D. (1973). Some interface phenomena in parasitic protozoa and platyhelminths. *Can. J. Zool.*, **51**, 367–77.
330. SMYTH, J. D. (1976). *Introduction to Animal Parasitology*. (2nd Edition). Hodder and Stoughton, London.
331. SOUTHGATE, V. R. (1970). Observations on the epidermis of the miracidium and the formation of the tegument of the sporocyst of *Fasciola hepatica*. *Parasitology*, **61**, 177–90.
332. SOUTHWOOD, T. R. E. (1976). Bionomic strategies and population parameters. In *Theoretical Ecology : principles and applications*, MAY, R.

M. (ed). Blackwell Scientific Publications, Oxford.

333. SPERBER, C. (1952). A guide for the determination of European Naididae. *Zool. Bidr. Upps.*, **29**, 45–77.

334. STANDEN, O. D. (1951). The effects of temperature, light and salinity upon the hatching of the ova of *Schistosoma mansoni. Trans. R. Soc. trop. Med. Hyg.*, **45**, 225–41.

335. STARR, M. P. (1975a). A generalized scheme for classifying organismic associations. *Symp. Soc. exp. Biol.*, **29**, 1–20.

336. STARR, M. P. (1957b). *Bdellovibrio* as symbiont: the associations of bdellovibrios with other bacteria interpreted in terms of a generalized scheme for classifying organismic associations. *Symp. Soc. exp. Biol.*, **29**, 93–124.

337. STARR, M. P. and HUANG, J. C-C. (1972). Physiology of Bdellovibrios. *Adv. micro. Physiol.*, **8**, 215–61.

338. STEPHENSON, W. (1947). Physiological and histochemical observations on the adult liver fluke, *Fasciola hepatica* L. II. Feeding. *Parasitology*, **38**, 123–7.

339. STEWART, J. E. and ZWICKAR, B. M. (1972). Natural and induced bactericidal activities in the haemolymph of the lobster *Homarus americanus*: products of haemocyte-plasma interaction. *Can. J. Microbiol.*, **18**, 1499–509.

340. STEWART, W. D. P. and ROWELL, P. (1977). Modifications of nitrogen-fixing algae in lichen symbioses. *Nature*, **265**, 371–72.

341. STIREWALT, M. A. (1971). Penetration stimuli for schistosome cercariae. In *The Biology of Symbiosis*, CHENG, T. C. (ed). Butterworths, London.

342. TAKAHASHI, T., MORI, K. and SHIGETA, Y. (1961). Phototactic, thermotactic and geotactic responses of the miracidia of *Schistosoma japonicum. Jap. J. Parasit.*, **10**, 686–91.

343. TAYLOR, E. W. and THOMAS, J. N. (1968). Membrane (contact) digestion in the three species of tapeworms *Hymenolepis diminuta*, *Hymenolepis microstoma* and *Moniezia expansa*. *Parasitology*, **58**, 535–46.

344. TAYLOR, F. J. R. (1974). Implications and extensions of the serial endosymbiotic theory of the origin of eukaryotes. *Taxon*, **23**, 229–58.

345. TERRY, R. J. and SMITHERS, S. R. (1975). Evasion of the immune response by parasites. *Symp. Soc. exp. Biol.*, **29**, 453–65.

346. THREADGOLD, L. T. (1963). The tegument and associated structures of *Fasciola hepatica. Q. Jl microsc. Sci.*, **104**, 505–12.

347. TIMMS, A. R. and BUEDING, E. (1959). Studies of a proteolytic enzyme from *Schistosoma mansoni. Br. J. Pharmac. Chemother.*, **14**, 68–73.

348. TINKER, P. B. H. (1975). Effects of vesicular-arbuscular mycorrhizas on higher plants. *Symp. Soc. exp. Biol.*, **29**, 325–49.

349. TRENCH, R. K. (1975). Of 'leaves that crawl': functional chloroplasts in animal cells. *Symp. Soc. exp. Biol.*, **29**, 229–65.

350. TRIMBLE III, J. J. and LUMSDEN, R. D. (1975). Cytochemical characterization of tegument membrane-associated carbohydrates in *Taenia crassiceps* larvae. *J. Parasit.*, **61**, 665–76.

351. TRUDGILL, D. L. (1967). The effect of environment on sex determination in *Heterodera rostochiensis*. *Nematologica*, **13**, 263–72.

352. TURTON, J. A. (1971). Distribution and growth of *Hymenolepis diminuta* in the rat, hamster and mouse. *Z. Parasit.*, **37**, 315–29.

353. TYRELL, D. A. J. (1976). *Interferon and its Clinical Potential.* Heinemann Medical, London.

354. UGLEM, G. L. and BECK, S. M. (1972). Habitat specificity and correlated

aminopeptidase activity in the acanthocephalans *Neoechinorhynchus cristatus* and *N. crassus*. *J. Parasit.*, **58**, 911–20.

355. UGLEM, G. L., PAPPAS, P. W. and READ, C. P. (1973). Surface aminopeptidase in *Moniliformis dubius* and its relation to amino acid uptake. *Parasitology*, **67**, 185–95.

356. URQUHART, G. M., MURRAY, M. and JENNINGS, F. W. (1972). The immune response to helminth infection in trypanosome-infected animals. *Trans. R. Soc. trop. Med. Hyg.*, **66**, 669–70.

357. VAN DER LAND, J. and DIENSKE, H. (1968). Two new species of *Gyrocotyle* (Monogenea) from chimaerids (Holocephali). *Zoöl. Meded., Leiden*, **43**, 97–105.

358. VAN DER LAND, J. and TEMPLEMAN, W. (1968). Two new species of *Gyrocotyle* (Monogenea) from *Hydrolagus affinis* (Brito Capello) (Holocephali). *J. Fish. Res. Bd Can.*, **28**, 335–41.

359. VAN MEIRVENNE, N., JANSENNS, P. G. and MAGNUS, E. (1975). Antigenic variation in syringe passaged populations of *Trypanosoma (Trypanozoon) brucei* I. Rationalization of the experimental approach. *Annls Soc. belge Méd. trop.*, **55**, 1–23.

360. VAN MEIRVENNE, N., JANSSENS, P. G., MAGNUS, E., LUMSDEN, W. H. R. and HERBERT, W. J. (1975). Antigenic variation in syringe passaged *Trypanosoma (Trypanozoon) brucei*. II. Comparative study of two antigenic type collections. *Annls Soc. belge Méd. trop.*, **55**, 25–30.

361. VARLEY, G. C., GRADWELL, G. R. and HASSELL, M. P. (1973). *Insect Population Biology*. Blackwell Scientific Publications, Oxford.

362. VICKERMAN, K. (1969). On the surface coat and flagellar adhesion in trypanosomes. *J. Cell Sci.*, **5**, 163–94.

363. VICKERMAN, K. (1971). Morphological and physiological considerations of extracellular blood protozoa. In *Ecology and Physiology of Parasites*, FALLIS, A. M. (ed). Toronto University Press, Toronto.

364. VICKERMAN, K. (1972). The host-parasite interface of parasitic protozoa. Some problems posed by ultrastructure studies. *Symp. Brit. Soc. Parasit.*, **10**, 71–92.

365. VICKERMAN, K. (1974) The ultrastructure of pathogenic flagellates. In Trypanosomiasis and Leishmaniasis with special reference to Chagas' Disease. *Ciba Fdn Symp.*, **20**, (New Series), 171–98.

366. VIENS, P., TARGETT, G. A. T., WILSON, V. C. L. C. and EDWARDS, C. T. (1972). The persistence of *Trypanosoma (Herpetosoma) musculi* in the kidneys of immune mice. *Trans. R. Soc. trop. Med. Hyg.*, **66**, 669–70.

367. VIGLIERCHIO, D. R. and MJUGE, S. G. (1975). Auxin inctivation systems of nemic origin. *Nematologica*, **21**, 471–75.

368. VON BRAND, T. (1973). *Biochemistry of Parasites*. (2nd Edition). Academic Press, London.

369. WARD, S. (1976). The use of mutants to analyse the sensory nervous system of *Caenorhabditis elegans*. In *The Organization of Nematodes*, CROLL, N. A. (ed). Academic Press, London.

370. WARD, S., THOMSON, N., WHITE, J. and BRENNER, S. (1975). Electron microscopical reconstruction of the anterior sensory anatomy of the nematode *Caenorhabditis elegans*. *J. comp. Neurol.*, **160**, 313–38.

371. WEBSTER, J. M. (1975). Aspects of the host-parasite relationship of plant-parasitic nematodes. *Adv. Parasit.*, **13**, 225–56.

372. WEIDNER, E. (1972). Ultrastructural study of microsporidian invasion into cells. *Z.* Parasit., **40**, 227–42.

373. WHITFIELD, P. J. (1971). The locomotion of the acanthor of *Moniliformis*

dubius (Archiacanthocephala). *Parasitology*, **62**, 35–47.

374. WHITFIELD, P. J., ANDERSON, R. M. and BUNDY, D. A. P. (1977). Experimental investigations on the behaviour of the cercariae of an ectoparasitic digenean *Transversotrema patialense*: general activity patterns. *Parasitology*, **75**, 9–30.

375. WHITFIELD, P. J., ANDERSON, R. M. and MOLONEY, N. A. (1975). The attachment of cercariae of an ectoparasitic digenean, *Transversotrema patialense*, to the fish host; behavioural and ultrastructural aspects. *Parasitology*, **70**, 311–29.

376. WHITFIELD, P. J. and WELLS, J. (1973). Observations on the ectoparasitic digenean *Transversotrema patialense*. *Parasitology*, **67**, 27–28.

377. WHITTAKER, R. H. and FEENY, P. P. (1971). Allelochemics: chemical interactions between species. *Science*, **171**, 757–69.

378. WILLIAMS, H. H. (1968). *Phyllobothrium piriei* sp. nov. (Cestoda: Tetraphyllidea) from *Raja naevus* with a comment on its habitat and mode of attachment. *Parasitology*, **58**, 929–37.

379. WILLIAMS, H. H., MCVICAR, A. and RALPH, R. (1970). The alimentary canal of fish as an environment for helminth parasites. *Symp. Brit. Soc. Parasit.*, **8**, 43–77.

380. WILSON, R. A. (1969). Fine structure of the tegument of the miracidium of *Fasciola hepatica* (L). *J. Parasit.*, **55**, 124–34.

381. WILSON, R. A. (1970). Fine structure of the nervous system and specialized nerve endings in the miracidium of *Fasciola hepatica*. *Parasitology*, **60**, 399–410.

382. WRIGHT, C. A. (1971). *Flukes and Snails*. Unwin University Books, London.

383. WRIGHT, R. D. and LUMSDEN, R. D. (1969). Ultrastructure of the tegumentary pore-canal system of the acanthocephalan *Moniliformis dubius*. *J. Parasit.*, **55**, 993–1003.

384. ZAHL, P. A. and MCLAUGHLIN, J. J. A. (1959). Studies in marine biology IV. On the role of algal cells in the tissues of marine invertebrates. *J. Protozool.*, **6**, 344–52.

385. ZUSSMAN, R. A., BAUMAN, P. M. and PETRUSKA, J. C. (1970). The role of ingested hemoglobin in the nutrition of *Schistosoma mansoni*. *J. Parasit.*, **56**, 75–79.

Index